11.00

STRUCTURAL ANALYSIS
OF SHELLS

STRUCTURAL ANALYSIS OF SHELLS

E. H. BAKER
Professor of Mechanical Engineering, California State Polytechnic College at San Luis Obispo

L. KOVALEVSKY
Member of the Technical Staff North American Rockwell Corp. Space Division

F. L. RISH
Member of the Technical Staff North American Rockwell Corp. Space Division

McGRAW HILL BOOK COMPANY

New York   St. Louis   San Francisco   Düsseldorf   Johannesburg
Kuala Lumpur   London   Mexico   Montreal   New Delhi
Panama   Rio de Janeiro   Singapore   Sydney   Toronto

Library of Congress Cataloging in Publication Data

Baker, E H
  Structural analysis of shells.

  Includes bibliographical references.
  1. Shells (Engineering) I. Kovalevsky, L.,
joint author. II. Rish, F. L., joint author.
III. Title.
TA660.S5B34   624'.1776   78–130678
ISBN 0–07–003354–4

Copyright © 1972 by McGraw-Hill, Inc. All Rights Reserved. Printed in the United States of America. No part of this publication may be reproduced, stored in a retrieval system, or transmitted, in any form or by any means, electronic, mechanical, photocopying, recording, or otherwise, without the prior written permission of the publisher.

1234567890 HDMM 75432

The editors for this book were William G. Salo, Jr., and Frank Purcell, the designer was Naomi Auerbach, and its production was supervised by Stephen J. Boldish. It was set in Imprint Old Face 101.

It was printed by Halliday Lithograph Corporation and bound by The Maple Press Company.

CONTENTS

Preface xi

1. INTRODUCTION TO THE THEORY OF SHELLS 1

1-1 General 1
1-2 Linear Shell Theory 2
1-3 Geometry of Shells 2
1-4 External Loadings 3
1-5 Internal Stresses 3
1-6 Condition of Equilibrium 5
1-7 Membrane Theory for Shells of Revolution 6
1-8 Bending Theory for Axisymmetrically Loaded Shells of Revolution 11
1-9 Elastic Laws 14
    Geometrical Relations between Deformations 14
    Hooke's Law 17
    Internal Loads 18
1-10 Classification of Shell Theories 20
    First-order-approximation Shell Theory 21
    Second-order-approximation Theories for Shells of Revolution 22
    Special Shell Theories 22
    Membrane Theory of Shells 22
1-11 Nonlinear Shell Theory 23
    REFERENCES 24

2. FORCE METHOD 25

2-1 Introduction 25
2-2 Geometrical Considerations of Shell Segments 26
2-3 Membrane Solutions 27
2-4 Bending Theory 28
2-5 Comparison of Membrane and Bending Theories for Nonshallow Shell 29
2-6 Force Method of Solution 29
2-7 Procedure for Force Method 30
2-8 Restraints of Boundaries 32
2-9 Interaction Between Shells of Various Geometries 32
      Breakdown for Complicated Shell Geometry 33
      Interaction between Two Shell Elements 33
      Summary 36
2-10 Determination of Stresses and Deformations at any Point 36
      Numerical Example 38
      REFERENCES 39

3. PRIMARY SOLUTIONS 40

3-1 Introduction 40
3-2 Determination of Membrane Internal Forces 40
3-3 Determination of Membrane Displacements 42
3-4 Any Shape of Meridian 44
3-5 Spherical Shells 45
3-6 Conical Shells 50
3-7 Cylindrical Shells 55
3-8 Elliptical Shells 59
3-9 Cassini Shells 61
3-10 Toroidal Shells 61
3-11 Other Geometries of Shells 69
3-12 Irregular Shell 71
      REFERENCES 74

4. SECONDARY SOLUTIONS 75

4-1 Introduction 75
4-2 Spherical Shells (Open, Closed) 76
      Open Spherical Shell 79
      Distortions 82
4-3 Conical Shells 86
      Open Conical Shell—Loading at Lower Edge 87
      Open Conical Shell—Loading at Upper Edge 89
4-4 Cylindrical Shells 95
      Long Cylinders 95
      Short Cylinders with Uniform Wall Thickness without Abrupt Discontinuity 96
4-5 Approximate Method 99
      Displacements 100
      Interaction 100
      The Maximum Values 101
4-6 Some Practical Considerations 102
      REFERENCES 105

5. SPECIAL SOLUTIONS 106

5-1 Introduction 106
5-2 Hampe's Solutions 107
    Cylinders with Uniform Thickness 107
    Cylinders with Rotationally Symmetric Discontinuities in Geometry or Loading 121
    Spherical Shell, Any Fixity at the Lower Boundary 124
    Definition of F-Factors 126
    Approximate Method for Determination of Location and Maximum Stresses in Cylinders 135
5-3 Circular plates 141
    Primary Solutions 141
    Secondary Solutions 141
    Special Cases 141
5-4 Circular Rings 151
    REFERENCES 151

6. MULTISHELL STRUCTURES 153

6-1 Introduction 153
6-2 Equation of Deformations of Shell Elements 154
6-3 Equilibrium of Junctions 156
6-4 Interaction Procedure 158
    Example 159
6-5 Final Sets of Equations for Corrective Loadings 162
    Structural System Combined from Statically Determinate Elements 162
    Structural System Combined from Statically Indeterminate Elements 163
6-6 Application of Tables 164
    REFERENCES 180

7. SHELLS WITH COMPOSITE OR STIFFENED WALLS 181

7-1 Introduction 181
7-2 Extensional and Bending Stiffness 182
7-3 Primary Solutions 183
7-4 Secondary Solutions 184
    Cylinders 185
    Cylinders, Transverse Shear Distortion Included 186
    Cylinders, Influence of Axial Force Included 188
    Spheres and Cones 189
7-5 Stiffened Shells 190
7-6 Sandwich Shells 192
    Introduction 192
    Modes of Failure of Sandwich Elements 193
    Types of Sandwich Cores 194
    Analysis of Sandwich Shells 194
    REFERENCES 195

8. UNSYMMETRICALLY LOADED SHELLS 196

8-1 Introduction 196
8-2 Shells of Revolution 196
8-3 Shells of Beam Systems 199
    Cantilever Cylindrical Shell 199
    Simple and Fixed-beam Cylindrical Shell 201
    Continuous Cylindrical Shell under Deadweight 203

8-4 Curved Panels (Barrel Vaults) 205
       REFERENCES 208

9. ALLOWABLE STRESSES AND MARGIN OF SAFETY 209

9-1 Introduction 209
9-2 Allowable Stress 210
9-3 Margin of Safety 210
9-4 Failure Theories 210
9-5 Practical Applications 214
9-6 Stress Ratios and Interaction Curves 214
9-7 A Theoretical Approach to Interaction 217
       Conclusion 218
       REFERENCES 219

10. STABILITY OF UNSTIFFENED SHELLS 220

10-1 General 220
10-2 Curved Cylindrical Panels 222
       Introduction 222
       Axial Compression, Curved Panels 222
       Shear, Curved Panels 225
       Bending, Curved Panels 227
       External Pressure, Curved Panels 228
       Combined Loading, Curved Panels 229
10-3 Cylinders 229
       Axial Compression, Thin-walled Cylinders 229
       Shear or Torsion, Unstiffened Cylinders 232
       Bending, Unstiffened Cylinders 234
       External Pressure, Unstiffened Cylinders 236
       Combined Loading, Unstiffened Cylinders 238
       Cylinders with an Elastic Core, General 241
       Axial Compression, Cylinders with Elastic Core 241
       External Pressure, Cylinders with Elastic Core 242
       Torsion, Cylinders with Elastic Core 244
       Combined Axial Compression and Lateral Pressure 245
10-4 Cones 246
       Axial Compression, Unstiffened Cones 246
       Shear or Torsion, Unstiffened Cones 248
       Bending, Unstiffened Cones 249
       Lateral and Axial External Pressure, Unstiffened Cones 251
       Combined Loading, Unstiffened Cones 253
10-5 Spherical Shells 253
       Uniform External Pressure, Spherical Caps 253
       Concentrated Load at the Apex, Spherical Caps 254
       Uniform External Pressure and Concentrated Load at the Apex, Spherical Caps 255
10-6 Other Shapes 256
       Uniform External Pressure, Complete Ellipsoidal Shells 256
       Uniform Internal Pressure, Complete Oblate Spheroidal Shells 258
       Internal Pressure, Ellipsoidal and Toroidal Bulkheads 258
       Uniform External Pressure, Complete Circular Toroidal Shells 260
       Axial Loading, Shallow Bowed-out Toroidal Segments 261
       Uniform External Pressure, Shallow Toroidal Segments 263
10-7 Inelastic Buckling 265
       General, Inelastic Buckling 265
       Plasticity Correction Factor 265
       Combined Loadings, Inelastic Buckling 275
       REFERENCES 276

11. STABILITY OF ORTHOTROPIC COMPOSITE SHELLS 279

11-1   General   279
11-2   Elastic Constants   280
           Orthotropic Layered Shells   282
           Sandwich Shells   286
           Integrally Stiffened Waffle Shells   288
           Special Cases   291
11-3   Cylinders   293
           Axial Compression, Orthotropic Cylinders   293
           Torsion, Orthotropic Cylinders   296
           Bending, Orthotropic Cylinders   299
           External Pressure, Orthotropic Cylinders   300
11-4   Cones   303
           Axial Compression, Orthotropic Cones   303
           Torsion, Orthotropic Cones   304
           Bending, Orthotropic Cones   304
           Lateral and Axial External Pressure, Orthotropic Cones   305
           REFERENCES   305

12. STABILITY OF STIFFENED SHELLS 306

12-1   General   306
12-2   Frame- and Stringer-stiffened Cylinders   307
           Axial Compression, Frame- and Stringer-stiffened Cylinders   307
           Bending, Frame- and Stringer-stiffened Cylinders   317
12-3   Frame-stiffened Cylinders   317
           Lateral and Axial External Pressure, Frame-stiffened Cylinders   317
           REFERENCES   322

13. STABILITY OF SANDWICH SHELLS 325

13-1   General   325
13-2   Local Instability   326
           Intracell Buckling   326
           Face-sheet Wrinkling   328
13-3   Cylinders   330
           Axial Compression, Sandwich Cylinders   330
           Shear or Torsion, Sandwich Cylinders   331
           Bending, Sandwich Cylinders   335
           External Pressure, Sandwich Cylinders   335
13-4   Cones   340
           Axial Compression, Sandwich Cones   340
           Torsion, Sandwich Cones   340
           Bending, Sandwich Cones   340
           Lateral and Axial External Pressure, Sandwich Cones   341
13-5   Doubly Curved Shells   341
           REFERENCES   342

Index   343

PREFACE

The purpose of this book is to provide instructions, procedures, and solutions for the static analysis of aerospace, civil, and mechanical engineering shell structures. This book also provides an introduction to and reference for the theory of shells.

To a great extent, much of the material from which this book was developed was obtained from the "Shell Analysis Manual," NASA CR 912. The "Shell Analysis Manual" was prepared for the National Aeronautics and Space Administration, Manned Spacecraft Center, Houston, Texas, by North American Rockwell Corporation, Space Division, Downey, California, under Contract NAS9-4387, for which Mr. Herbert C. Kavanaugh, Jr., was the NASA technical monitor.

Generally, the information presented in this book is a condensation of material published by U.S. Government agencies, universities, scientific and technical journals, text books, aerospace industries, including North American Rockwell Corporation, and foreign publications. Particular credit is given to the following publishers who granted permission to use their publications.

American Institute of Aeronautics and Astronautics, New York, New York.

American Rocket Society Journal, Vol. 31, No. 2, February 1961, pp. 237–246, "Stability of Orthotropic Cylindrical Shells Under Combined Loading" by T. E. Hess.

Journal of the Aerospace Sciences, Vol. 29, No. 5, May 1962, pp. 505–511, "Elastic Stability of Orthotropic Shells" by H. Becker and G. Gerard.

American Concrete Institute, Detroit, Michigan.

Journal of the American Concrete Institute, No. 2, Vol. 27, October 1955, "Line Load and Temperature Moments in Shells of Rotation Built into Cylinders" by M. G. Salvadori.

American Society of Civil Engineers, New York, New York.

Journal of the Structural Division, Vol. 82, July 1956, "Bending Stresses in Edge Stiffened Domes" by M. G. Salvadori and Shermann.

R. Oldenbourg Verlage, München, West Germany.

Drang and Swang by L. Föppl, 1941, 1944, and 1947.

Springer Verlag, Berlin, West Germany.

Stresses in Shells by W. Flügge, 1957.

Wilhelm Ernst and Sohn Verlag, Berlin, West Germany.

Beton Kalender by G. Worch, 1943 and 1958.

We also found the books *Elementary Statics of Shells* by A. Pflüger and *Statik Rotationssymmetrischer Flächentragwerke* by E. Hampe to be of great benefit.

Detailed derivations of formulas are limited because it is not believed to serve the purpose of this book. Numerous references to more detailed discussions are given.

The book has been developed primarily from existing material in the field of shells. The original works are referenced in the bibliography.

Chapter 2 outlines the force method for shells and simpler multishells which are combined from not more than two shell elements.

Chapter 3 presents the primary solutions needed for the force method for many shell geometries for many loadings.

Chapter 4 presents the secondary solutions for the same purpose.

Chapter 5 presents some special cases such as cylinders and spheres with different boundary conditions. Also the solution of interaction for a cylinder with abrupt change of wall thickness.

Chapter 6 finally presents the force method for complicated multishells with more than two shell elements.

Chapter 7 treats composite shells, reducing them to the same methods which were explained previously.

Chapter 8 presents the special cases of unsymmetrical shells (unsymmetrical due to geometry or loading).

Chapter 9 treats allowables and margins of safety for the biaxial state of stress as occurs in a shell structure. This chapter concludes the static analysis of multishells.

Chapter 10 is the first chapter in which stability is presented. The monocoque shells are discussed and formulas are presented.

Chapter 11 continues the stability analysis of shells, treating orthotropic shells in general.

Chapter 12 presents in more detail stability of stiffened shells.

Chapter 13 presents stability of sandwich shells.

This book was written by engineers for engineers and for the personal usage of the authors who participated in writing of this document and whose names are listed in alphabetical order. It is the authors' hope that this book will be not only useful to the practicing engineers but also for the students who would like to extend their knowledge of shell analysis. To help them, primarily, the introductory chapter is included which contains basic derivations which are needed for good understanding of shell analysis. An experienced engineer can simply omit this introductory chapter.

E. H. Baker
L. Kovalevsky
F. L. Rish

Chapter 1

INTRODUCTION TO THE THEORY OF SHELLS

1-1 General

The most common shell theories are those based on linear elasticity concepts. Linear shell theories adequately predict stresses and deformations for shells exhibiting small elastic deformations, that is, deformations for which it is assumed that the equilibrium-equation conditions for deformed elements are the same as if they were not deformed and Hooke's law applies.

The nonlinear theory of elasticity forms the basis for the finite-deflection and stability theories of shells. Large-deflection theories are often required when dealing with shallow shells, highly elastic membranes, and buckling problems. The nonlinear shell equations are considerably more difficult to solve and for this reason are more limited in use.

Development of more exact theoretical expressions does not necessarily assist in the solution of practical shell problems, since often the theoretical expressions can be solved only with great difficulty, and then only for special cases. The experimental approach is also limited because data are not available for every special case.

Practical difficulties in both theory and experiment have led to the development and application of applied engineering methods for the analysis of shells. While these methods are approximate and are valid only under specific conditions, they generally are very useful and give good accuracy for the analysis of practical engineering shell structures.

1-2 Linear Shell Theory

The theory of small deflections of thin elastic shells is based upon the equations of the mathematical theory of linear elasticity. The geometry of shells (i.e., one dimension much smaller than the other dimensions) does not warrant, in general, the consideration of the complete three-dimensional elasticity equations. In fact, the consideration of the complete elasticity equations leads to expressions and equations which are so complicated that it becomes impossible to obtain solutions for shell problems of practical interest.

Fortunately, however, sufficiently accurate analyses of thin shells can be obtained using simplified versions of the general elasticity equations. In the development of thin-shell theories, simplification is accomplished by reducing the shell problem to the study of the deformations of the middle (or reference) surface of the shell. In all cases, one begins with the governing equations in the three-dimensional theory of elasticity and attempts to reduce the system of equations, involving three independent space variables, to a new system involving only two space variables. These two variables are more conveniently taken as coordinates on the middle surface of the shell.

Shell theories of varying degrees of accuracy may be derived, depending upon the degree to which the elasticity equations are simplified. The approximations necessary for the development of an adequate theory of shells have been the subject of considerable controversy among investigators in the field. A brief discussion of the approximations is presented in Sec. 1-11. The theory presented in Secs. 11-8 and 11-9 is a first-order-approximation shell theory for axisymmetrically loaded shells of revolution.

1-3 Geometry of Shells

Before shell theory is discussed, the geometry of an arbitrary shell in three-dimensional space is defined. The geometry of a shell is entirely defined by specifying the form of the middle surface and the thickness of the shell at each point. To describe the form of the middle surface, it is necessary to present some of the important geometrical properties of a surface. A more detailed presentation of the theory of surfaces can be found in books on tensor analysis and differential geometry.

In the engineering application of thin shells, a shell whose reference surface is in the form of a surface of revolution has extensive usage. This discussion is restricted to surfaces of revolution. A surface of revolution is obtained by rotation of a plane curve about an axis lying in the plane of the curve. This curve is called the meridian, and its plane is the meridian plane. The intersections of the surface with planes perpendicular to the axis of rotation are parallel circles and are called parallels.

For such a shell the lines of principal curvature are its meridians and parallels. The following nomenclature is given in Fig. 1-1.

ϕ = angle between the axis of the shell and the shell normal at the point under consideration on the middle surface of the shell
θ = angle between r and any defined line ξ

The radii of curvature of a shell of revolution are

R_ϕ = radius of curvature of meridian
R_θ = length of the normal between any point on the middle surface and the axis of rotation
r = radius of curvature of the parallel
R_ϕ and R_θ = principal radii of curvature of the surface

The following geometrical relation is of fundamental importance:

$$r = R_\theta \sin \phi$$

1-4 External Loadings

The external loads consist of body forces that act on the element and surface forces that act on the upper and lower surfaces of the shell element.

All loadings under consideration at any point on the shell can be resolved into three components in the x, y, and z directions. The x direction is parallel to the tangent to the meridian. The y direction is parallel to the tangent to the parallel circle, and the z direction is normal to the surface of the shell. For example: The deadweight p (weight of shell per unit area) for a shell of revolution can be resolved into load per unit area in the x, y, and z directions, respectively, in the following manner (Fig. 1-2):

$$p_x = p \sin \phi \qquad p_y = 0 \qquad p_z = p \cos \phi$$

1-5 Internal Stresses

The external forces are resisted by internal forces, or stresses, which are in equilibrium with the external forces. It is convenient to investigate the stresses along a meridian and parallel, which are defined by the angles ϕ and θ.

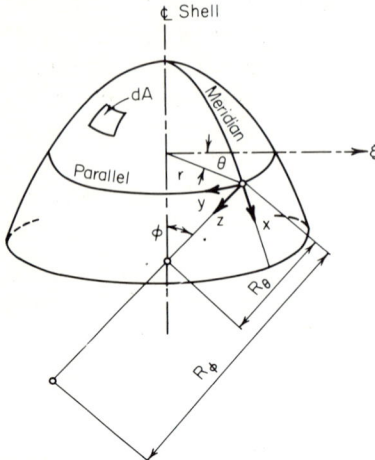

figure 1-1 *Shell of revolution.*

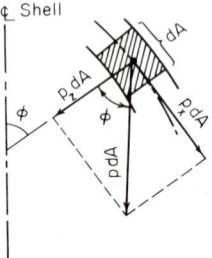

figure 1-2 *Loading components from deadweight.*

The internal forces consist of membrane forces, transverse shears, bending moments and twisting moments.

1. The membrane forces, which act in the plane of the surface of shell, are shown in Fig. 1-3.

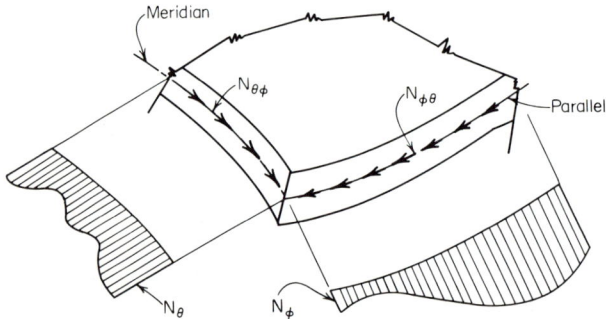

figure 1-3 *Membrane forces.*

N_ϕ, N_θ = normal inplane forces per unit length (load/unit length)
$N_{\phi\theta}$, $N_{\theta\phi}$ = inplane shear forces per unit length (load/unit length)
These forces can vary along the meridian and parallel (see Fig. 1-3).

2. The transverse shear forces per unit length Q_ϕ and Q_θ are shown in Fig. 1-4.

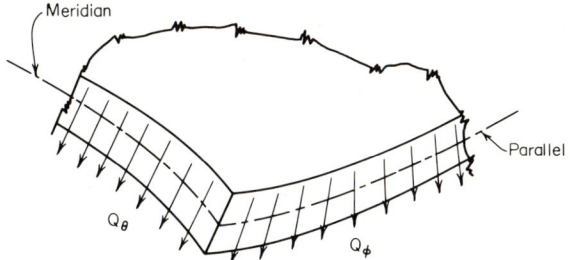

figure 1-4 *Transverse shear forces.*

3. Bending moments M_ϕ and M_θ per unit length and twisting moments $M_{\phi\theta}$ and $M_{\theta\phi}$ per unit length are shown in Fig. 1-5.

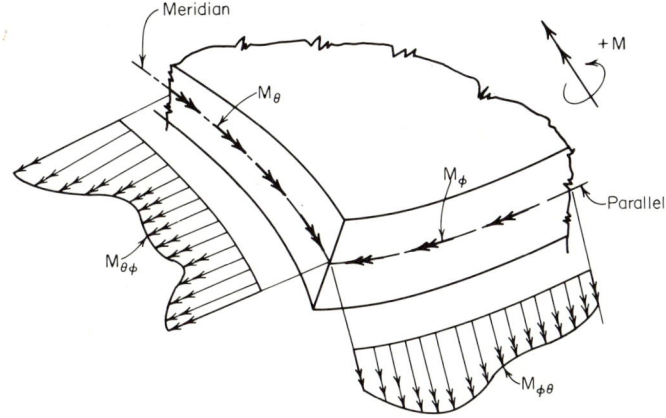

figure 1-5 *Bending and twisting moments.*

The positive directions of all stresses under 1, 2, and 3 are shown in the corresponding figures. All positive loadings act in the positive direction of the system of coordinates.

In the preceding section, all internal forces are replaced by statically equivalent forces and moments.

1-6 Condition of Equilibrium

The conditions for equilibrium of the shell element under external and internal loads will be determined. The equations arising by virtue of the demands of equilibrium and the compatibility of deformations will be derived by considering an individual differential shell element. These equations are relations between differential quantities or between

6 Structural Analysis of Shells

differential changes in the internal forces and therefore are called differential equations. If a differential element is imagined separated from the loaded shell, it is stressed by 10 internal components which must be in equilibrium with the external loads.

$$N_\phi, N_\theta, N_{\phi\theta}, N_{\theta\phi}, Q_\phi, Q_\theta, M_\phi, M_\theta, M_{\phi\theta}, M_{\theta\phi}$$

To determine these components, there are known only six equilibrium equations:

$$\sum F_x = 0 \qquad \sum M_x = 0$$
$$\sum F_y = 0 \qquad \sum M_y = 0 \qquad (1\text{-}1)$$
$$\sum F_z = 0 \qquad \sum M_z = 0$$

where $\sum F_i$ is the sum of the force in the i direction ($i = x, y, z$) and $\sum M_i$ is the sum of the moments about the i axis. This problem is four times internally statically indeterminate.

1-7 Membrane Theory for Shells of Revolution

Consider a truss structure, which is physically many times internally statically indeterminate. This complicated problem can be simplified by assuming all joints of the truss are pinned. This means that each member of the truss is stressed only axially. End moments and shears are zero, and the truss is analyzed as an internally statically determinate structure.

Similar assumptions may be introduced in the shell equations:

$$M_\phi = M_\theta = M_{\phi\theta} = M_{\theta\phi} = Q_\phi = Q_\theta = 0$$

Consequently, only four unknowns remain:

$$N_\phi, N_\theta, N_{\phi\theta}, N_{\theta\phi}$$

which are called the membrane forces. If a shell theory includes only the membrane forces in the analyses, it is called a membrane theory. Certain restrictions in the use of membrane theory will be discussed in Chap. 2.

Figure 1-6 shows a differential element of the shell whose area may be expressed

$$dA = r \, d\theta \, R_\phi \, d\phi$$

Figure 1-7 shows all forces in equilibrium which may act on a differential element in the membrane theory. The components of the external

Introduction to the Theory of Shells 7

loading are designated by X, Y, and Z, which act in the x, y, and z directions, respectively, and are in units of force.

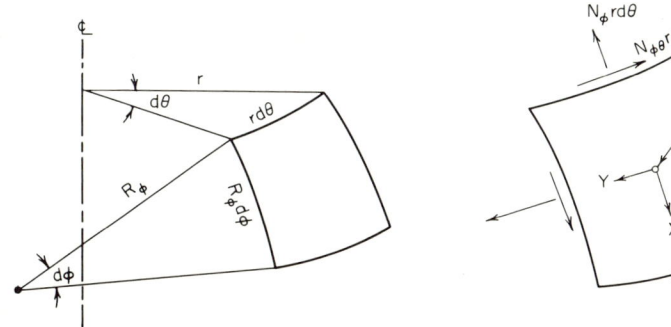

figure 1-6 *Differential element of shell.*   **figure 1-7** *Forces on differential element.*

The forces are shown on one end only. On the opposite end the forces will be differentially changed:

(a) $N_\phi r\, d\theta$ with change of ϕ becomes $N_\phi r\, d\theta + \dfrac{\partial}{\partial \phi}(N_\phi r\, d\theta)\, d\phi$

(b) $N_\theta R_\phi\, d\phi$ with change of θ becomes $N_\theta R_\phi\, d\phi + \dfrac{\partial}{\partial \theta}(N_\theta R_\phi\, d\phi)\, d\theta$

(c) $N_{\theta\phi} R_\phi\, d\phi$ with change of θ becomes $N_{\theta\phi} R_\phi\, d\phi + \dfrac{\partial}{\partial \theta}(N_{\theta\phi} R_\phi\, d\phi)\, d\theta$

(d) $N_{\phi\theta} r\, d\theta$ with change of ϕ becomes $N_{\phi\theta} r\, d\theta + \dfrac{\partial}{\partial \phi}(N_{\phi\theta} r\, d\theta)\, d\phi$

The loading components are

(e) $Z = p_z r\, d\theta\, R_\phi\, d\phi$

(f) $X = p_x r\, d\theta\, R_\phi\, d\phi$

(g) $Y = p_y r\, d\theta\, R_\phi\, d\phi$

The forces acting on the differential element must be in static equilibrium; therefore, each Eq. 1-1 must be satisfied.

(a) $\sum M_x = 0$

(b) $\sum M_y = 0$

8   Structural Analysis of Shells

These equations are always satisfied, since there are no forces in Fig. 1-7 which would produce moments about the x and y axis.

(c) $$\sum M_z = 0$$

$$\sum M_z = N_{\phi\theta} r \, d\theta \, R_\phi \, d\phi - N_{\theta\phi} R_\phi \, d\phi \, r \, d\theta = 0$$

where the values of higher order are neglected.

This leads to the relation

$$N_{\phi\theta} = N_{\theta\phi} \tag{1-2}$$

(d) $$\sum F_x = 0$$

$$-N_\phi r \, d\theta + N_\phi r \, d\theta + \frac{\partial}{\partial \phi}(N_\phi r \, d\theta) \, d\phi - N_{\theta\phi} R_\phi \, d\phi$$

$$+ N_{\theta\phi} R_\phi \, d\phi + \frac{\partial}{\partial \theta}(N_{\theta\phi} R_\phi \, d\phi) \, d\theta - N_\theta R_\phi \, d\phi \, d\theta \cos\phi + p_x r \, d\theta \, R_\phi \, d\phi = 0$$

This leads to the following expression:

$$\frac{\partial}{\partial \phi}(N_\phi r \, d\theta) \, d\phi + \frac{\partial}{\partial \theta}(N_{\theta\phi} R_\phi \, d\phi) \, d\theta - N_\theta R_\phi \, d\phi \, d\theta \cos\phi + p_x r \, d\theta \, R_\phi \, d\phi = 0 \tag{1-3}$$

Additional explanation of the origin of the third term in Eq. 1-3 is necessary. Figure 1-8a shows a meridian section, and Fig. 1-8b shows a hoop section of a shell element where higher-order terms have been neglected. Figure 1-8b shows why there is a force $U = N_\theta R_\phi \, d\phi \, d\theta$ which is directed toward the shell axis and is contained in the hoop section. It is clear from Fig. 1-8a that this force U has a component in the x direction

$$U_x = N_\theta R_\phi \, d\phi \, d\theta \cos\phi$$

which has been included in the derivations of Eq. 1-3.

(e) $$\sum F_y = 0$$

which may be written

$$-N_\theta R_\phi \, d\phi + N_\theta R_\phi \, d\phi + \frac{\partial}{\partial \theta}(N_\theta R_\phi \, d\phi) \, d\theta - N_{\phi\theta} r \, d\theta$$

$$+ N_{\phi\theta} r \, d\theta + \frac{\partial}{\partial \phi}(N_{\phi\theta} r \, d\theta) \, d\phi + N_{\theta\phi} R_\phi \frac{r \, d\theta}{R_\theta \tan\phi} + p_y R_\phi \, d\phi \, r \, d\theta = 0$$

This leads to the following expression:

$$\frac{\partial}{\partial \theta}(N_\theta R_\phi \, d\phi) \, d\theta + \frac{\partial}{\partial \phi}(N_{\phi\theta} r \, d\theta) \, d\phi + N_{\theta\phi} R_\phi \frac{r \, d\theta}{R_\theta \tan\phi} + p_y R_\phi \, d\phi \, r \, d\theta = 0 \tag{1-4}$$

An explanation of the origin of the third term in Eq. 1-4 is necessary. Figure 1-9 shows the projection of a shell element in the direction of the normal to the shell surface, and two of the membrane shear forces acting on the element.

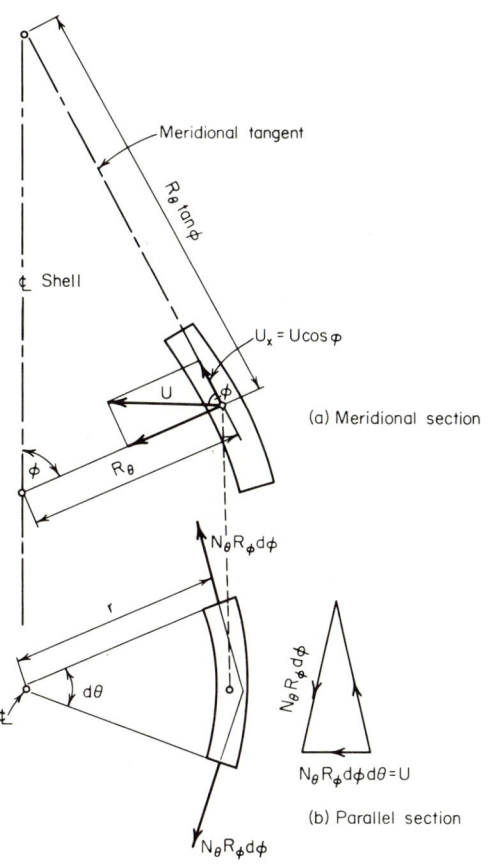

figure 1-8 *Sections of shell of revolution with forces.*

It can be seen that there is a component of force in the y direction of magnitude $N_{\theta\phi} R_\phi \, d\phi \, d\epsilon$ if higher-order terms are neglected. The angle $d\epsilon$ is the angle included between the shearing vectors.

From Fig. 1-10,
$$r \, d\theta = R_\theta \tan \phi \, d\epsilon$$

Consequently
$$d\epsilon = \frac{r \, d\theta}{R_\theta \tan \phi}$$

10 Structural Analysis of Shells

Therefore, the component of force in the y direction, which has been included in the derivation of Eq. 1-4, is

$$N_{\theta\phi} R_\phi \, d\phi \, \frac{r \, d\theta}{R_\theta \tan \phi}$$

(f) $\sum F_z = 0$

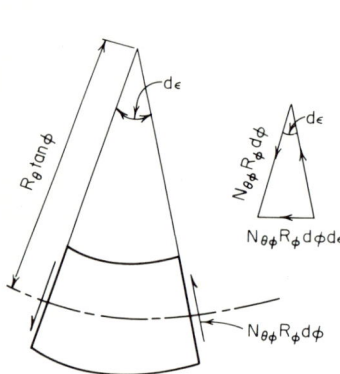

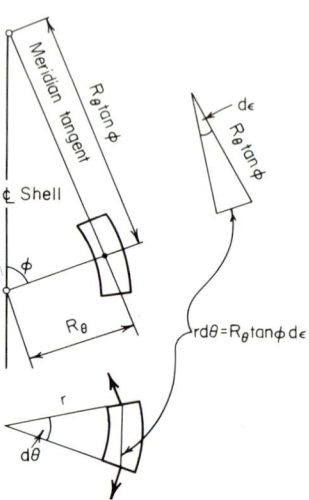

figure 1-9 *Shell element with shear forces.*

figure 1-10 *Geometrical meaning of term $r \, d\theta$.*

Equilibrium in the z direction (Fig. 1-11) is satisfied by the following equation:

$$N_\phi r \, d\theta \, d\phi + N_\theta R_\phi \, d\phi \, d\theta \sin \phi + p_z R_\phi \, d\phi \, r \, d\theta = 0 \qquad (1\text{-}5)$$

The second term is U_z from Fig. 1-8a. The first term is obtained in a manner similar to Fig. 1-8b.

In this way three equations express the condition of equilibrium in the x, y, and z directions.

It is noted that R_ϕ and R_θ depend only on the position of the point on the meridian ϕ, but not on θ because the shell is a shell of revolution.

Equations 1-3, 1-4, and 1-5 can be simplified by dividing all terms with $d\theta \, d\phi$ and by noting that

$$r = R_\theta \sin \phi \qquad N_{\phi\theta} = N_{\theta\phi}$$

Finally we arrive at the following system of equations, which consist of two differential equations and one algebraic equation:

$$\frac{\partial}{\partial \phi}(N_\phi R_\theta \sin \phi) + \frac{\partial N_{\phi\theta}}{\partial \theta} R_\phi - N_\theta R_\phi \cos \phi + p_x R_\phi R_\theta \sin \phi = 0$$

$$\frac{\partial N_\theta}{\partial \theta} R_\phi + \frac{\partial}{\partial \phi}(N_{\phi\theta} R_\theta \sin \phi) + N_{\phi\theta} R_\phi \cos \phi + p_y R_\phi R_\theta \sin \phi = 0 \qquad (1\text{-}6)$$

$$N_\phi R_\theta + N_\theta R_\phi + p_z R_\phi R_\theta = 0$$

According to membrane theory, Eqs. 1-6 may be used to find the membrane forces N_ϕ, N_θ, $N_{\phi\theta}$ for any loading condition. The solution of these equations is discussed in Chap. 3.

It also should be noted that the stresses do not depend upon the stiffness properties of the material since the structure is statically determinate.

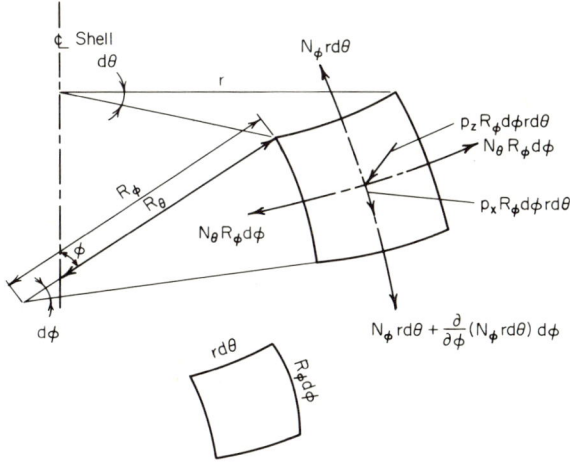

figure 1-11 *Membrane forces on element of shell of revolution.*

1-8 Bending Theory for Axisymmetrically Loaded Shells of Revolution

The membrane theory is very useful for many practical cases, but the results are not always satisfactory. The membrane theory fails to predict accurate stresses whenever bending is involved because the membrane theory does not consider bending. In this section the bending theory of shells will be described. This theory includes the bending resistance of shells. Because of the complexity of the bending theory only the special case of shells of revolution subjected to axisymmetrical loads is presented.

12   **Structural Analysis of Shells**

This special case includes a large percentage of the problems encountered in actual practice. More general bending theories are presented in many references, but most solutions to these nonaxisymmetric shell equations can be obtained only by large computer programs.

Figure 1-11 shows the membrane forces on a differential element which has been separated from the loaded shell. In addition to the forces indicated in Fig. 1-11, the transverse shear forces and moments as shown in Fig. 1-12 must be included in bending theory.

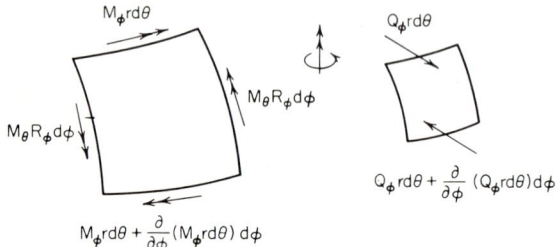

figure 1-12   *Bending and shear loads.*

It can be noted that, because of axial symmetry,

$$M_{\theta\phi} = 0 \quad M_{\phi\theta} = 0 \quad Q_\theta = 0 \quad \frac{\partial}{\partial \theta}(M_\theta R_\phi \, d\phi) \, d\theta = 0 \quad p_y = 0$$

The differential element must be in equilibrium. Consequently, six equations of equilibrium must be satisfied.

(a) $$\sum F_x = 0$$

$$\frac{\partial}{\partial \phi}(N_\phi r \, d\theta) \, d\phi - N_\theta R_\phi \, d\phi \cos\phi - Q_\phi r \, d\theta \, d\phi + p_x R_\phi r \, d\theta = 0 \quad (1\text{-}7)$$

All terms in the above equation are from the membrane theory except the third term, the meaning of which is depicted in Fig. 1-13.

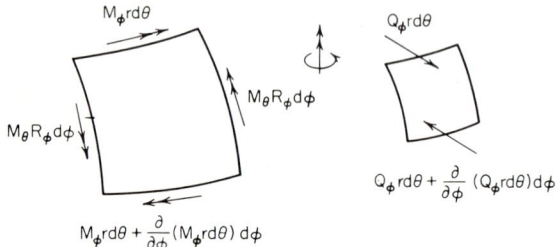

figure 1-13   *Geometrical meaning of term $Q_\phi r \, d\theta \, d\phi$.*

The increment $(\partial/\partial\phi)(Q_\phi r \, d\theta) \, d\phi$ is negligible, compared with $Q_\phi r \, d\theta$

(b) $$\sum F_y = 0$$

Introduction to the Theory of Shells 13

This condition is satisfied by itself because of the symmetry, and no equation is contributed.

(c) $$\sum F_z = 0$$

$$N_\phi r\, d\theta\, d\phi + N_\theta R_\phi\, d\phi\, d\theta \sin\phi + \frac{\partial}{\partial \phi}(Q_\phi r\, d\theta)\, d\phi + p_z R_\phi\, d\phi\, r\, d\theta = 0 \quad (1\text{-}8)$$

The above equation differs from the similar equation in membrane theory only by the third term.

(d) $$\sum M_x = 0$$

Because of the symmetry this condition is satisfied by itself, and no useful equation is introduced.

(e) $$\sum M_y = 0$$

$$-M_\phi r\, d\theta + M_\phi r\, d\theta + \frac{\partial}{\partial \phi}(M_\phi r\, d\theta)\, d\phi - Q_\phi r\, d\theta\, R_\phi\, d\phi + M_\theta R_\phi\, d\phi\, \frac{r\, d\theta}{R_\theta \tan \phi} = 0$$

(1-9)

The differential increment of Q is negligible. In the above equation an additional term was added. The last term in this equation is the contribution of $M_\theta R_\phi\, d\phi$. This moment vector has a contribution to the sum moment about the y axis for the same reason $N_{\theta\phi}$ contributed to the sum of the forces in the y direction (Figs. 1-12 and 1-9).

(f) $$\sum M_z = 0$$

Because of the rotational symmetry, this equation is satisfied by itself, and no new equation results.

The equations of equilibrium can be modified and simplified by dividing out $d\phi\, d\theta$ and substituting with $R_\theta \sin \phi$. The equilibrium equations can finally be written as

$$\frac{d}{d\phi}(N_\phi R_\theta \sin \phi) - N_\theta R_\phi \cos \phi - Q_\phi R_\theta \sin \phi + p_x R_\phi R_\theta \sin \phi = 0$$

$$N_\phi R_\theta \sin \phi + N_\theta R_\phi \sin \phi + \frac{d}{d\phi}(Q_\phi R_\theta \sin \phi) + p_z R_\phi R_\theta \sin \phi = 0 \quad (1\text{-}10)$$

$$\frac{d}{d\phi}(M_\phi R_\theta \sin \phi) - Q_\phi R_\phi R_\theta \sin \phi + M_\theta R_\phi \cos \phi = 0$$

We have five unknowns:

$$N_\phi,\ N_\theta,\ Q_\phi,\ M_\phi,\ M_\theta$$

14 Structural Analysis of Shells

and only three equations (1-10), which means that the problem is statically indeterminate. Additional equations will be developed which include the elastic properties of the material.

1-9 Elastic Laws

Geometrical Relations between Deformations

The strains of the middle surface of the shell and the angle of rotation of the tangent to the meridian are now considered. Designate:

ϵ_ϕ = strain in the meridional direction of the shell middle surface (θ = const)

ϵ_θ = strain in the hoop direction of the shell middle surface (ϕ = const)

z = distance to any shell fiber from shell middle surface measured along a normal to the middle surface (on middle surface $z = 0$)

$\epsilon_{\phi_z}, \epsilon_{\theta_z}$ = strain in the meridional and hoop directions, respectively, of a surface which is parallel to the shell middle surface and located at distance z from the middle surface

β = rotation of a meridional tangent due to deformations

These deformations are not independent of each other.

Figure 1-14 shows an undeformed and a deformed shell element of meridian. The following changes occurred:

$$R_\phi \rightarrow R_\phi + \Delta R_\phi$$
$$R_\theta \rightarrow R_\theta + \Delta R_\theta$$
$$\phi \rightarrow \phi + \beta$$

A fiber of the shell element which is at a distance z from the neutral surface is shown by a thick line in Fig. 1-14. The length of this fiber before and after deformation is

$$(R_\phi - z)\, d\phi \rightarrow (R_\phi + \Delta R_\phi - z)\, d(\phi + \beta)$$

The corresponding strain is

$$\begin{aligned}
\epsilon_{\phi_z} &= \frac{(R_\phi + \Delta R_\phi - z)\, d(\phi + \beta) - (R_\phi - z)\, d\phi}{(R_\phi - z)\, d\phi} \\
&= \frac{R_\phi\, d\phi + \Delta R_\phi\, d\phi - z\, d\phi + R_\phi\, d\beta + \Delta R_\phi\, d\beta - z\, d\beta - R_\phi\, d\phi + z\, d\phi}{(R_\phi - z)\, d\phi} \\
&= \frac{\Delta R_\phi\, d\phi + R_\phi\, d\beta + \Delta R_\phi\, d\beta - z\, d\beta}{(R_\phi - z)\, d\phi} \\
&= \frac{\Delta R_\phi\, d\phi}{(R_\phi - z)\, d\phi} + \frac{(R_\phi - z)\, d\beta}{(R_\phi - z)\, d\phi} + \frac{\Delta R_\phi\, d\beta}{(R_\phi - z)\, d\phi}
\end{aligned}$$

The last term is negligible. Therefore

$$\epsilon_{\phi_z} = \frac{\Delta R_\phi}{R_\phi - z} + \frac{d\beta}{d\phi} \tag{1-11}$$

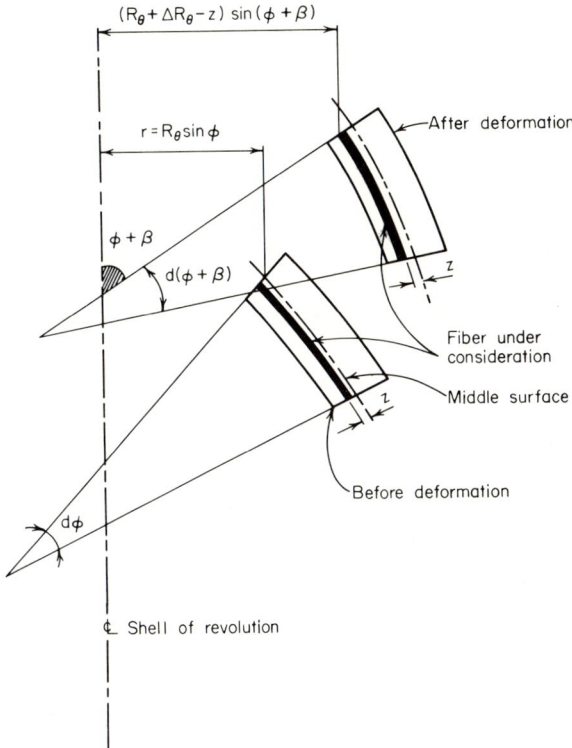

figure 1-14 *Meridional section.*

For the middle surface $z = 0$:

$$\epsilon_\phi = \frac{\Delta R_\phi}{R_\phi} + \frac{d\beta}{d\phi} \tag{1-12}$$

From (1-11) and (1-12) ΔR_ϕ can be eliminated, and the following relation is obtained:

$$\epsilon_{\phi_z} = \frac{R_\phi}{R_\phi - z}\epsilon_\phi - \frac{z}{R_\phi - z}\frac{d\beta}{d\phi} \tag{1-13}$$

Similarly the strain in the hoop direction is

$$\epsilon_{\theta_z} = \frac{(R_\theta + \Delta R_\phi - z)\sin(\phi + \beta)\,d\theta - (R_\theta - z)\sin\phi\,d\theta}{(R_\theta - z)\sin\phi\,d\theta} \tag{1-14}$$

16 Structural Analysis of Shells

The deformation can be assumed small; consequently $\cos \beta \approx 1$, $\sin \beta \approx \beta$. Therefore,

$$\sin(\phi + \beta) = \sin \phi \cos \beta + \cos \phi \sin \beta = \sin \phi + \beta \cos \phi$$

$$\Delta R_\theta \beta \approx 0$$

Consequently the formula for ϵ_{θ_z} reduces to

$$\epsilon_{\theta_z} = \frac{\Delta R_\theta}{R_\theta - z} + \beta \cot \phi$$

$$\epsilon_\theta = \frac{\Delta R_\theta}{R_\theta} + \beta \cot \phi$$

and after elimination of ΔR_θ:

$$\epsilon_{\theta_z} = \frac{R_\theta}{R_\theta - z} \epsilon_\theta - \frac{z}{R_\theta - z} \beta \cot \phi \tag{1-15}$$

An additional relation of importance is obtained from geometry (Fig. 1-15):

$$\cos \phi = \frac{dr}{R_\phi \, d\phi}$$

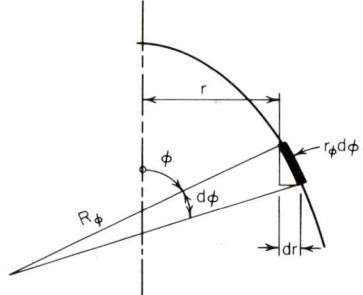

figure 1-15 *Geometrical explanation of equation* $\cos \phi = dr/R_\phi \, d\phi$.

After deformation the values ϕ, r, and $R_\phi \, d\phi$ become

$$\phi \to \phi + \beta$$

$$r \to (1 + \epsilon_\theta) r$$

$$R_\phi \, d\phi \to (1 + \epsilon_\phi) R_\phi \, d\phi$$

The radius r and the circumference of the parallels are elongated by the same ϵ_θ; consequently the following equation can be modified:

$$\cos(\phi + \beta) = \frac{1}{(1 + \epsilon_\phi) R_\phi} \frac{d}{d\phi} (1 + \epsilon_\theta) r$$

$$(1 + \epsilon_\phi)(\cos \phi \cos \beta - \sin \phi \sin \beta) = \frac{r}{R_\phi} \frac{d}{d\phi} (1 + \epsilon_\theta) + \frac{1 + \epsilon_\theta}{R_\phi} \frac{dr}{d\phi}$$

Considering that the deformations are small and also $r = R_\theta \sin \phi$,

$$-\beta \sin \phi + \epsilon_\phi \cos \phi = \frac{r}{R_\phi} \frac{d\epsilon_\theta}{d\phi} + \frac{\epsilon_\theta}{R_\phi} \frac{dr}{d\phi} = \frac{R_\theta}{R_\phi} \frac{d\epsilon_\theta}{d\phi} \sin \phi + \epsilon_\theta \cos \phi$$

$$\boxed{\beta = (\epsilon_\phi - \epsilon_\theta) \cot \phi - \frac{R_\theta}{R_\phi} \frac{d\epsilon_\theta}{d\phi}} \qquad (1\text{-}16)$$

Hooke's Law

Hooke's law relates the strains ϵ_{ϕ_z}, ϵ_{θ_z}, with the corresponding stresses, σ_ϕ and σ_θ

$$\epsilon_{\phi_z} = \frac{1}{E} (\sigma_\phi - \mu \sigma_\theta) \qquad \epsilon_{\theta_z} = \frac{1}{E} (\sigma_\theta - \mu \sigma_\phi) \qquad (1\text{-}17)$$

or

$$\sigma_\phi = \frac{E}{1 - \mu^2} (\epsilon_{\phi_z} + \mu \epsilon_{\theta_z})$$

$$\sigma_\theta = \frac{E}{1 - \mu^2} (\epsilon_{\theta_z} + \mu \epsilon_{\phi_z}) \qquad (1\text{-}18)$$

where E is Young's modulus of elasticity and μ is Poisson's ratio for the shell material. The influence of the stress and strain in the third direction is small and will not be considered.

Consequently the equations for stresses can be rewritten

$$\sigma_\phi = \frac{E}{1 - \mu^2} \left(\epsilon_\phi + \mu \epsilon_\theta - \frac{z}{R_\phi} \frac{d\beta}{d\phi} - \mu \frac{z}{R_\theta} \beta \cot \phi \right)$$

$$\sigma_\theta = \frac{E}{1 - \mu^2} \left(\epsilon_\theta + \mu \epsilon_\phi - \frac{z}{R_\theta} \beta \cot \phi - \mu \frac{z}{R_\phi} \frac{d\beta}{d\theta} \right) \qquad (1\text{-}19)$$

18 Structural Analysis of Shells

Internal Loads

The internal forces N_i and M_i can be determined by integrating the stress through the shell thickness where $i = \phi, \theta$. Figure 1-16 shows the necessary geometrical relationship. Then

$$N_\phi r \, d\theta = \int_{-t/2}^{+t/2} \sigma_\phi (R_\theta - z) \frac{r}{R_\theta} \, dz \, d\theta$$

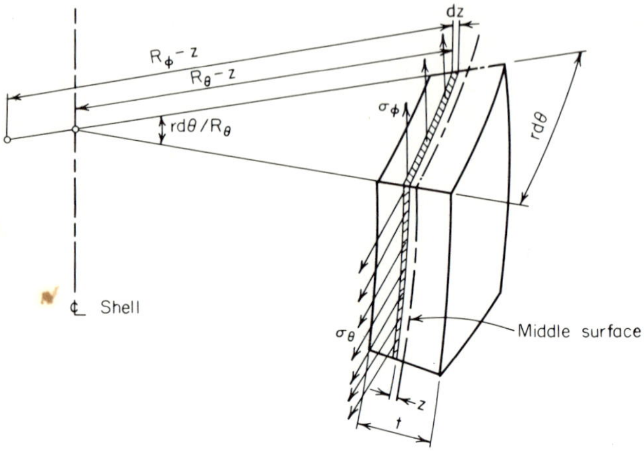

figure 1-16 *Stresses in shell element.*

N_ϕ has the dimensions (force/unit length); z is small in comparison with R_θ and may be neglected.

Consequently, the stress resultant can be expressed as follows:

$$N_\phi = \int_{-t/2}^{+t/2} \sigma_\phi \, dz \tag{1-20}$$

Similarly

$$N_\theta = \int_{-t/2}^{+t/2} \sigma_\theta \, dz$$

$$M_\phi = \int_{-t/2}^{+t/2} \sigma_\phi z \, dz$$

$$M_\theta = -\int_{-t/2}^{+t/2} \sigma_\theta z \, dz$$

Note the minus sign in the last equation. Positive σ and positive z results in negative M. From Fig. 1-12, this is self-explanatory (direction of moment vector).

Introduction to the Theory of Shells 19

The expressions for σ_ϕ and σ_θ from Eq. 1-19 can be entered into Eq. 1-20 for the stress resultants, and the integration with respect to z can be performed remembering that

$$\int_{-t/2}^{+t/2} dz = t \qquad \int_{-t/2}^{+t/2} z\, dz = 0 \qquad \int_{-t/2}^{+t/2} z^2\, dz = \frac{t^3}{12}$$

With this definition the following formulas will finally be obtained:

$$\boxed{\begin{aligned} N_\phi &= B(\epsilon_\phi + \mu\epsilon_\theta) \\ N_\theta &= B(\epsilon_\theta + \mu\epsilon_\phi) \\ M_\phi &= -D(\chi_\phi + \mu\chi_\theta) \\ M_\theta &= D(\chi_\theta + \mu\chi_\phi) \end{aligned}} \qquad (1\text{-}21)$$

where

$$\chi_\phi = \frac{1}{R_\phi}\frac{d\beta}{d\phi}$$

$$\chi_\theta = \frac{\beta \cot \phi}{R_\theta}$$

and B and D are defined as

$$B = \frac{Et}{1-\mu^2} = \text{extensional rigidity (extensional stiffness)}$$

$$D = \frac{Et^3}{12(1-\mu^2)} = \text{flexural rigidity (flexural stiffness)}$$

The stresses may be obtained in terms of the stress resultants from the following equations:

$$\sigma_\phi = \frac{E}{1-\mu^2}\left(\frac{N_\phi}{B} + \frac{M_\phi}{D}z\right) = \frac{N_\phi}{t} + \frac{M_\phi}{t^3/12}z \qquad (1\text{-}22)$$

$$\sigma_\theta = \frac{E}{1-\mu^2}\left(\frac{N_\theta}{B} - \frac{M_\theta}{D}z\right) = \frac{N_\theta}{t} - \frac{M_\theta}{t^3/12}z \qquad (1\text{-}23)$$

We can see that theoretically the problem of an axisymmetrically loaded shell of revolution can now be solved using bending theory. We have now eight unknowns:

$$N_\phi,\, N_\theta,\, Q_\phi,\, M_\phi,\, M_\theta,\, \epsilon_\phi,\, \epsilon_\theta \text{ and } \beta$$

20   *Structural Analysis of Shells*

To find the unknowns, we have also eight equations:

(1-10) 3
(1-16) 1
(1-21) 4
            ―
            8

Theoretically, with these eight equations and the boundary conditions, we can solve for the stresses and deflections at any point in the shell. However, practically speaking, this task is almost insurmountable for the general case, and solutions have been obtained only for specific cases. These solutions are presented in the following chapters. Some approximations which have been made in order to obtain solutions for different shells will be discussed in more detail later for the specific cases presented.

1-10   Classification of Shell Theories

In the preceding sections the basic relations for shells were developed either from statics or from purely geometrical considerations. As in the theory of elasticity, a relationship for connecting the geometric and static phenomena is presented by the introduction of a generalized Hooke's law.

The physical hypothesis expressed by these relations is sufficient for the description of the state of deformation or stress in the shell. To be able to establish a connection between forces, moments, and deformation components of the middle surface, it is necessary to know how either the stresses or strains vary across the shell thickness. This situation arises from attempts to reduce the shell problem from a three-dimensional elasticity problem to a two-dimensional one.

The problem can be resolved to one of arbitrarily choosing quantities to represent the state of deformation in the shell. The introduction of certain assumptions permits the evaluation of stress-resultant equations, thereby rendering approximate relationships between force and deformations.

The selection of the proper form of these approximations has been the subject of considerable controversy among the many investigators in the field. As a result, a large number of general and specialized thin-shell theories exist, developed within the framework of linear elasticity. It will be desirable in the subsequent discussion to consider the most commonly encountered theories and classify them according to the assumptions on which they are based.

For the purpose of discussion, the various linear shell theories will be classified into four basic categories:

1. First-order-approximation shell theory
2. Second-order-approximation shell theory
3. Specialized theories for shells
4. Membrane shell theory

The order of a particular approximate theory will be established by the order of the terms in the thickness coordinate that are retained in the strain and constitutive equations.

In the case of thin shells, the simplified bending theories of shells are (in general) based on Love's first-approximation and second-approximation shell theories. Although some theories do not strictly adhere to Love's original approximations, they can be considered as modifications thereof and as either first- or second-order-approximation theories. Linear membrane theory is understood to be the limiting case corresponding to a zero-order approximation or momentless state.

First-order-approximation Shell Theory

Love was the first investigator to present a successful approximation shell theory based on classical elasticity. To simplify the strain-displacement relationships and, consequently, the constitutive relations, Love (Ref. 1-1) introduced the following assumptions, known as first approximations and commonly termed the Kirchhoff-Love hypothesis:

1. The shell thickness t is negligibly small in comparison with the least radius of curvature R_{min} of the middle surface; i.e., $t/R_{min} \ll 1$ (therefore, terms $z/R \ll 1$).
2. Linear elements normal to the unstrained middle surface remain straight during deformation and suffer no extensions.
3. Normals to the undeformed middle surface remain normal to the deformed middle surface.
4. The component of stress normal to the middle surface is small compared with other components of stress and may be neglected in the stress-strain relationships.
5. Strains and displacements are small so that quantities containing second- and higher-order terms are neglected in comparison with first-order terms in the strain equations.

The last assumption is consistent with the formulation of the classical theory of linear elasticity. The other assumptions will be used to simplify the elasticity relations.

Assumption 2 of Love's first approximation is analogous to Navier's hypothesis in elementary beam theory which requires that plane sections remain plane. The bending theory presented in Secs. 1-8 and 1-9 is a first order of approximation.

Second-order-approximation Theories for Shells of Revolution

The second-order-approximation theory of Flügge (Ref. 1-2) and Byrne (Ref. 1-3) retains the z/R terms in comparison with unity in the stress-resultant equations and in the strain-displacement relations. Flügge-Byrne-type equations for a general shell are discussed by Kempner (Ref. 1-4), who obtains them as a special case of a unified thin-shell theory. Applications of this second-order-approximation theory have generally been restricted to circular cylindrical shapes, for which case solutions are obtained in Refs. 1-3 and 1-5. In the latter reference, the Flügge-Byrne-type equations are considered as a standard against which simplified first-order-approximation theories are compared.

Second-order-approximation equations are derived by Vlasov (Ref.1-6) directly from the general three-dimensional linear elasticity equations for a thick shell. The assumption $\epsilon_z = \gamma_{\phi z} = \gamma_{\phi\theta} = 0$ is made, where $\gamma_{\phi z}$ and $\gamma_{\phi\theta}$ are transverse shear strains, and the remaining strains are represented by the first three terms of their series expansion. The assumption of zero normal strain as well as zero transverse shear strains permits a rapid transition from the three-dimensional theory to the two-dimensional equations of shell theory, but it should not be interpreted in its strict sense as implying a state of plane strain. Rather, it is a convenient assumption equivalent to the basic Kirchhoff-Love hypothesis that normal lines remain normal and their extensions are negligible. An excellent discussion of this assumption is given by Novozhilov (Ref. 1-7).

Special Shell Theories

Many theories have been developed to include features of special types of shells. Examples are (1) shallow-shell theory, (2) shell theory including shear deformation, and (3) Geckeler's approximation for symmetrically loaded shells. Most of these theories are based on Love's first-order approximation.

Membrane Theory of Shells

The shell theories studied in the previous sections are generally referred to as "bending" theories of shells because this development includes the consideration of the flexural behavior of shells. If, in the study of equilibrium of a shell, all moment expressions are neglected, the resulting theory is the so-called "membrane" theory of shells.

A shell can be considered to act as a membrane if flexural strains are zero or negligible compared with direct axial strains. It is apparent that two types of shells comply with this definition of a membrane: (1) shells with bending stiffness sufficiently small so that they are physically incapable of resisting bending and (2) shells that are flexurally stiff but loaded and supported in a manner that avoids the introduction of bending strains. The state of stress in a membrane is referred to as a "momentless" state of stress. For an absolutely flexible shell, since it offers no resistance to bending, only a momentless state of stress is possible. For shells with finite stiffness, such a state of stress is only one of the possible stress conditions, and for a momentless state, several supplementary conditions relating to the shape of the shell, the character of the load applied, and the support of its edges must be fulfilled.

Thin shells adapt themselves badly to bending, so that relatively small bending moments generate considerable stresses and deflections. Therefore, the pure bending stress condition is to be avoided and is technically disadvantageous to shells. The momentless state of stress condition is a desirable feature in the design of shell structures because it offers the advantage of uniform utilization of the strength capabilities of the shell material, in most cases using less material and thus resulting in less weight. The study of membrane theory is considerably simpler than that of bending theory and for this reason historically preceded the latter theory. The first contributions to membrane theory were furnished by Lamé and Clapeyron early in the nineteenth century. These works considered symmetrical loading on shells of revolution. On the assumption that no moments could exist in the shell, the loading could produce only in-plane forces. On this basis, the calculation of the shell could be "statically determined" (i.e., the analysis could be performed solely with the force-equilibrium equations without the need of the deformation relations). Additional discussion of this subject may be found in References 1-7, 1-8, 1-9, and 1-10.

1-11 Nonlinear Shell Theory

The small-deflection equations presented earlier were formulated from the classical linear theory of elasticity. It is known that these equations, which are based on Hooke's law and the omission of nonlinear terms in both the equations for strain components and the equilibrium equations, have a unique solution in every case. In other words, linear shell theory determines a unique position of equilibrium for every shell with prescribed load and constraints.

In reality, however, the solution of a physical shell problem is not always unique. A shell under identical conditions of loading and constraints may have several possible positions of equilibrium. The incorrect inference to which linear shell theory leads can be explained by the approximations introduced in the development of the shell equations. In this development rotations were neglected in the expressions for strains and equilibrium in order that the equations could be linearized. It is essential in the investigation of the multiple equilibrium states of a shell to include these rotation terms.

A theory of shells that is free of this hypothesis can be thought of as being "geometrically nonlinear" and requires formulation on the basis of the nonlinear elasticity theory. Additionally, the shell may be "physically nonlinear" with respect to the stress-strain relations.

Theories based on nonlinear elasticity are required in analyzing the so-called "large" deformations of shells. "Large" or finite-deflection shell theories form the basis for the investigation of the stability of shells. In the case of stability, the effects of deformation on equilibrium cannot be ignored.

REFERENCES

1-1. Love, A. E. H.: *A Treatise on the Mathematical Theory of Elasticity*, 4th ed., Dover Publications, Inc., New York, 1944.
1-2. Flügge, W.: *Statik und Dynamik der Schalen*, Springer-Verlag OHG, Berlin, 1957.
1-3. Byrne, Ralph, Jr.: Theory of Small Deformations of a Thin Elastic Shell, Seminar Reports in Mathematics, University of California at Los Angeles, *Math.* N.S., vol. 2, no. 1, pp. 103–152, 1944.
1-4. Kempner, J.: *Unified Thin-shells Theory*, Symposium on the Mechanics of Plates and Shells for Industry Research Associates, Polytechnic Institute of Brooklyn, PIBAL no. 566, Mar. 9–11, 1960.
1-5. Kempner, J.: Remarks on Donnell's Equations, *J. Appl. Mech.*, vol. 22, no. 1, March, 1955.
1-6. Vlasov, V. Z.: *General Theory of Shells and Its Applications in Engineering*, NASA Technical Translations, NASA TTF-99, 1949.
1-7. Novozhilov, V. V.: *The Theory of Thin Shells*, Erven P. Noordhoff, Ltd., Groningen, Netherlands, 1959.
1-8. Goldenveiser, A. L.: *Theory of Elastic Thin Shells*, Pergamon Press, New York, 1961.
1-9. Timoshenko, S. P., and S. Woinowsky-Krieger: *Theory of Plates and Shells*, 2d ed., McGraw-Hill Book Company, New York, 1959.
1-10. Koiter, W. T.: The Theory of Thin Elastic Shells, *Proc. Symp. IUTAM Delft 24–28 August, 1959*, North-Holland Publishing Company, Amsterdam, 1960.

Chapter 2

FORCE METHOD

2-1 Introduction

Chapter 1 defined the structural shell and described several classical shell theories. The many characteristics of shells, such as the thickness-to-radius-of-curvature ratio, material behavior, type of construction (e.g., honeycomb sandwich or ring-stiffened shells), types of loadings, and other factors, all play a role in establishing which theory is applicable in any particular case. Furthermore, shallow as distinct from nonshallow shells require different approaches even though they fall into the same thin-shell theory.

The membrane theory, as will be shown in this chapter, has a significant limitation. The bending theories are more exact, but generally speaking, they are very long and too time-consuming to be used in everyday analysis. Both theories, however, can be combined into a practical engineering method, called the "force method," which eliminates limitations of both theories and makes it possible to analyze complicated shells in a relatively short time. The following definitions are made:

26 *Structural Analysis of Shells*

1. The force method is an analytical procedure in which the deflection relationships of the shell are expressed in terms of the redundant edge loads and/or moments. Solution for the unknown redundants leads to the solution for the statically loaded shell. In this text the solution is obtained by the superposition of the primary and secondary solutions, which are dependent on corrective edge loadings.

2. A primary solution is a solution of a shell in which loads are applied to the surface of the shell and the boundary conditions are consistent with the assumptions of membrane theory.

3. A secondary solution is a solution of a shell in which uniform loads (moments, shears) are applied only to the edges of the shell and the edges are assumed to be free. Secondary solutions are usually obtained with the bending theory.

4. Unit edge loadings are equally distributed loadings along the boundary of the shell and are of unit intensity. The unit loads are moments, vertical loads, and horizontal loads.

5. Corrective edge loadings are unit edge loadings multiplied by certain factors in order to close the discontinuity in deformations and to return the shell to the condition prescribed by the actual boundary conditions.

Before the procedure is presented in this chapter, some additional aspects, needed for a clear understanding of theory, will be discussed.

In this chapter, nonshallow shells of revolution are considered. The resulting differential equations for nonshallow shells have solutions which will be tabulated later and used for the solution of simple and complex rotationally symmetric loads. There are certain restraining conditions along the edges of the structural shell, called edge restraints. The edge restraints are enforced by additional shears and moments along the boundary. These loads introduce bending and shear distortions in shells in addition to the membrane state of strains which in the majority of cases would exist in the shell if the edge restraints were not present. The final state of stresses is the result of the superposition of corresponding stresses (due to the membrane loading and corrective edge loadings).

2-2 Geometrical Considerations of Shell Segments

A shell or a combination of shells such as that shown in Fig. 2-1, having the characteristics of (1) a nonshallow thin shell of revolution, (2) a rotationally symmetrically loaded shell, and (3) a rotationally symmetrical distribution of materials, will be treated. In addition, the described procedure is limited to "thin" shells.

A thin shell is defined as a shell that conforms to the Navier hypothesis and the Bernoulli-Euler theory of bending. A basic assumption in this theory is that a normal plane section before bending remains a normal plane section after bending, without extension. A characteristic of a nonshallow shell is that for many loadings the bending moments exist only in the neighborhood of the edge of the shell or in the area where a concentrated load is applied.

Novozhilov (Ref. 2-1) recommends the criterion that a thin shell be defined as a shell where the ratio t/R (where t is the thickness and R is the average radius of curvature) can be neglected in comparison with unity. If this ratio is not small in comparison with unity, the theory of the so-called "thick" shell may have to be used. The division into thin and thick shells is still artificial and arbitrary unless those values which are negligible in comparison with unity are defined. For example, if it is assumed that the usual error of 5 percent is permissible, then the range of thin monocoque shells will generally be dictated by the relation $t/R < 1/20$. The great majority of monocoque shells commonly used in practice are in the $1/1,000 < t/R < 1/50$ range, which means that they belong to the thin-shell family. However, as was noted above, the division into thin and thick shells is arbitrary and depends on the degree of accuracy required for the solution of the problem, as well as the type of problem being investigated.

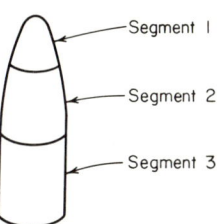

figure 2-1
Combination of shells.

Furthermore, only thin shells with small deflections in the elastic range will be discussed in this chapter (i.e., the deflection of the shell must be small in comparison with the wall thickness). After the analysis is conducted, the result should be qualified to ensure that the deflection is small in comparison with the wall thickness. Loads and material restrictions are such that the laws of the linear theory of elasticity are applicable. Stresses must be below the proportional limit of the material.

2-3 Membrane Solution

Generally, a shell is a statically indeterminate structure. There are six equations of equilibrium, which are derived from the three force and three moment equilibrium conditions (see Chap. 1), but 10 unknowns that make the problem internally statically indeterminate.

The elaborate calculation of statically indeterminate values in shell analysis may be bypassed with the help of an approximate method that can lead to useful results for many cases in practice. This method is called the membrane theory. Its justification and success are closely

28 Structural Analysis of Shells

connected with the interplay of forces in curved-surface structures, as explained in the introduction.

Of the 10 unknown forces acting on the differential element of a shell (two bending moments, two torsional moments, two normal shears, two inplane shears, and two inplane loads), only four are of any significance in many problems. Consequently, a simplified theory was developed which assumes that normal shears, bending moments, and twisting moments are negligibly small compared with other terms; hence they are set equal to zero.

The membrane theory for shells assumes that the basic resistance of the shell to load is by inplane tension, compression, and shear. The membrane theory is applicable only if boundary conditions are compatible with conditions of equilibrium such as the example shown in Fig. 2-2. It is noted in Fig. 2-3 that the concentrated loads normal to the middle surface are not compatible with the membrane theory because the membrane forces in the shell cannot react the concentrated load.

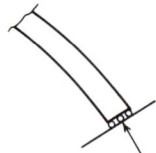

figure 2-2 *Boundary conditions compatible with the membrane theory.*

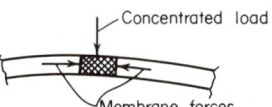

figure 2-3 *Disequilibrium due to concentrated load.*

2-4 Bending Theory

The bending theory, in which all stresses including vertical shear, bending, and twisting are considered, is more general and exact than the membrane theory. Unfortunately, as shown in the discussion in Chap. 1, this method is much more elaborate. However, in certain instances, this theory can be simplified when applied to rotationally symmetric geometries subjected to rotationally symmetric loads.

For a better understanding of the shell-bending action, the following analogies can be given: A plate supported along the edges and loaded perpendicularly to the plate surfaces is actually a two-dimensional equivalent of a beam supported at the ends and loaded perpendicularly to the beam axis. In this case the plate, like the beam, resists loads by two-dimensional bending and shear. Beams resist loads by one-dimensional bending and shear. The plate is a two-dimensional surface. A shell is also a surface but is three-dimensional. Bending is resisted by the shell in a similar manner to the plate, except that for the plate, bending is the main mechanism for resistance and for a shell it is only secondary (Ref. 2-2).

Theoretically, if one of the bending theories is used, any shell with any boundary condition may be solved. Unfortunately, because of the complexity of such analysis, the process in most cases is very difficult. It requires the solution of the systems of differential equations, which is a complicated problem in itself. Solutions can be obtained only for certain loadings and geometries. A method is needed for the practical engineer which will make possible the solution of complicated shell problems in a relatively short time. The next section will describe such a method, which is based on both membrane and bending theory, using the tabulated existing solutions for different cases of shell geometries and loadings, obtained by membrane and bending theories. In this way the derivations are bypassed.

2-5 Comparison of Membrane and Bending Theories for Nonshallow Shell

The bending theory is more general than the membrane theory because it permits the use of all possible boundary conditions. To compare the two theories, assume a nonshallow spherical shell with built-in edges subjected to some axisymmetrical loading. Assume further that two solutions are obtained, one using just membrane theory, another the bending theory. When the results of the analysis are compared, the following conclusions can be made:

1. The stresses and deformations are almost identical for all locations of the shell, with the exception of a narrow strip on the shell surface which is adjacent to the boundary. This narrow strip is generally no wider than $\sqrt{Rt}$, where R is the radius and t is the thickness of the spherical shell.

2. Except for the strip along the boundary, all bending moments, twisting moments, and vertical shears are negligible; this causes the entire solution to be practically identical to the membrane solution.

3. Disturbances along the supporting edge are very significant; however, the local bending and shear decrease rapidly along the meridian and may become negligible outside the narrow strip described in item 1.

The above comparison leads directly to the force method.

2-6 Force Method of Solution

Since the bending and membrane theories give practically the same results except for a strip adjacent to the boundary, the simple membrane

30   Structural Analysis of Shells

theory can be used to solve the problem shown in Fig. 2-4*b*; then, at the edges, the corrective moments and shears as shown in Fig. 2-4*c* can be

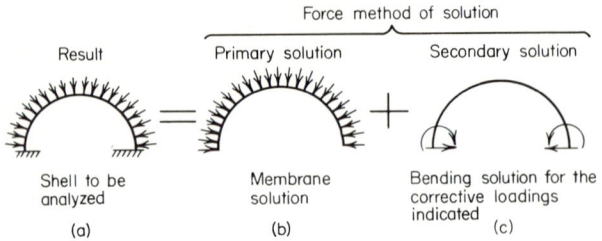

figure 2-4 *Substitution for bending theory by unit-loads solution.*

applied to bring the displaced edge of the shell into the position prescribed by boundary conditions. This is the unit-edge-force method of solving the given problem. The bending theory is used to solve the problem shown in Fig. 2-4*c*, but solutions are already available for many geometries subjected to this type of loading.

In the next section the superposition of primary and secondary solutions is clarified. It is important that only the bending theory is needed for two solutions of very special loadings at the boundaries of the shell which consist of moment and shear both axisymmetrically distributed along the boundary. Consequently, once the solutions are obtained, they can be used later without any special derivation. The results obtained from application of both theories can be superimposed, and the final results (Fig. 2-4*a*) will be almost identical to those obtained by using the exact bending theory.

2-7 Procedure for Force Method

The solution of a shell of revolution under axisymmetrical loading can be conducted in a simplified way, as described below (see also Ref. 2-3).

1. Assume that the shell under consideration is a free membrane. Obtain a solution for this membrane for the given external loads (primary loads). For many cases such solutions exist and are tabulated. Find the overall stresses and distortions of the edge and other points on the shell surface which are of interest. This is the primary solution. The primary solutions for frequently used loadings and geometries are collected and tabulated in Chap. 3. The category of primary solution is usually obtained from membrane theory, with the exception of a few cases obtained by other methods when the membrane theory fails to provide the answer.

2. Apply the following axisymmetrical edge loadings:
   a. Moment, M, in inch-pounds per inch along the edge
   b. Horizontal shear, H, in pounds per inch along the edge
   c. Vertical shear, V, in pounds per inch along the edge

These loadings should be of such magnitude as to be able to return the distorted edge of the membrane into a position prescribed by the nature of the supports (edge condition). The amount of applied corrective loading depends on the magnitude of edge deformations due to the primary solution. The exact magnitude will be determined by the interaction procedure explained in Sec. 2-9. However, to start the interaction process, formulas will be necessary for deformations at the boundaries due to the following:

   a. Unit edge moment
   b. Unit edge horizontal shear
   c. Unit edge vertical shear

These solutions will be referred to as unit edge influences or as secondary solutions, and are presented in Chaps. 4 and 6. They are presented in terms of algebraic formulas for the moments, shears, inplane forces, and deformations (displacements and rotations) due to the $M = H = V = 1$ acting along the free boundary of the shell (see Fig. 2-4c). These influences can be obtained at any point on the shell with these formulas. For interaction, however, only the deformations at the boundary are needed.

3. Having the primary and unit edge solutions, the corrective loadings M, H, and V can be obtained using the interaction procedure described in Sec. 2-9. All stresses and distortions at any point of the shell due to the loadings M, H, and V can now be determined using solutions presented in Chap. 4. V may sometimes be neglected.

4. Superposition of stresses and distortions obtained by the primary solution and corrective-loadings solution leads to the final solution.

It is noted that the most complicated step in the force method is the interaction between the structural elements. For this purpose the formulas for the stresses and deformations at the boundary due only to the action of $H = M = 1$ and the primary loading are needed. These formulas are a special case of the more general formulas for the stresses and deformations at any point of the shell. Chapters 3 and 4 will present the tabulated formulas (for different geometries and loadings) for the stresses and deformations at any point of shell. In Chap. 6 the more complicated interactions will be considered. The corresponding formulas for the special case of influence coefficients at the boundaries only will be presented again as special cases.

These formulas are shorter and better suited for the interplay of many interacting structural elements and in many cases are adequate because usually the maximum value of stresses appears at the boundaries. If the analyst is interested only in the stresses at the boundaries, the general formulas for the stresses and deformations at any point of shell are not required, and the simplified formulas given at the end of Chap. 6 can be used.

2-8 Restraints of Boundaries

The boundaries of shells sometimes are restrained elastically, which means that the edge supports permit rotation and horizontal and vertical deflections depending linearly upon the edge reaction. The special cases of restraints are so-called "free edges," which are not restricted at all and may rotate and deflect because of the application of the loading, and so-called "fixed" or "built-in" edges which are totally restricted against any deformations. These cases are only approximations of what may actually exist physically, but very often free or fixed edges are assumed in order to simplify the analysis. Almost always in reality, the boundaries are "elastically built in" or "elastically restricted." For example, a dome with a built-in ring along the boundary has restricted rotations and deflections along the boundaries. This restriction is imposed by the ring, which is deformable under the loading. Without those restrictions the edges would be "free," and under the external loading the deformations would be greater. A shell element which is built into a more complicated shell is similarly restrained by the next shell element, which is nothing but an elastic support with respect to the shell under consideration. The restraints for the "fixed boundaries" can be specified as follows:

$$\text{Rotation along boundaries } \beta = 0$$
$$\text{Horizontal deflection along boundaries } \delta_H = 0$$
$$\text{Vertical deflection along boundaries } \delta_V = 0$$

The conditions along the free or unrestrained boundaries are $M = 0$, $H = 0$, and $V = 0$. At the boundary β, δ_H, and δ_V usually are not zero. The values of deformations for an elastically restrained boundary are somewhere in between the above cases. A simply supported boundary is a special case in which the deflection is zero in a prescribed direction.

2-9 Interaction between Shells of Various Geometries

Usually, for purposes of analysis, structures are represented by a system of simple members that mechanically interact with each other. A shell can be regarded as one of these possible members. For example, pressure vessels and aerospace vehicles contain bulkhead and cylinder combinations, attached together. Consequently, a discontinuity relationship exists for each of these shell elements because of the interaction between them. Analytical methods are required to determine stresses and deflections, including the effects of the interaction.

Breakdown for Complicated Shell Geometry

Complicated shell configurations usually can be broken down into simple elements. Very often the combination of shells, circular plates, and rings must be considered. Usual shapes include spherical, elliptical, conical, conoidal, toroidal, or compound (irregular) shapes of bulkhead. Figure 2-5, for example, illustrates a compound bulkhead which consists of a spherical transition and conical shell. The analyst can consider such a system as an irregular shell and use some approximation, or he can calculate it as a compound shell, using the method of interaction.

In this section, the interaction method, which is applicable not only to monocoque shells but also to sandwich and orthotropic shells, is presented. The interacting elements are often made from different materials, and the loading can vary from element to element. The most frequently used loadings are internal or external pressure, axial tension or compression loads, and thermally induced loads.

Interaction between Two Shell Elements

The method of interaction, which is a significant step in the force method, is now described. For simplicity, the interaction between two structural elements only is chosen. The more general case of interaction of several elements is described in Chap. 6.

A shell system consisting of an internally pressurized bulkhead and cylinder is selected as an example (see Fig. 2-6). The bulkhead is

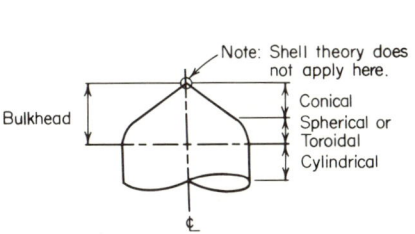

figure 2-5 *Compound bulkhead.*

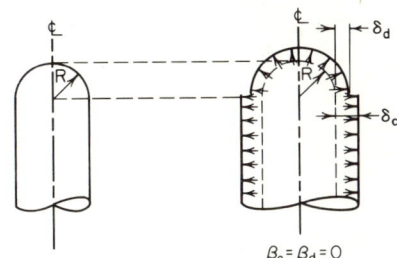

figure 2-6 *Cylindrical shell with spherical bulkhead.*

considered as a single element of some defined shape and is not subdivided into separate portions. Assume the pressurized container is separated into two main parts, the cylindrical shell and the dome, as shown in Fig. 2-6. Stresses and deformations introduced by internal pressure can be determined for each part separately, using membrane analysis. The membrane analysis (primary solution) yields the radial displacements $\Delta r = \delta_c$ and rotation β_c for the cylinder along the discontinuity line and $\Delta r = \delta_d$ and β_d for the dome along the discontinuity line. Since the structure is separated into two elements,

$$\delta_c \neq \delta_d$$

$$\beta_c \neq \beta_d$$

Consequently, there exists the discontinuity:

1. In displacement $\delta_c - \delta_d$
2. In slope $\beta_c - \beta_d$

To close this gap, unknown forces H and M will be introduced around the junction to hold the two pieces together. The displacement and rotation of the edge of the cylinder due to unit values of H and M are defined as follows:

$$\delta_{Hc}, \beta_{Hc} \quad \text{and} \quad \delta_{Mc}, \beta_{Mc}$$

The displacement and rotation of the dome edge due to unit values of M and H are

$$\delta_{Hd}, \beta_{Hd} \quad \text{and} \quad \delta_{Md}, \beta_{Md}$$

These unit deformations and unit loadings at the junctions are presented in Fig. 2-7. The value of any influence (deflection, rotation, etc.) at a

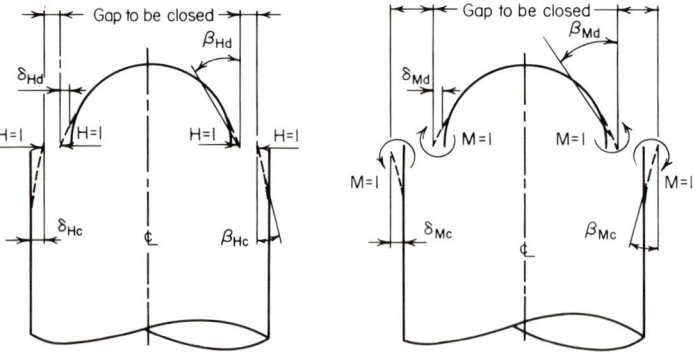

figure 2-7 *Unit deformations, unit loadings, and remaining gaps to be closed by corrective loadings M and H.*

certain point caused by the application of some unit loading (load, moment, deformation, etc.) at some point is defined as an influence coefficient at this point. Consequently, the terms δ and β are the influence coefficients at the edge of the shell due to H and M.

Generally the sign convention is arbitrary. Any rule of signs may be adopted if it is logical and is used consistently. The following sign convention is used for this illustrative example:

1. Horizontal deflection δ is positive outward.
2. Shears are positive if they cause deflection outward.
3. Moments are positive if they cause tension on the inside fibers of the shell.
4. Rotations of the meridian are positive if the rotation is in a clockwise direction, observed by watching the parallel circle in the increasing θ direction.

To close the gap, the following equations can be written:

$$\delta_d + \delta_{Hd}H + \delta_{Md}M = \delta_c - \delta_{Hc}H + \delta_{Mc}M$$

$$\beta_d - \beta_{Hd}H - \beta_{Md}M = \beta_c - \beta_{Hc}H + \beta_{Mc}M$$

which leads to the following set of equations:

$$(\delta_{Hd} + \delta_{Hc})H + (\delta_{Md} - \delta_{Mc})M = \delta_c - \delta_d$$
$$(\beta_{Hc} - \beta_{Hd})H + (-\beta_{Md} - \beta_{Mc})M = \beta_c - \beta_d \quad (2\text{-}1)$$

where H and M are the corrective edge shear and moment necessary to assure continuity of deflection and slope at the junction. All coefficients δ and β are known (usually tabulated). For many shell geometries and loadings of practical cases, these coefficients will be given in later sections of this book as algebraic formulas. Thus, the following can be indicated:

$$\delta_{Hc} + \delta_{Hd} = \delta_H \qquad \beta_{Hc} - \beta_{Hd} = \beta_H$$
$$\delta_{Mc} - \delta_{Md} = \delta_M \qquad -\beta_{Mc} - \beta_{Md} = \beta_M \quad (2\text{-}2)$$
$$\delta_c - \delta_d = \delta \qquad \beta_c - \beta_d = \beta$$

As illustrated, 12 coefficients are known in general. In the special case of interaction of a pressurized cylindrical shell and a spherical dome with a tangent intersection, the number of coefficients is reduced to 10 because

$$\beta_c = 0 \quad \text{and} \quad \beta_d = 0$$

Finally, Eq. 2-1 is reduced to a system of two equations with the two unknowns H and M:

$$\delta_H H + \delta_M M = \delta$$
$$\beta_H H + \beta_M M = \beta$$
(2-3)

The determinants of the above system are as follows:

$$D = \begin{vmatrix} \delta_H & \delta_M \\ \beta_H & \beta_M \end{vmatrix} \quad D_1 = \begin{vmatrix} \delta & \delta_M \\ \beta & \beta_M \end{vmatrix} \quad D_2 = \begin{vmatrix} \delta_H & \delta \\ \beta_H & \beta \end{vmatrix}$$

The statically indeterminate values of H and M are determined:

$$H = \frac{D_1}{D} \quad M = \frac{D_2}{D}$$

It is noted that one cut through the shell leads to two algebraic equations with two unknowns.

It is also noted that in addition to M and H, there is an axial force due to the reaction of the bulkhead distributed around the junction between the cylinder and dome, but the effect of this force on the displacement, due to M and H, is negligible if R/t is not very large. In Chap. 6 this force will be considered.

Summary

This section has presented an interaction process, to compute H and M. It can be concluded that the problem of interaction is reduced to the problem of finding the rotation β and the displacement $\Delta r = \delta$ of interacting structural elements due to the primary loadings and the secondary loadings $M = H = 1$ around the junction. The rotations and displacements will then be introduced into a set of linear equations, as shown in Eq. 2-3. The corrective loads M and H can be found when the set of equations is solved.

2-10 Determination of Stresses and Deformations at Any Point

Stresses and deformations due to the original external loading (primary load) can be obtained from the membrane solution. Using Sec. 2-9, the corrective edge loadings H and M can be determined, in accordance with the boundary conditions. For the next step of the analysis we need to know the stresses and deformations not only along the boundary but also at any location on the shell. Consequently, the δ and β must be given for all locations on the shell due to the action of $H = M = 1$

along the boundaries. These influences are known for many geometries and are tabulated in Chap. 4. For any point on the shell the corresponding values of stresses and deformations due to the unit edge loadings have to be multiplied by the actual edge loads of H and M as determined from the interaction procedure. Consequently, at every point of the shell we know:

1. The stresses and deformation of the membrane solution due to primary loading
2. The stresses and deformations due to M and H

Now in accordance with Fig. 2-4, the superposition of the solutions under 1 and 2 leads to the final solution of the stresses and deformations at any point of the shell. From the above discussion, the following generalization can be made.

For a shell of revolution the rotations and translations, designated by $\bar{\alpha}_i$, are the only deformations. These deformations may be expressed in terms of the moments and shears (M, H, acting at the boundaries) and the primary loading p. The influence coefficients at the boundary are not dependent on M, H, or p but on the elastic characteristics and geometry of shell only, because in corresponding formulas M, H, and p are multiplying factors.

The deformation at every point of shell can be determined if the influence coefficients α_{ij} at the point under consideration i are known.

Assume that there are the points $1, 2, \ldots, i, \ldots$ of interest on the shell surface and a set of forces $P_1, P_2, \ldots, P_i, \ldots$ acting on the same shell. Then the deformation at any point i can be expressed as follows:

$$\bar{\alpha}_i = \alpha_{i1} P_1 + \alpha_{i2} P_2 + \cdots + \alpha_{ij} P_j + \cdots$$

Loadings P_i can be moments, concentrated loads, distributed loads, etc. In the case of an axisymmetrical shell P_i represents the corrective edge loadings. In addition at any point i, the influence of the primary loading, say α_{i0}, must be added. Consequently, for any point i the following relation can be established:

$$\bar{\alpha}_i = \sum \alpha_{ij} P_j + \alpha_{i0}$$

In matrix form, the deformations are

$$\begin{bmatrix} \bar{\alpha}_1 \\ \bar{\alpha}_2 \\ \bar{\alpha}_3 \\ \vdots \end{bmatrix} = \begin{bmatrix} \alpha_{11} & \alpha_{12} & \alpha_{13} & \cdots \\ \alpha_{21} & \alpha_{22} & \alpha_{23} & \cdots \\ \alpha_{31} & \alpha_{32} & \alpha_{33} & \cdots \\ \vdots & \vdots & \vdots & \end{bmatrix} \begin{bmatrix} P_1 \\ P_2 \\ P_3 \\ \vdots \end{bmatrix} + \begin{bmatrix} \alpha_{10} \\ \alpha_{20} \\ \alpha_{30} \\ \vdots \end{bmatrix}$$

Result   Secondary solutions   Primary solution

38 Structural Analysis of Shells

From this relation the unknown deformations are determined. Symbolically the above relation can be written

$$\{\bar{\alpha}_i\} = [\alpha_{ij}]\{P_j\} + \{\alpha_{i0}\}$$

$[\alpha_{ij}]$ is known as the flexibility matrix.

It is not necessary for the analyst to use the matrix form, but it is convenient because of its compactness.

Similarly the stress resultants at any point on the shell

$$(N_\theta, N_\phi, M_\theta, M_\phi) = \bar{N}$$

can be determined:

$$\{\bar{N}\} = [b_{ij}]\{P_j\} + \{N_{i0}\}$$

where the first term on the right side is the contribution from the secondary solution and the second term is the contribution of the primary solution.

The above set of equations may be rewritten:

$$\{\bar{N}\} - \{N_{i0}\} = \{\bar{N} - N_{i0}\} = [b_{ij}]\{P_j\}$$

Numerical Example

This approach can be illustrated with an example. For a given shell of revolution loaded by the axisymmetrical loading as shown in Fig. 2-8

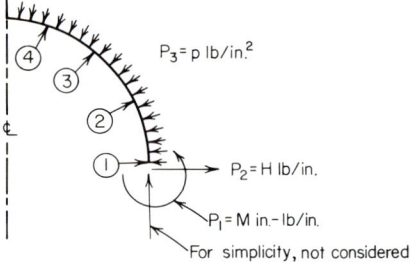

figure 2-8 *Loads on shell.*

determine the deformations (β = rotation, δ = horizontal translation) due to the loading P_i, $i = 1, 2, 3$ at the points (1), (2), (3), and (4), where

$$P_1 = M, \text{ in.-lb/in.}$$
$$P_2 = H, \text{ lb/in.}$$
$$P_3 = p, \text{ lb/in.}^2$$

Then,

$$\bar{\beta}_1 = (\beta_{11}P_1 + \beta_{12}P_2) + \beta_{10}$$
$$\bar{\delta}_1 = (\delta_{11}P_1 + \delta_{12}P_2) + \delta_{10}$$
for point (1)

$$\bar{\beta}_2 = (\beta_{21}P_1 + \beta_{22}P_2) + \beta_{20}$$
$$\bar{\delta}_2 = (\delta_{21}P_1 + \delta_{22}P_2) + \delta_{20}$$
for point (2)

Similarly for points (3) and (4).

Or expressed as a matrix:

$$\begin{bmatrix} \bar{\beta}_1 \\ \bar{\delta}_1 \\ \bar{\beta}_2 \\ \bar{\delta}_2 \\ \bar{\beta}_3 \\ \bar{\delta}_3 \\ \bar{\beta}_4 \\ \bar{\delta}_4 \end{bmatrix} = \begin{bmatrix} \beta_{11} & \beta_{12} \\ \delta_{11} & \delta_{12} \\ \beta_{21} & \beta_{22} \\ \delta_{21} & \delta_{22} \\ \beta_{31} & \beta_{32} \\ \delta_{31} & \delta_{32} \\ \beta_{41} & \beta_{42} \\ \delta_{41} & \delta_{42} \end{bmatrix} \begin{bmatrix} P_1 \\ P_2 \end{bmatrix} + \begin{bmatrix} \beta_{10} \\ \delta_{10} \\ \beta_{20} \\ \delta_{20} \\ \beta_{30} \\ \delta_{30} \\ \beta_{40} \\ \delta_{40} \end{bmatrix}$$

where $\bar{\beta}_i$, $\bar{\delta}_i$ = final deformations at points i

β_{ij}, δ_{ij} = flexibility coefficients

β_{i0}, δ_{i0} = deformations due to the primary loading

REFERENCES

2-1. Novozhilov, V. V.: *The Theory of Thin Shells*, Erven P. Noordhoff, Ltd., Groningen, Netherlands, 1959.
2-2. Borg, S. F., and J. J. Gennaro: *Advanced Structural Analysis*, D. Van Nostrand Company, Inc., Princeton, N.J.
2-3. Baker, E. H., A. P. Cappelli, L. Kovalevsky, F. L. Rish, and R. M. Verette: *Shell Analysis Manual*, NASA CR-912, April, 1968.

Chapter 3

PRIMARY SOLUTIONS

3-1 Introduction

The theoretical background of membrane theory was reviewed in Chap. 1. The set of differential equations 1-6 is adequate to determine the stresses for any membrane shell which is a surface of revolution loaded with an axisymmetric loading. In this chapter a collection of solutions for membrane shells is presented for use as primary solutions in the force method outlined in Chap. 2.

The formulas for membrane forces are valid for any type of shell construction. The formulas for deformations are valid only if the shell has uniform extensional stiffness. For types of shell construction other than homogeneous isotropic, the deformations can be obtained by replacing Et in the formulas presented in this chapter by $B(1 - \mu^2)$, where $B = B_x = B_y$ (see Chap. 7). The geometry of each shell considered is described, and the limitations of the analysis are indicated.

3-2 Determination of Membrane Internal Forces

The forces acting on the sides of the shell element are denoted by the symbols indicated in Fig. 3-1.

θ = angle in horizontal plane, which locates any point on the shell
ϕ = angle in vertical plane (measured from axis of rotation)
R_ϕ = radius of curvature of meridian at any point
R_θ = radial distance between point on the shell and the axis of rotation
$\left. \begin{array}{l} Z = p_z \\ X = p_\phi \end{array} \right\}$ = radial and meridional components of the loading, which acts on the differential element (because of assumed axisymmetrical loading, the component in circumferential direction is zero)
$\left. \begin{array}{l} N_\phi \\ N_\theta \end{array} \right\}$ = loads on meridional and circumferential side of the differential element

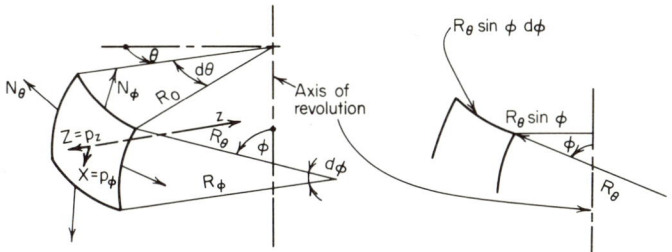

figure 3-1 *Shell-element forces.*

In general, shells of revolution are loaded by a distributed pressure on the surface in combination with a vertical loading at the vertex or around a hole at the vertex. The following solution is given for axisymmetric membrane shells loaded with an internal (or external) pressure:

$$\frac{N_\phi}{R_\phi} + \frac{N_\theta}{R_\theta} = p_z \qquad (3\text{-}1)$$

where

$$N_\phi = \frac{1}{R_\theta \sin^2 \phi} \left[\int R_\phi R_\theta (p_z \cos \phi - p_\phi \sin \phi) \sin \phi \, d\phi + C \right] \qquad \text{(Ref. 3-4)}$$

The resultant of the vertical loads applied above the circle $\phi = \phi_0$ is $2\pi C$. The angle ϕ_0 defines the opening in the shell of revolution as shown in Fig. 3-2. If the shell is closed, the resultant vertical loading degenerates to a concentrated load P at the vertex of the shell.

$$2\pi C = -P$$

where P = vertical load, lb.

If P is the only applied load, the meridional and circumferential forces are given by

$$N_\phi = \frac{P}{2\pi R_\theta \sin^2 \phi} \quad \text{and} \quad N_\theta = \frac{P}{2R_\phi \sin^2 \phi}$$

These loads may be superimposed on any other loads on the shell. If the shell is closed, local bending in the shell in addition to membrane loads will occur in the region of application of the load.

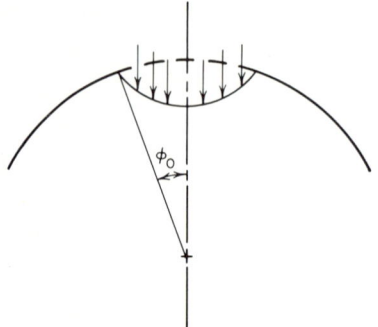

figure 3-2 *Partial loading above circle $\phi = \phi_0$.*

3-3 Determination of Membrane Displacements

It was shown that in a membrane analysis the determination of stress resultants in an axisymmetrically loaded shell of revolution is a statically determinate problem, and the membrane forces N_θ and N_ϕ are easily obtained. The displacements are needed, however, for the interaction procedure between two or more shells, which was explained in Chap. 2.

In the symmetrical deformation of a shell, a small displacement of a point due to arbitrary loading can be resolved into two components:

u in the direction of the tangent to the meridian (positive in direction of increase of ϕ)

w in the direction of the normal to the middle surface (positive in direction of increase of θ)

The strain components ϵ_ϕ and ϵ_θ are expressed in terms of the forces N_ϕ and N_θ (from Eq. 1-21):

$$\epsilon_\phi = \frac{1}{Et}(N_\phi - \mu N_\theta)$$
$$\epsilon_\theta = \frac{1}{Et}(N_\theta - \mu N_\phi)$$
(3-2)

where $E =$ Young's modulus of elasticity
$t =$ thickness of the shell
$\mu =$ Poisson's ratio

The next step is to make use of the following differential equation (Ref. 3-6):

$$\frac{du}{d\phi} - u \cot \phi = \frac{1}{Et}[N_\phi(R_\phi + \mu R_\theta) - N_\theta(R_\theta + \mu R_\phi)] = f(\phi)$$

where ϕ is the angle which locates any point on the shell middle surface along the meridian with respect to the axis of revolution.

Then the general solution for u is

$$u = \sin\phi \left[\int \frac{f(\phi)}{\sin\phi} d\phi + K \right] \qquad \text{(Ref. 3-6)}$$

where K is the constant of integration to be determined from the condition at the support. The displacement w is found from the equation (Ref. 3-6)

$$w = u \cot\phi - R_\theta \epsilon_\theta$$

because $\epsilon_\theta = (u/R_\theta) \cot\phi - w/R_\theta$, substituting the value ϵ_θ from Eq. 3-2.

Having u and w, the corresponding displacements, Δr and h, in the horizontal and vertical directions, respectively, can be found using simple trigonometric relations in connection with Fig. 3-3.

$$\Delta r = w \sin\phi + u \cos\phi$$
$$h = -w \cos\phi + u \sin\phi$$

If $\phi = 90°$, $w = -R_\theta \epsilon_\theta$, $u = (\epsilon_\theta - w/R_\theta)(R_\theta/\cot\phi) = 0$.

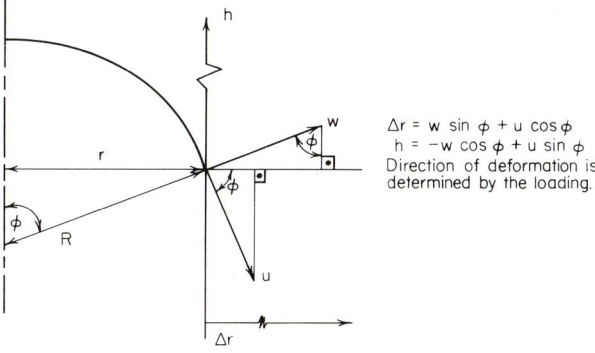

figure 3-3 *Geometric relations between displacements.*

For preliminary design for the range $45° \leq \phi \leq 90°$ the expression $w \approx -R_\theta \epsilon_\theta$ can be used as an approximation.

In practical analysis the values u and w are seldom used. Instead the horizontal movement δ and rotation β are used.

$$\delta = r_0 \epsilon_\theta = \frac{1}{Eh}(N_\theta - \mu N_\phi) r_0 = \frac{R_\theta \sin\phi}{Et}(N_\theta - \mu N_\phi)$$

$$\beta = \frac{dw}{R_\phi \, d\phi} = \frac{\cot\phi}{R_\phi} \frac{du}{d\phi} - \frac{d}{R_\phi \, d\phi}\left[\frac{R_\theta}{Et}(N_\theta - \mu N_\phi) \right]$$

At the edge $u = 0$.

$$\beta = \frac{\cot \phi}{R_\phi Et}\left[N_\phi(R_\phi + \mu R_\theta) - N_\theta(R_\theta + \mu R_\phi) - \frac{d}{R_\phi d\phi}\left(\frac{\delta}{\sin \phi}\right)\right]$$

Finally Eqs. 1-6 and 3-1 can be replaced by the simpler relation

$$N_\phi = -\frac{\text{total loading}}{2\pi a \sin \phi_i}$$

$$N_\phi + N_\theta + p_z R = 0$$

The total loading is defined as the loading above the circumference under consideration, as per Fig. 3-4. The value a is indicated on the same figure.

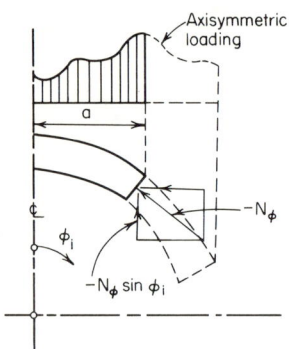

figure 3-4 *Arbitrary partial axisymmetric loading.*

3-4 Any Shape of Meridian

Table 3-1 presents a summary of the equations of linear membrane theory in a convenient form for the general case of a shell of revolution loaded axisymmetrically by a pressure loading.

TABLE 3-1 Axisymmetrically Loaded Shell of Revolution. Summary of Equations for Linear Membrane Theory.

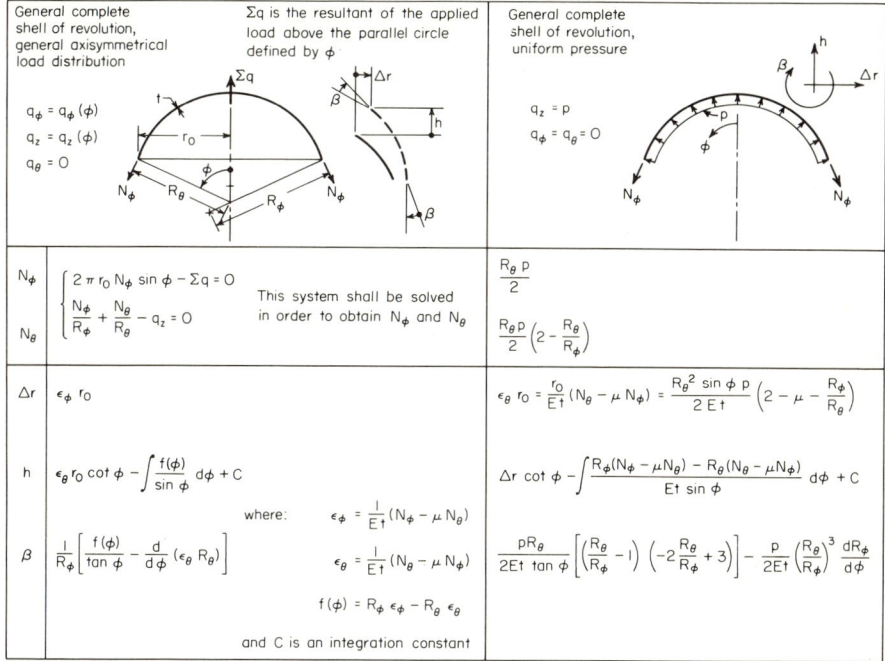

NOTE: Membrane forces are positive if they cause tension in the shell.
Positive directions for deformations are indicated.
q_z positive if oriented outward.

3-5 Spherical Shells

This section presents the solutions for nonshallow spherical shells loaded by axisymmetrical loading. Both closed and open spherical shells are considered.

Listed below are the loading cases under consideration, which are divided into the circumferential, meridional, and normal components X, Y, Z. Additional designations are indicated in the figures that correspond to each loading case. Spherical shells which satisfy the relation

$$\cot \phi = \frac{\cos \phi}{\sin \phi} \approx \frac{1}{\phi}$$

46 *Structural Analysis of Shells*

are not in the scope of this section and fall into the category of shallow shells. In order for a membrane state of stress to exist, the boundaries of the shell must be free to rotate and deflect normal to the shell middle surface. Formulas for deformation are valid only if the shell has uniform extensional stiffness.

The loading cases in Table 3-2 are considered. In Tables 3-3 and 3-4 the membrane stresses are positive if they cause tension in the shell.

Δr is positive for an increase in r.

β is positive for rotation in the clockwise direction by looking in the positive direction of meridian (with increasing angle θ).

TABLE 3-2 Loadings under Consideration

1. Dead weight	2. Uniformly distributed loading over the base (on projected area)	3. Hydrostatic pressure loading	4. Uniform loading in Z direction (pressurization)	5. Load per unit of length of the upper shell edge
$X = q \sin \phi$, $Y = 0$	$X = p \cos \phi \sin \phi$	p = Specific weight of liquid	$X = Y = 0$	$X = Y = Z = 0$
$Z = q \cos \phi$	$Y = 0$	$X = Y = 0$	$Z = p$	p = lb/in.
	$Z = p \cos^2 \phi$	$Z = p[f + R(1 - \cos \phi)]^*$		

*For reversed spherical shell, f is the distance from the surface of liquid to the apex of the reversed shell, and $Z = [f - R(1 - \cos \phi)] p$.

NOTE: The corresponding formulas for stresses and deformations are listed in tabular form for each loading case for the closed and open spherical shell.

TABLE 3-3 Stress and Deformation in Spherical Shell Membrane Solution (Ref. 3-1)

	Case A	Case B	Case C	Case D
	(full hemisphere, angle ϕ, ϕ_1)	(angle ϕ_1, ϕ)	(angles ϕ_1, ϕ_2, ϕ)	(angles ϕ_1, ϕ_2, ϕ)

1. Deadweight loading

N_ϕ	$-\dfrac{Rq}{1+\cos\phi}$	$\dfrac{Rq}{1+\cos\phi}$	$-\dfrac{Rq}{\sin^2\phi}(\cos\phi_2-\cos\phi)$	$\dfrac{Rq}{\sin^2\phi}(\cos\phi_2-\cos\phi)$
N_θ	$-Rq\left(\cos\phi-\dfrac{1}{1+\cos\phi}\right)$	$Rq\left(\cos\phi-\dfrac{1}{1+\cos\phi}\right)$	$-Rq\left[\cos\phi-\dfrac{1}{\sin^2\phi}(\cos\phi_2-\cos\phi)\right]$	$Rq\left[\cos\phi-\dfrac{1}{\sin^2\phi}(\cos\phi_2-\cos\phi)\right]$
Δr	$\dfrac{R^2 q}{Et}\sin\phi\left[-\cos\phi+\dfrac{1+\mu}{\sin^2\phi}(1-\cos\phi)\right]$	$-\dfrac{R^2 q}{Et}\sin\phi\left[-\cos\phi+\dfrac{1+\mu}{\sin^2\phi}(1-\cos\phi)\right]$	$\dfrac{R^2 q}{Et}\sin\phi\left[-\cos\phi+\dfrac{1+\mu}{\sin^2\phi}(\cos\phi_2-\cos\phi)\right]$	$-\dfrac{R^2 q}{Et}\sin\phi\left[-\cos\phi+\dfrac{1+\mu}{\sin^2\phi}(\cos\phi_2-\cos\phi)\right]$
β	$-\dfrac{Rq}{Et}(2+\mu)\sin\phi$	$\dfrac{Rq}{Et}(2+\mu)\sin\phi$	$-\dfrac{Rq}{Et}(2+\mu)\sin\phi$	$\dfrac{Rq}{Et}(2+\mu)\sin\phi$

2. Uniformly distributed loading over the base (on projected area)

N_ϕ	$-\dfrac{pR}{2}$	$\dfrac{pR}{2}$	$-\dfrac{pR}{2}\left(1-\dfrac{\sin^2\phi_2}{\sin^2\phi}\right)$	$\dfrac{pR}{2}\left(1-\dfrac{\sin^2\phi_2}{\sin^2\phi}\right)$
N_θ	$-\dfrac{pR}{2}\cos 2\phi$	$\dfrac{pR}{2}\cos 2\phi$	$-\dfrac{pR}{2}\left(2\cos^2\phi-1+\dfrac{\sin^2\phi_2}{\sin^2\phi}\right)$	$\dfrac{pR}{2}\left(2\cos^2\phi-1+\dfrac{\sin^2\phi_2}{\sin^2\phi}\right)$
Δr	$\dfrac{R^2 p}{Et}\sin\phi\left[-\cos^2\phi+\dfrac{(1+\mu)}{2}\right]$	$-\dfrac{R^2 p}{Et}\sin\phi\left(\dfrac{1+\mu}{2}-\cos^2\phi\right)$	$\dfrac{R^2 p}{Et}\sin\phi\left[-\cos^2\phi+\dfrac{1+\mu}{2}\left(1-\dfrac{\sin^2\phi_2}{\sin^2\phi}\right)\right]$	$-\dfrac{R^2 p}{Et}\sin\phi\left[-\cos^2\phi+\dfrac{1+\mu}{2}\left(1-\dfrac{\sin^2\phi_2}{\sin^2\phi}\right)\right]$
β	$-\dfrac{Rp}{Et}(3+\mu)\sin\phi\cos\phi$	$\dfrac{Rp}{Et}(3+\mu)\sin\phi\cos\phi$	$-\dfrac{Rp}{Et}(3+\mu)\sin\phi\cos\phi$	$\dfrac{Rp}{Et}(3+\mu)\sin\phi\cos\phi$

3. Hydrostatic pressure loading

N_ϕ	$-\dfrac{pR^2}{6}\left(-1+3\dfrac{f}{R}-\dfrac{2\cos^2\phi}{1+\cos\phi}\right)$	$-\dfrac{pR^2}{6}\left(-1-3\dfrac{f}{R}-\dfrac{2\cos^2\phi}{1+\cos\phi}\right)$	$-\dfrac{pR^2}{6}\left[3\left(1+\dfrac{f}{R}\right)\left(1-\dfrac{\sin^2\phi_2}{\sin^2\phi}\right)-2\dfrac{\cos^3\phi_2-\cos^3\phi}{\sin^2\phi}\right]$	$-\dfrac{pR^2}{6}\left[3\left(1-\dfrac{f}{R}\right)\left(1-\dfrac{\sin^2\phi_2}{\sin^2\phi}\right)-\dfrac{2(\cos^3\phi_2-\cos^3\phi)}{\sin^2\phi}\right]$

TABLE 3-3 Continued

N_θ	$-\frac{pR^2}{6}\left(-1+3\frac{t}{R}-\frac{4\cos^2\phi-6}{1+\cos\phi}\right)$	$\frac{pR^2}{6}\left(+1+3\frac{t}{R}+\frac{4\cos^2\phi-6}{1+\cos\phi}\right)$	$-\frac{pR^2}{6}\left[3\left(1-\frac{t}{R}\right)\left(1+\frac{\sin^2\phi_2}{\sin^2\phi}\right)-6\cos\phi\right.$ $\left.+\frac{2(2\cos^3\phi+\cos^3\phi_2)-6\cos\phi}{\sin^2\phi}\right]$
Δr	$-\frac{pR^3}{6Et}\sin\phi\left[3\left(1+\frac{t}{R}\right)(1-\mu)\right.$ $\left.-6\cos\phi-\frac{2(1+\mu)}{\sin^2\phi}(\cos^3\phi-1)\right]$	$-\frac{pR^3}{6Et}\sin\phi\left[3\left(1-\frac{t}{R}\right)(1-\mu)\right.$ $\left.-6\cos\phi-\frac{2(1+\mu)}{\sin^2\phi}(\cos^3\phi-1)\right]$	$-\frac{pR^3}{6Et}\sin\phi\left\{3\left(1-\frac{t}{R}\right)\left[1-\mu+(1+\mu)\frac{\sin^2\phi_2}{\sin^2\phi}\right]\right.$ $\left.-6\cos\phi+2(1+\mu)\frac{\cos^3\phi_2-\cos^3\phi}{\sin^2\phi}\right\}$
β	$-\frac{pR^2}{Et}\sin\phi$	$-\frac{pR^2}{Et}\sin\phi$	$-\frac{pR^3}{Et}\sin\phi$

4. Uniform loading in normal direction (pressurization)

N_ϕ	$-\frac{pR}{2}$	$-\frac{pR}{2}$	$-\frac{pR}{2}\left(1-\frac{\sin^2\phi_2}{\sin^2\phi}\right)$
N_θ	$-\frac{pR}{2}$	$-\frac{pR}{2}$	$-\frac{pR}{2}\left(1+\frac{\sin^2\phi_2}{\sin^2\phi}\right)$
Δr	$-\frac{pR^2}{2Et}(1-\mu)\sin\phi$	$-\frac{pR^2}{2Et}(1-\mu)\sin\phi$	$-\frac{pR^2}{Et}\sin\phi\left[1-\frac{1+\mu}{2}\left(1-\frac{\sin^2\phi_2}{\sin^2\phi}\right)\right]$
β	0	0	0

5. Lantern loading

N_ϕ	No case	No case	$\frac{p\sin\phi_2}{\sin^2\phi}$
N_θ			$-\frac{p\sin\phi_2}{\sin^2\phi}$
Δr			$-\frac{pR(1+\mu)}{Et}\frac{\sin\phi_2}{\sin\phi}$
β			0

TABLE 3-4 Hydrostatic-pressure Loading over Portion of Spherical Shell (Ref. 3-3)

Shell	Loading	N_ϕ	N_θ
	$Z = \rho(R - R\cos\phi - f)$	Points above the liquid level: 0 Points below the liquid level: $-\rho \dfrac{R^2}{6}\left\{\dfrac{f}{R}\left[\dfrac{1}{\sin^2\phi}\dfrac{f}{R}\left(3-\dfrac{f}{R}\right)-3\right]\right.$ $\left. +1 - \dfrac{2\cos^2\phi}{1+\cos\phi}\right\}$	Points above the liquid level: 0 $-\rho R^2\left(1-\cos\phi-\dfrac{f}{R}\right)-N_\phi$
$\phi' = 180° - \phi$	$Z = -\rho(R - R\cos\phi' - f)$	Points above the liquid level: $\rho\dfrac{f^2}{6}\left(3-\dfrac{f}{R}\right)\dfrac{1}{\sin^2\phi'}$ Points below the liquid level: $\rho\dfrac{R^2}{6}\left(3\dfrac{f}{R}-1+\dfrac{2\cos^2\phi'}{1+\cos\phi'}\right)$	$-\rho\dfrac{f^2}{6}\left(3-\dfrac{f}{R}\right)\dfrac{1}{\sin^2\phi'}$ $\rho R^2\left(\dfrac{f}{R}-1+\cos\phi'\right)-N_\phi$

NOTE: Membrane forces are positive if causing tension in the shell.

3-6 Conical Shells

This section presents the solutions for nonshallow conical shells loaded by axisymmetrical loading. Closed and open conical shells are considered. The loading cases under consideration are categorized by the circumferential, meridional, and normal components X, Y, Z.

Additional designations are indicated in the figures that correspond to each loading case. In order for the shell to be classified as a nonshallow shell, the angle α_0 must be larger than $45°$.

In order for a membrane state of stress to exist at the boundaries of the shell, the boundary must be free to rotate and deflect normal to the middle shell surface. Formulas for deformations are valid only if the shell has uniform extensional stiffness. All formulas presented are based on small-deflection membrane theory.

The loading cases in Table 3-5 are considered. In the following table, Table 3-6, the sign convention is the same as in Sec. 3-5.

TABLE 3-5 Loadings under Consideration

	Geometry	Components
(a)	Loading dead weight	$X = q \sin \alpha_0$ $Y = 0$ $Z = q \cos \alpha_0$ q = weight of shell per unit area
(b)	Uniformly distributed loading over the base	$X = p \cos \alpha_0 \sin \alpha_0$ $Y = 0$ $Z = p \cos^2 \alpha_0$ p is in lb/in.^{2}
(c)	Hydrostatic pressure loading	ρ = specific weight of liquid $X = 0$ $Y = 0$ $Z = \rho(f + x \sin \alpha_0)$ Case A $Z = \rho(f - x \sin \alpha_0)$ Case B

52 *Structural Analysis of Shells*

TABLE 3-5 Continued

		Geometry	Components
(d)		Uniform normal pressure p (psi)	$X = 0$ $Y = 0$ $Z = p$
(e)		Equally distributed loading along the opening edge (lantern load)	p is in lb/in.
(f)		Hydrostatic pressure over portion of shell	p = specific weight of liquid $X = 0$ $Y = 0$ $Z = \rho(x \sin \alpha_0 - f)$ for (A) $Z = p_z = \rho(f - x \sin \alpha_0)$ for (B)

Corresponding formulas and deformations are shown in Table 3-6 for each loading case as indicated for closed and open conical shell.

TABLE 3-6   Membrane Solutions for Conical Shell   (Ref. 3-1)

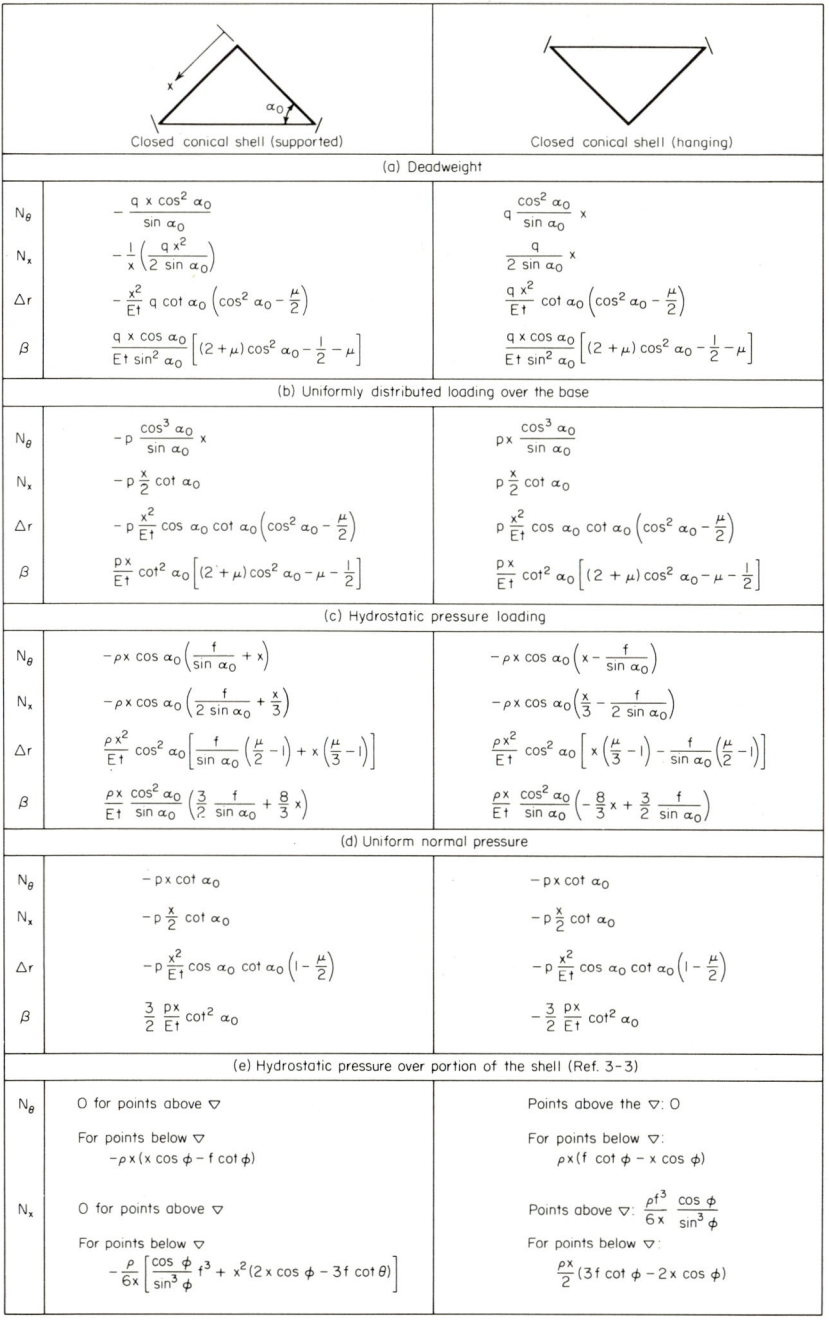

	Closed conical shell (supported)	Closed conical shell (hanging)
	(a) Deadweight	
N_θ	$-\dfrac{q\,x\,\cos^2\alpha_0}{\sin\alpha_0}$	$q\,\dfrac{\cos^2\alpha_0}{\sin\alpha_0}\,x$
N_x	$-\dfrac{1}{x}\left(\dfrac{q\,x^2}{2\sin\alpha_0}\right)$	$\dfrac{q}{2\sin\alpha_0}\,x$
Δr	$-\dfrac{x^2}{Et}\,q\,\cot\alpha_0\left(\cos^2\alpha_0 - \dfrac{\mu}{2}\right)$	$\dfrac{q\,x^2}{Et}\cot\alpha_0\left(\cos^2\alpha_0 - \dfrac{\mu}{2}\right)$
β	$\dfrac{q\,x\,\cos\alpha_0}{Et\,\sin^2\alpha_0}\left[(2+\mu)\cos^2\alpha_0 - \dfrac{1}{2} - \mu\right]$	$\dfrac{q\,x\,\cos\alpha_0}{Et\,\sin^2\alpha_0}\left[(2+\mu)\cos^2\alpha_0 - \dfrac{1}{2} - \mu\right]$
	(b) Uniformly distributed loading over the base	
N_θ	$-p\,\dfrac{\cos^3\alpha_0}{\sin\alpha_0}\,x$	$p\,x\,\dfrac{\cos^3\alpha_0}{\sin\alpha_0}$
N_x	$-p\,\dfrac{x}{2}\cot\alpha_0$	$p\,\dfrac{x}{2}\cot\alpha_0$
Δr	$-p\,\dfrac{x^2}{Et}\cos\alpha_0\cot\alpha_0\left(\cos^2\alpha_0 - \dfrac{\mu}{2}\right)$	$p\,\dfrac{x^2}{Et}\cos\alpha_0\cot\alpha_0\left(\cos^2\alpha_0 - \dfrac{\mu}{2}\right)$
β	$\dfrac{p\,x}{Et}\cot^2\alpha_0\left[(2+\mu)\cos^2\alpha_0 - \mu - \dfrac{1}{2}\right]$	$\dfrac{p\,x}{Et}\cot^2\alpha_0\left[(2+\mu)\cos^2\alpha_0 - \mu - \dfrac{1}{2}\right]$
	(c) Hydrostatic pressure loading	
N_θ	$-\rho\,x\cos\alpha_0\left(\dfrac{f}{\sin\alpha_0} + x\right)$	$-\rho\,x\cos\alpha_0\left(x - \dfrac{f}{\sin\alpha_0}\right)$
N_x	$-\rho\,x\cos\alpha_0\left(\dfrac{f}{2\sin\alpha_0} + \dfrac{x}{3}\right)$	$-\rho\,x\cos\alpha_0\left(\dfrac{x}{3} - \dfrac{f}{2\sin\alpha_0}\right)$
Δr	$\dfrac{\rho\,x^2}{Et}\cos^2\alpha_0\left[\dfrac{f}{\sin\alpha_0}\left(\dfrac{\mu}{2}-1\right) + x\left(\dfrac{\mu}{3}-1\right)\right]$	$\dfrac{\rho\,x^2}{Et}\cos^2\alpha_0\left[x\left(\dfrac{\mu}{3}-1\right) - \dfrac{f}{\sin\alpha_0}\left(\dfrac{\mu}{2}-1\right)\right]$
β	$\dfrac{\rho\,x\,\cos^2\alpha_0}{Et\,\sin\alpha_0}\left(\dfrac{3}{2}\dfrac{f}{\sin\alpha_0} + \dfrac{8}{3}x\right)$	$\dfrac{\rho\,x\,\cos^2\alpha_0}{Et\,\sin\alpha_0}\left(-\dfrac{8}{3}x + \dfrac{3}{2}\dfrac{f}{\sin\alpha_0}\right)$
	(d) Uniform normal pressure	
N_θ	$-p\,x\,\cot\alpha_0$	$-p\,x\,\cot\alpha_0$
N_x	$-p\,\dfrac{x}{2}\cot\alpha_0$	$-p\,\dfrac{x}{2}\cot\alpha_0$
Δr	$-p\,\dfrac{x^2}{Et}\cos\alpha_0\cot\alpha_0\left(1 - \dfrac{\mu}{2}\right)$	$-p\,\dfrac{x^2}{Et}\cos\alpha_0\cot\alpha_0\left(1 - \dfrac{\mu}{2}\right)$
β	$\dfrac{3}{2}\dfrac{p\,x}{Et}\cot^2\alpha_0$	$-\dfrac{3}{2}\dfrac{p\,x}{Et}\cot^2\alpha_0$
	(e) Hydrostatic pressure over portion of the shell (Ref. 3-3)	
N_θ	0 for points above ▽ For points below ▽ $-\rho x(x\cos\phi - f\cot\phi)$	Points above the ▽: 0 For points below ▽: $\rho x(f\cot\phi - x\cos\phi)$
N_x	0 for points above ▽ For points below ▽ $-\dfrac{\rho}{6x}\left[\dfrac{\cos\phi}{\sin^3\phi}f^3 + x^2(2x\cos\phi - 3f\cot\theta)\right]$	Points above ▽: $\dfrac{\rho f^3}{6x}\dfrac{\cos\phi}{\sin^3\phi}$ For points below ▽: $\dfrac{\rho x}{2}(3f\cot\phi - 2x\cos\phi)$

54 Structural Analysis of Shells

TABLE 3-6 Continued

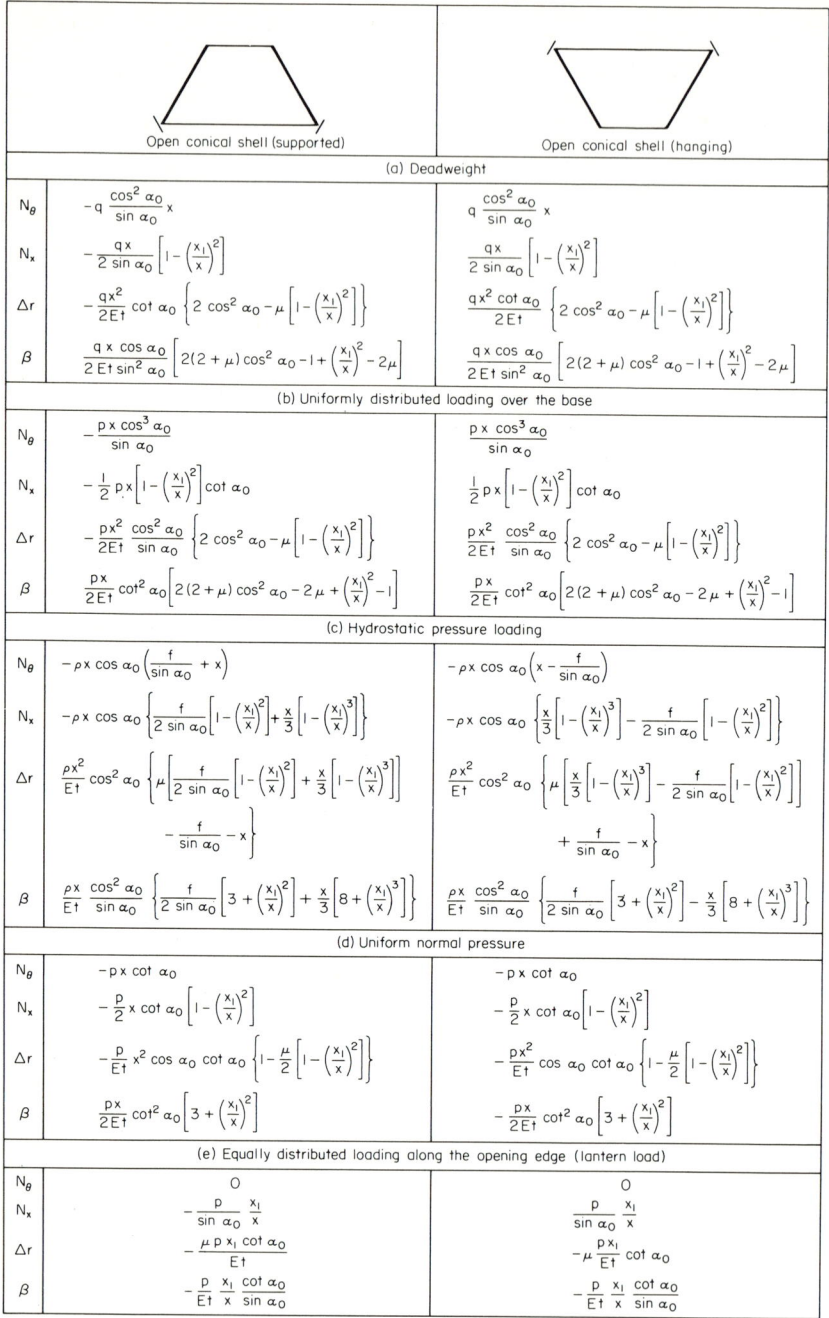

	Open conical shell (supported)	Open conical shell (hanging)
	(a) Deadweight	
N_θ	$-q\dfrac{\cos^2\alpha_0}{\sin\alpha_0}x$	$q\dfrac{\cos^2\alpha_0}{\sin\alpha_0}x$
N_x	$-\dfrac{qx}{2\sin\alpha_0}\left[1-\left(\dfrac{x_1}{x}\right)^2\right]$	$\dfrac{qx}{2\sin\alpha_0}\left[1-\left(\dfrac{x_1}{x}\right)^2\right]$
Δr	$-\dfrac{qx^2}{2Et}\cot\alpha_0\left\{2\cos^2\alpha_0-\mu\left[1-\left(\dfrac{x_1}{x}\right)^2\right]\right\}$	$\dfrac{qx^2\cot\alpha_0}{2Et}\left\{2\cos^2\alpha_0-\mu\left[1-\left(\dfrac{x_1}{x}\right)^2\right]\right\}$
β	$\dfrac{q\,x\cos\alpha_0}{2Et\sin^2\alpha_0}\left[2(2+\mu)\cos^2\alpha_0-1+\left(\dfrac{x_1}{x}\right)^2-2\mu\right]$	$\dfrac{q\,x\cos\alpha_0}{2Et\sin^2\alpha_0}\left[2(2+\mu)\cos^2\alpha_0-1+\left(\dfrac{x_1}{x}\right)^2-2\mu\right]$
	(b) Uniformly distributed loading over the base	
N_θ	$-\dfrac{px\cos^3\alpha_0}{\sin\alpha_0}$	$\dfrac{px\cos^3\alpha_0}{\sin\alpha_0}$
N_x	$-\dfrac{1}{2}px\left[1-\left(\dfrac{x_1}{x}\right)^2\right]\cot\alpha_0$	$\dfrac{1}{2}px\left[1-\left(\dfrac{x_1}{x}\right)^2\right]\cot\alpha_0$
Δr	$-\dfrac{px^2}{2Et}\dfrac{\cos^2\alpha_0}{\sin\alpha_0}\left\{2\cos^2\alpha_0-\mu\left[1-\left(\dfrac{x_1}{x}\right)^2\right]\right\}$	$\dfrac{px^2}{2Et}\dfrac{\cos^2\alpha_0}{\sin\alpha_0}\left\{2\cos^2\alpha_0-\mu\left[1-\left(\dfrac{x_1}{x}\right)^2\right]\right\}$
β	$\dfrac{px}{2Et}\cot^2\alpha_0\left[2(2+\mu)\cos^2\alpha_0-2\mu+\left(\dfrac{x_1}{x}\right)^2-1\right]$	$\dfrac{px}{2Et}\cot^2\alpha_0\left[2(2+\mu)\cos^2\alpha_0-2\mu+\left(\dfrac{x_1}{x}\right)^2-1\right]$
	(c) Hydrostatic pressure loading	
N_θ	$-\rho x\cos\alpha_0\left(\dfrac{f}{\sin\alpha_0}+x\right)$	$-\rho x\cos\alpha_0\left(x-\dfrac{f}{\sin\alpha_0}\right)$
N_x	$-\rho x\cos\alpha_0\left\{\dfrac{f}{2\sin\alpha_0}\left[1-\left(\dfrac{x_1}{x}\right)^2\right]+\dfrac{x}{3}\left[1-\left(\dfrac{x_1}{x}\right)^3\right]\right\}$	$-\rho x\cos\alpha_0\left\{\dfrac{x}{3}\left[1-\left(\dfrac{x_1}{x}\right)^3\right]-\dfrac{f}{2\sin\alpha_0}\left[1-\left(\dfrac{x_1}{x}\right)^2\right]\right\}$
Δr	$\dfrac{\rho x^2}{Et}\cos^2\alpha_0\left\{\mu\left[\dfrac{f}{2\sin\alpha_0}\left[1-\left(\dfrac{x_1}{x}\right)^2\right]+\dfrac{x}{3}\left[1-\left(\dfrac{x_1}{x}\right)^3\right]\right]-\dfrac{f}{\sin\alpha_0}-x\right\}$	$\dfrac{\rho x^2}{Et}\cos^2\alpha_0\left\{\mu\left[\dfrac{x}{3}\left[1-\left(\dfrac{x_1}{x}\right)^3\right]-\dfrac{f}{2\sin\alpha_0}\left[1-\left(\dfrac{x_1}{x}\right)^2\right]\right]+\dfrac{f}{\sin\alpha_0}-x\right\}$
β	$\dfrac{\rho x}{Et}\dfrac{\cos^2\alpha_0}{\sin\alpha_0}\left\{\dfrac{f}{2\sin\alpha_0}\left[3+\left(\dfrac{x_1}{x}\right)^2\right]+\dfrac{x}{3}\left[8+\left(\dfrac{x_1}{x}\right)^3\right]\right\}$	$\dfrac{\rho x}{Et}\dfrac{\cos^2\alpha_0}{\sin\alpha_0}\left\{\dfrac{f}{2\sin\alpha_0}\left[3+\left(\dfrac{x_1}{x}\right)^2\right]-\dfrac{x}{3}\left[8+\left(\dfrac{x_1}{x}\right)^3\right]\right\}$
	(d) Uniform normal pressure	
N_θ	$-px\cot\alpha_0$	$-px\cot\alpha_0$
N_x	$-\dfrac{p}{2}x\cot\alpha_0\left[1-\left(\dfrac{x_1}{x}\right)^2\right]$	$-\dfrac{p}{2}x\cot\alpha_0\left[1-\left(\dfrac{x_1}{x}\right)^2\right]$
Δr	$-\dfrac{p}{Et}x^2\cos\alpha_0\cot\alpha_0\left\{1-\dfrac{\mu}{2}\left[1-\left(\dfrac{x_1}{x}\right)^2\right]\right\}$	$-\dfrac{px^2}{Et}\cos\alpha_0\cot\alpha_0\left\{1-\dfrac{\mu}{2}\left[1-\left(\dfrac{x_1}{x}\right)^2\right]\right\}$
β	$\dfrac{px}{2Et}\cot^2\alpha_0\left[3+\left(\dfrac{x_1}{x}\right)^2\right]$	$-\dfrac{px}{2Et}\cot^2\alpha_0\left[3+\left(\dfrac{x_1}{x}\right)^2\right]$
	(e) Equally distributed loading along the opening edge (lantern load)	
N_θ	0	0
N_x	$-\dfrac{p}{\sin\alpha_0}\dfrac{x_1}{x}$	$\dfrac{p}{\sin\alpha_0}\dfrac{x_1}{x}$
Δr	$-\dfrac{\mu p\,x_1\cot\alpha_0}{Et}$	$-\mu\dfrac{p\,x_1}{Et}\cot\alpha_0$
β	$-\dfrac{p}{Et}\dfrac{x_1}{x}\dfrac{\cot\alpha_0}{\sin\alpha_0}$	$-\dfrac{p}{Et}\dfrac{x_1}{x}\dfrac{\cot\alpha_0}{\sin\alpha_0}$

3-7 Cylindrical Shells

The primary solutions for cylindrical shells with various axisymmetrical loading conditions are presented in this section. All solutions are based on membrane theory. In order for a membrane state of stress to exist, the boundaries of the shell must be free to deflect normal to the shell middle surface and to rotate. Formulas for deformations are valid only if the shell has uniform extensional stiffness. The nondimensional coordinate and other designations, shown in Fig. 3-5, are adopted with the positive direction as shown.

$$\xi = \frac{x}{L} \qquad \text{where} \quad 0 \leqslant \xi \leqslant 1$$

In addition, the following factors are defined:

$$B = \frac{Et}{1 - \mu^2}$$

$$k = \frac{\sqrt[4]{3(1 - \mu^2)}}{\sqrt{Rt}}$$

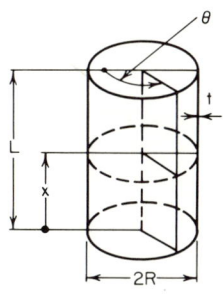
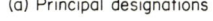
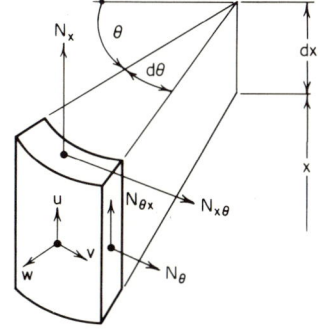

(a) Principal designations (b) Differential elements

figure 3-5 *Cylindrical shell.*

The following loadings are considered:

1. Linear loading, as a result of the superposition of uniform and triangular loadings (Fig. 3-6)

$$Y = X = 0 \qquad Z = p_v(1 + \lambda_p - \xi)$$

56 *Structural Analysis of Shells*

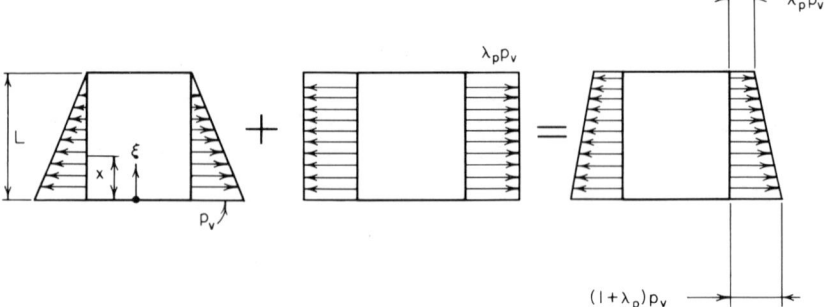

figure 3-6 *Linear loadings.*

2. Trigonometric loading as a result of superposition (Fig. 3-7)

$$Y = X = 0 \qquad Z = p(\xi) = -p_0(\sin \alpha\xi + \lambda_p \cos \beta\xi)$$

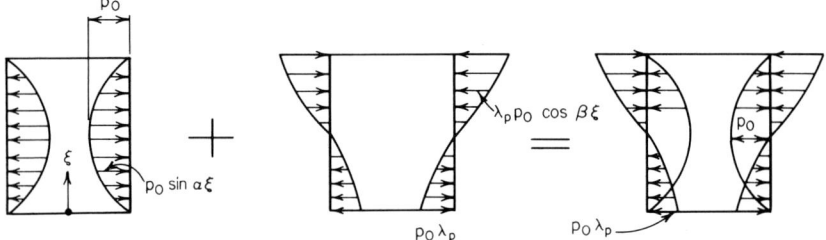

figure 3-7 *Trigonometrical loadings.*

3. Exponential loading (Fig. 3-8)

$$Z = p(\xi) = -p_0 \exp(-\alpha\xi)$$

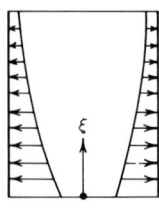

figure 3-8 *Exponential-loading case.*

Also the special case:
$$Z = p(\xi) = -\sum_i p_i \exp(-\alpha\xi)$$
$$= p \sum_i \lambda_{pi} \exp(-\alpha_i \xi)$$
$$X = Y = 0$$

λ_i is obtained as in previous cases.

4. Linear loading as per Fig. 3-9 (deadweight loading)

p_{x0} = weight of shell per unit area
$$X = p_x(\xi) = p_{x0}(1 - \xi)$$
$$Y = Z = 0$$

5. The constant tangential load Y in the circumferential direction (Fig. 3-10)

p_y = load per unit area
$$Y = p_y$$
$$X = Z = 0$$

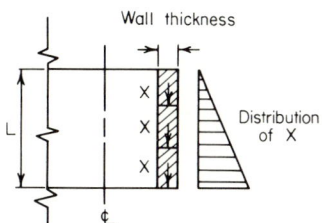

figure 3-9 *Deadweight loading.*

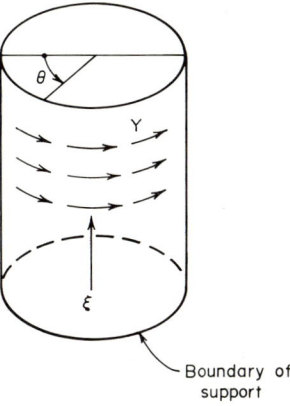

figure 3-10 *Circumferential loading.*

Membrane solutions for the above loadings are presented in Table 3-7. In Table 3-7 the adopted sign convention is as follows:

The membrane stresses N_θ and N_x are positive if causing tension in the shell.

$N_{x\theta}$ is positive if acting in the direction of positive θ.
$N_{\theta x}$ is positive if acting in the positive x direction.
u is positive in the direction of the positive x (or ξ) coordinate.
v is positive in the direction of positive θ.
w is positive if directed outward.

TABLE 3-7 Membrane Solutions for Cylindrical Shells (Ref. 3-1)

Item	Loading	Stresses			Deformations		
		N_θ	N_x	$N_{x\theta}$	u	v	w
1	Linear loading	$p_v R(1+\lambda_p - \xi)$	Zero	Zero	$\frac{1}{Et}\left[-\mu p_v RL\xi\left(1+\lambda_p-\frac{\xi}{2}\right)\right]$	Zero	$\frac{1}{Et}\left[p_v R^2(1+\lambda_p-\xi)\right]$
2	Trigonometrical loading	$p_0 R(\sin\alpha\xi + \lambda_p \cos\beta\xi)$	Zero	Zero	$\frac{\mu}{Et}p_0 RL\left(\frac{\cos\alpha\xi}{\alpha}-\lambda_p\frac{\sin\beta\xi}{\beta}\right)$	Zero	$\frac{1}{Et}p_0 R^2(\sin\alpha\xi + \lambda_p\cos\beta\xi)$
3	Exponential loading	$pR\sum_i \lambda_{pi}\exp(-\alpha_i\xi)$	Zero	Zero	$\frac{1}{Et}\mu pRL\sum_i \frac{\lambda_{pi}}{\alpha_i}\exp(-\alpha_i\xi)$	Zero	$\frac{1}{Et}pR^2\sum_i \lambda_{pi}\exp(-\alpha_i\xi)$
4	Linear loading. Loading applied to segments	Zero	$p_{xo}L\left(\frac{1}{2}-\xi+\frac{\xi^2}{2}\right)$	Zero	$\frac{1}{Et}\left[p_{xo}L^2\xi\left(\frac{1}{2}-\frac{\xi}{2}+\frac{\xi^2}{6}\right)\right]$	Zero	$\frac{1}{Et}\mu p_{xo}\alpha L\left(-\frac{1}{2}+\xi-\frac{\xi^2}{2}\right)$
5	Circumferential loading. Loads reacted at lower base	Zero	Zero	$p_y L(1-\xi)$	Zero	$\frac{1}{Et}\left[2(1+\mu)p_y L^2\left(\xi-\frac{\xi^2}{2}\right)\right]$	Zero

3-8 Elliptical Shells

This section presents the solutions for elliptical shells loaded by axisymmetrical loadings. Only closed elliptical shells are considered. The loadings under consideration are presented in Table 3-8. The boundaries of the elliptical shell must be free to rotate and deflect normal to the shell middle surface. Formulas for deformations are valid only if the shell has uniform extensional stiffness.

The formulas for stresses and deformations are presented in Table 3-8 and are obtained with small-deflection membrane theory (see also Figs. 3-11 to 3-13).

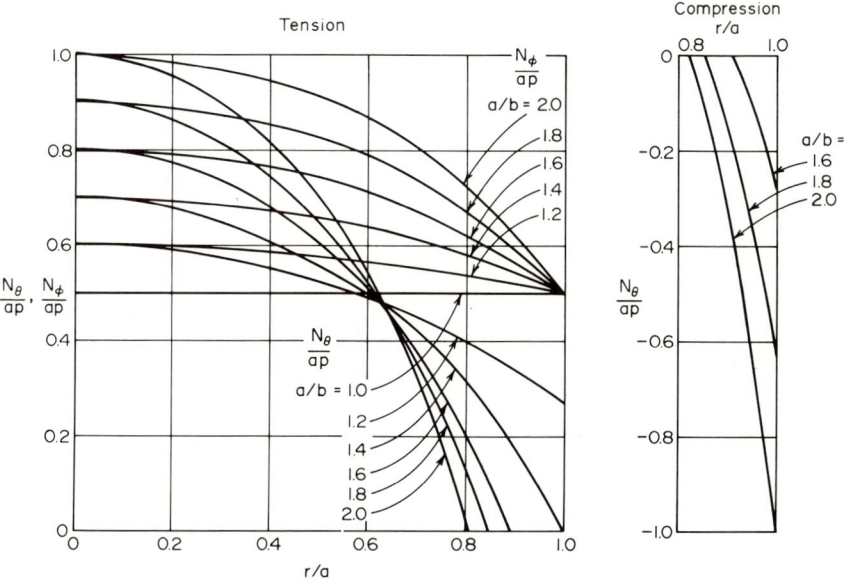

figure 3-11 *Membrane force parameters for ellipsoidal shells under uniform pressure (case 1).*

The membrane forces are positive if causing tension in the shell. Deformations are positive as indicated in the tables.

TABLE 3-8 Elliptical Membrane Shell

Loading case	Loading	Forces		Displacements*	
		N_ϕ	N_θ		
1	Uniform pressure $Z = p$ $X = Y = 0$	$\dfrac{R_\theta p}{2}$	$\dfrac{R_\theta p}{2}\left(2 - \dfrac{R_\theta}{R_\phi}\right)$	Δr	$\dfrac{pR_\theta^2 \sin\phi}{2Et}\left(2 - \mu - \dfrac{R_\theta}{R_\phi}\right)$
				h	$\dfrac{pa^2}{4Et}\left[(3\eta^2 + 1 - 2\mu)\dfrac{\cos\phi}{1 + K^2\sin^2\phi} - 2\eta^2 \cos\phi \right.$ $\left. + \dfrac{K}{\eta}(\eta^2 - \tfrac{1}{2} + \mu)\ln\dfrac{\eta + K\cos\phi}{\eta - K\cos\phi}\right]$
		The stresses are plotted in nondimensional form according to following equation: $\dfrac{N_\phi}{ap} = \dfrac{a}{2b}\left[1 - \left(\dfrac{r}{a}\right)^2\left[1 - \left(\dfrac{b}{a}\right)^2\right]\right]^{1/2}$ $\dfrac{N_\theta}{ap} = \dfrac{N_\phi}{ap}\left[2 - \dfrac{1}{1 - \left(\dfrac{r}{a}\right)^2\left[1 - \left(\dfrac{b}{a}\right)^2\right]}\right]$		β	$\dfrac{pR_\theta}{2Et\tan\phi}\left(\dfrac{R_\theta}{R_\phi} - 1\right)\left(\dfrac{R_\theta}{R_\phi} + 3\right)$
2	(Dead weight) $\dfrac{\sqrt{a^2 - b^2}}{a} = \epsilon$	$-\dfrac{p}{2}\dfrac{\sqrt{a^2\tan^2\phi + b^2}}{a^2\sin\phi\tan\phi}\left[a^2 - \dfrac{b^2\sqrt{1+\tan^2\phi}}{b^2 + a^2\tan^2\phi}\right.$ $\left. + \dfrac{b^2}{\epsilon}\ln\dfrac{(1+\epsilon)\sqrt{b^2+a^2\tan^2\phi}}{b(\epsilon + \sqrt{1+\tan^2\phi})}\right]$	$p\left[\dfrac{(b^2 + a^2\tan^2\phi)^{3/2}}{2\tan^2\phi\sqrt{b^2 + a^2\tan^2\phi}}\left(\dfrac{1}{\epsilon a^2}\right)\ln\dfrac{(1+\epsilon)\sqrt{b^2 + a^2\tan^2\phi}}{b(\epsilon + \sqrt{1+\tan^2\phi})}\right.$ $\left. + \dfrac{1}{b^2}\dfrac{\sqrt{1+\tan^2\phi}}{b^2 + a^2\tan^2\phi} - \dfrac{a^2}{\sqrt{b^2 + a^2\tan^2\phi}}\right]$		See Ref. 3-3
3		$-\dfrac{p}{2}\dfrac{a^2\sqrt{1+\tan^2\phi}}{\sqrt{b^2+a^2\tan^2\phi}}$	$-\dfrac{p}{2}\dfrac{a^2}{b^2}\dfrac{b^2 - a^2\tan^2\phi}{\sqrt{b^2+a^2\tan^2\phi}\sqrt{1+\tan^2\phi}}$		See Ref. 3-3

* In this case h represents vertical displacement (upwards +) if Δr is horizontal (positive outwards); $\eta = \dfrac{a}{b}$; $K = \sqrt{\eta^2 - 1}$.

3-9 Cassini Shells

The discrepancy of the hoop forces of the cylindrical shell and the bulkheads loaded by internal pressure may be avoided by choosing one family of Cassinian curves as a meridian shape. Its equation is

$$(r^2 + n^2z^2)^2 + 2a^2(r^2 - n^2z^2) = 3a^4$$

where n is a number > 1, r and z are variables along r and z lines, and $a = \max r$ as per Fig. 3-14 (Ref. 3-4).

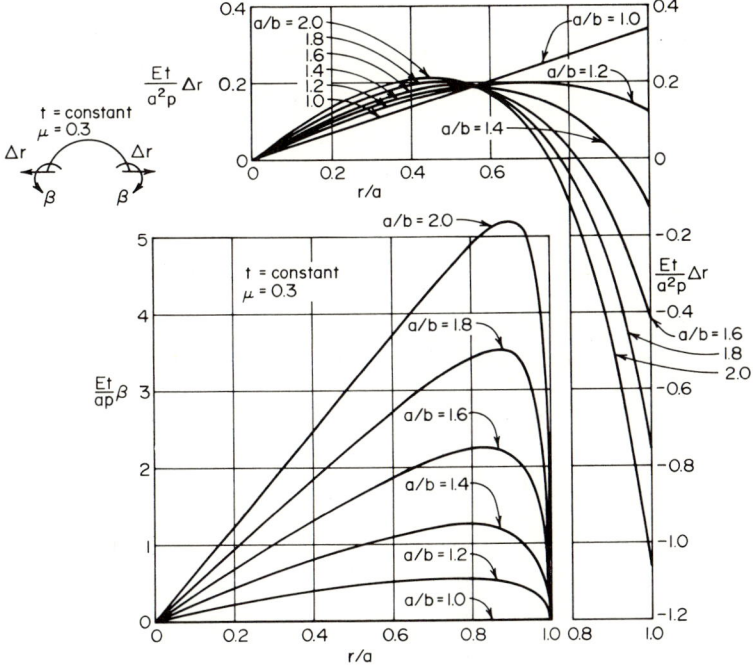

figure 3-12 *Displacement and rotation parameters for ellipsoidal shells under uniform pressure.*

The formulas for stresses and deformations are presented in Table 3-9 for $n = 1$ and $n = 2$.

The formulas presented are based on the membrane and small-deflection theory (Ref. 3-4). Consequently, the boundaries of the shell must be free to rotate and deflect normal to the middle shell surface. Formulas for deformations are valid only if the shell has uniform extensional stiffness. The only loading that is considered is a uniform pressure p.

The membrane forces for the case $n = 2$ are plotted (Fig. 3-15) in nondimensional form according to the following equations:

$$\frac{N_\phi}{ap} = \frac{2}{5(4K+3)} [5(16K^4 + 24K^3 - 7K^2 + 8K + 3)]^{1/2}$$

$$\frac{N_\theta}{ap} = \frac{4(64K^5 + 144K^4 + 44K^3 + 85K^2 - 36K + 23)}{(4K+3)^2 [5(16K^4 + 24K^3 - 7K^2 + 8K + 3)]^{1/2}}$$

where K is given in Table 3-9.

$$K = \left[1 - \frac{15}{16}\left(\frac{r_0}{a}\right)^2\right]^{1/2}$$

Additional graphs for Δr, y, and β are given in Figs. 3-16 and 3-17. The curves of cases 1 and 2 are not the same curves, but both belong to the Cassinian family of curves.

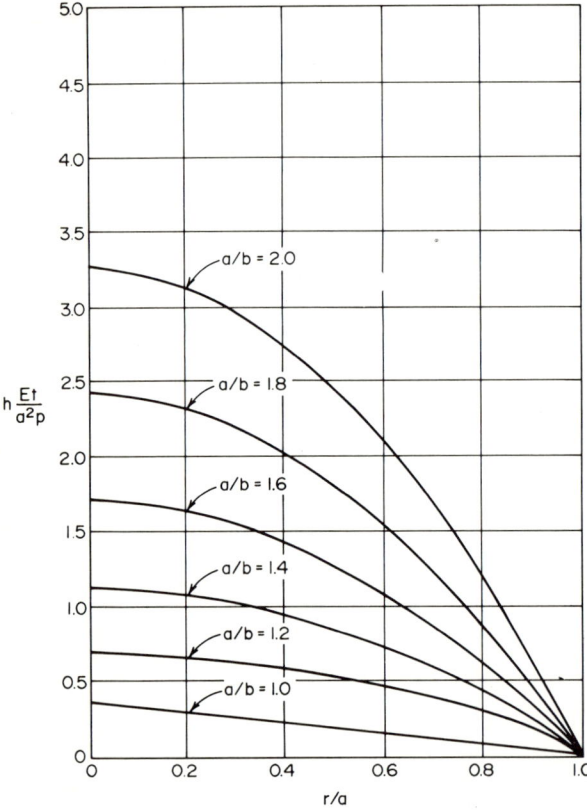

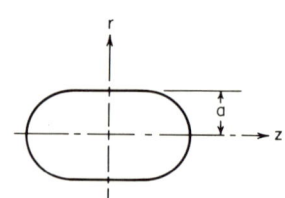

figure 3-13 *Displacement parameter for ellipsoidal shells under uniform pressure.*

figure 3-14 *Cassini's curve.*

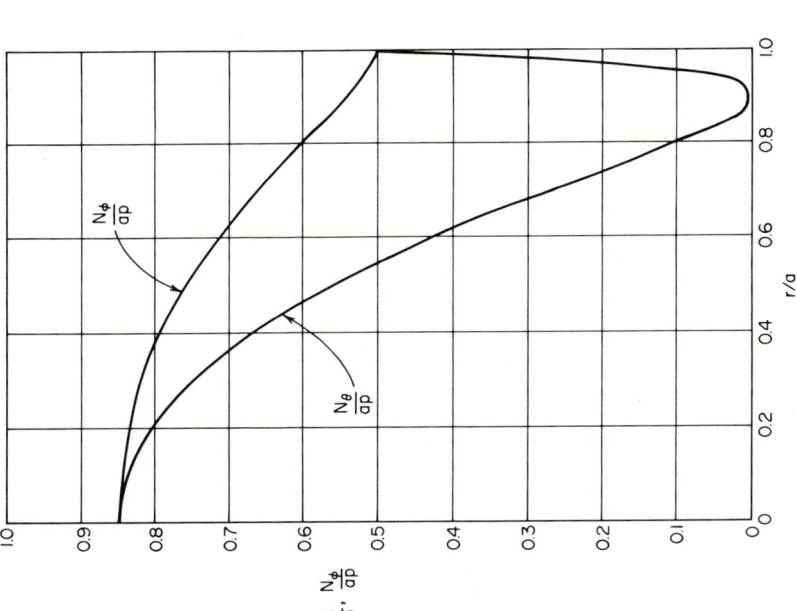

figure 3-15 *Membrane force parameters for Cassini shells under uniform pressure (case 2).*

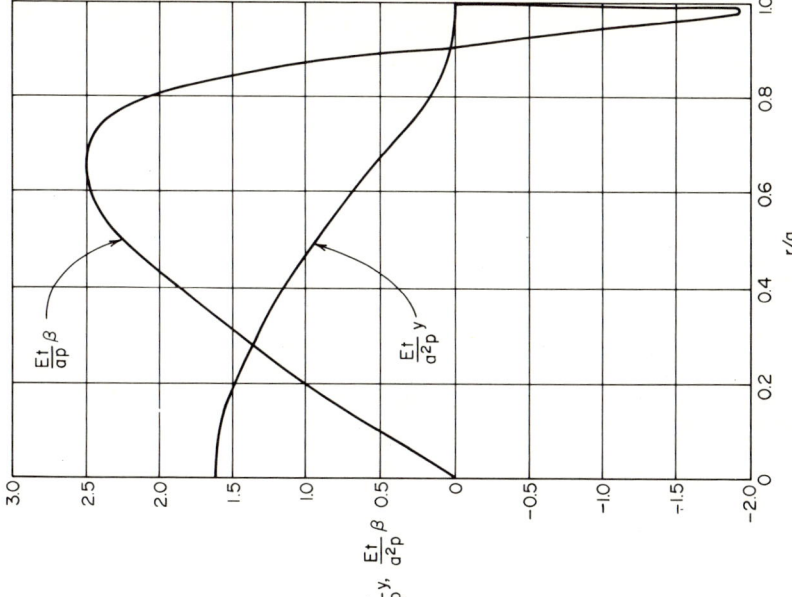

figure 3-16 *Displacement and rotation parameters for Cassini domes under uniform pressure (case 1).*

64 *Structural Analysis of Shells*

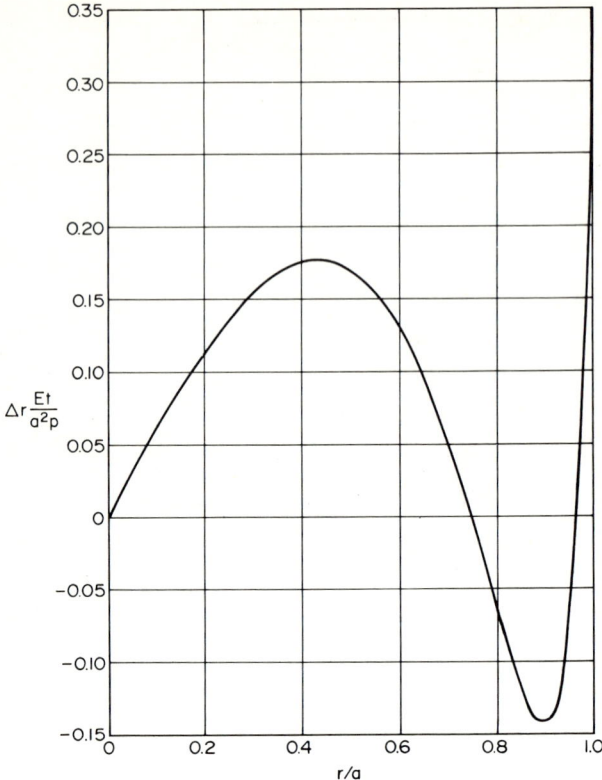

figure 3-17 *Displacement parameter for Cassinian domes under uniform pressure (case 1).*

TABLE 3-9 Cassini Shells—Membrane Solutions

		Case 1	Case 2
Loading and shape		$n > 1$	Special case: $n = 2$, $b = 0.743a$
Forces	N_ϕ	$-pa \dfrac{[r^2(a^2 + n^2 z^2) + n^4 z^2(a^2 - r^2)]^{1/2}}{a^2 + r^2 + n^2 z^2}$	$\dfrac{R_\theta p}{2}$
Forces	N_θ	$N_\phi \left[2 - \dfrac{3n^2 a^4}{(a^2 + r^2 + n^2 z^2)} \dfrac{a^2 - r^2 + n^2 z^2}{[r^2(a^2 + n^2 z^2) + n^4 z^2(a^2 - r^2)]} \right]$	$\dfrac{R_\theta p}{2}\left(2 - \dfrac{R_\theta}{R_\phi}\right)$
Displacements	Δr		$\dfrac{pR_\theta^2 \sin\phi}{2Et}\left(2 - \mu - \dfrac{R_\theta}{R_\phi}\right)$
Displacements	h	For determination of Δr, h, and β, see Sec. 3·3	$\Delta r \cot\phi - \int \dfrac{R_\phi(N_\phi - \mu N_\theta) - R_\theta(N_\theta - \mu N_\phi)}{Et \sin\phi} d\phi + C$
Displacements	β		$\dfrac{pR_\theta}{2Et \tan\phi}\left[\left(\dfrac{R_\theta}{R_\phi} - 1\right)\left(\dfrac{R_\theta}{R_\phi} + 3\right) - R_\theta^2 \sin^4\phi\, U\right]$
Remarks		Ref. 3-4	$U = \dfrac{2160}{a^2 K(3 + 4K)^4}$ $K = \left[1 - 0.94\left(\dfrac{r}{a}\right)^2\right]^{1/2}$

3-10 Toroidal Shells

This section presents some known solutions for closed and open toroidal shells. The loading and systems under consideration are shown in Tables 3-10 and 3-11. The solutions are based on membrane small-deflection theory; consequently, the boundaries must be free to rotate and deflect normal to the shell middle surface. Formulas for deformations are valid only if the shell has uniform extensional stiffness. The walls of the shells are assumed to be of uniform thickness.

TABLE 3-10 Toroidal-shell Primary Solutions

Loading case	Loading and shell	Forces		Displacements		
		N_ϕ	N_θ	Horizontal displacement Δr	Vertical displacement h	Rotation β
1	Toroidal segment	$\dfrac{pR}{2}\left(\dfrac{2b + R\sin\phi}{b + R\sin\phi}\right)$	$\dfrac{pR}{2}$	$\dfrac{pR^2}{2Et}\left[\dfrac{b}{R}(1-2\mu) + (1-\mu)\sin\phi\right]$	(See footnote*)	(See footnote*)
2	Toroidal segment (See footnote †)	$\dfrac{pR}{2}\left[\dfrac{b}{R\sin\phi} + 1\right]$	$\dfrac{pR}{2}\left[1 - \dfrac{b^2}{R^2\sin^2\phi}\right]$	$\dfrac{pR^2}{2Et}\left[\dfrac{b}{R}(1-2\mu) + (1-\mu)\sin\phi \right. \\ \left. + \dfrac{b^2(1+\mu)}{R^2\sin\phi} - \dfrac{b^3}{R^3\sin^2\phi}\right]$	$\dfrac{pR^2}{2Et}\left[(1-\mu)\cos\phi + \dfrac{b}{R}\left(1 + \dfrac{2b^2}{3R^2}\right)\cot\phi \right. \\ \left. + \dfrac{b^2}{R^2}\left(\dfrac{3}{2}+2\mu\right)\dfrac{\cot\phi}{\sin\phi} - \dfrac{b^2}{R^2}\left(1-\dfrac{b}{3R}\right)\dfrac{\cot\phi}{\sin^2\phi}\right. \\ \left. - \dfrac{b^2}{2R^2}(1+2\mu)\ln\tan\dfrac{\phi}{2}\right]$	$-\dfrac{pR\cot\phi}{2Et\sin\phi}\left[\dfrac{b}{R}\left(\dfrac{2b}{R\sin\phi}-1\right)\left(\dfrac{b}{R\sin\phi}+1\right)\right]$
3	Toroidal segment (See footnote †)	$\dfrac{pR}{2}\left[\dfrac{b}{R\sin\phi} - 1\right]$	$\dfrac{pR}{2}\left[\dfrac{b^2}{R^2\sin^2\phi} - 1\right]$	$\dfrac{pR^2}{2Et}\left[-\dfrac{b}{R}(1-2\mu) + (1-\mu)\sin\phi \right. \\ \left. + \dfrac{b^2(1+\mu)}{R^2\sin\phi} + \dfrac{b^3}{R^3\sin^2\phi}\right]$	$\dfrac{pR^2}{2Et}\left[(1-\mu)\cos\phi + \dfrac{b}{R}\left(1+\dfrac{2b^2}{3R^2}\right)\cot\phi \right. \\ \left. + \dfrac{b^2}{R^2}\left(\dfrac{3}{2}+2\mu\right)\dfrac{\cot\phi}{\sin\phi} - \dfrac{b^2}{R^2}\left(1-\dfrac{b}{3R}\right)\dfrac{\cot\phi}{\sin^2\phi}\right. \\ \left. - \dfrac{b^2}{2R^2}(1+2\mu)\ln\tan\dfrac{\phi}{2}\right]$	$-\dfrac{pR\cot\phi}{2Et\sin\phi}\left[\dfrac{b}{R}\left(\dfrac{2b}{R\sin\phi}+1\right)\left(\dfrac{b}{R\sin\phi}-1\right)\right]

*The equations for the displacement h, and rotation β, as derived from the linear membrane theory, are discontinuous and incompatible at $\phi = 0°$ and $180°$.

† For approximate useful range: $35° \leq \phi \leq 90°$.

TABLE 3-11 Toroidal Shell (Ref. 3-3)

(a) Ring axis bisects the cross-section

N_θ and N_ϕ positive if causing tension in shell

No.	Shell	Loading condition	N_ϕ	N_θ
4		$p_x = p\sin\phi$ $p_z = p\cos\phi$	$-pR\dfrac{\cos\phi_0 - \cos\phi - (\phi - \phi_0)\sin\phi_0}{(\sin\phi - \sin\phi_0)\sin\phi}$	$-p\dfrac{R}{\sin^2\phi}\left[(\phi - \phi_0)\sin\phi_0 - (\cos\phi_0 - \cos\phi) + (\sin\phi - \sin\phi_0)\sin\phi\cos\phi\right]$
5		$p_x = p\sin\phi\cos\phi$ $p_z = p\cos^2\phi$	$-p\dfrac{R}{2}\left(1 - \dfrac{\sin^2\phi_0}{\sin^2\phi}\right)$	$-p\dfrac{R}{2}\left(\cos 2\phi + 2\sin\phi\sin\phi_0 - \dfrac{\sin^2\phi_0}{\sin^2\phi}\right)$
6		$p_x = p\sin\phi$ $p_z = p\cos\phi$ $\Sigma p = 2p\pi\left(R\phi_0\sin\phi_0 - 2\sin^2\dfrac{\phi_0}{2}\right)$	$-pR\dfrac{1-\cos\phi + \phi\sin\phi_0}{\sin\phi(\sin\phi + \sin\phi_0)}$	$-pR\left[\cos\phi - \dfrac{1-\cos\phi + \sin\phi_0}{\sin^2\phi} + \sin\phi_0\left(\cot\phi - \dfrac{\phi}{\sin^2\phi}\right)\right]$
7		$p_x = p\sin\phi\cos\phi$ $p_z = p\cos^2\phi$ $\Sigma p = p\pi R^2\sin^2\phi_0$	$-p\dfrac{R\sin\phi + 2\sin\phi_0}{2\sin\phi + \sin\phi_0}$	$-p\dfrac{R}{2}(\cos 2\phi - 2\sin\phi\sin\phi_0)$

Pointed dome · Central support · Toroid surface

TABLE 3-11 Continued

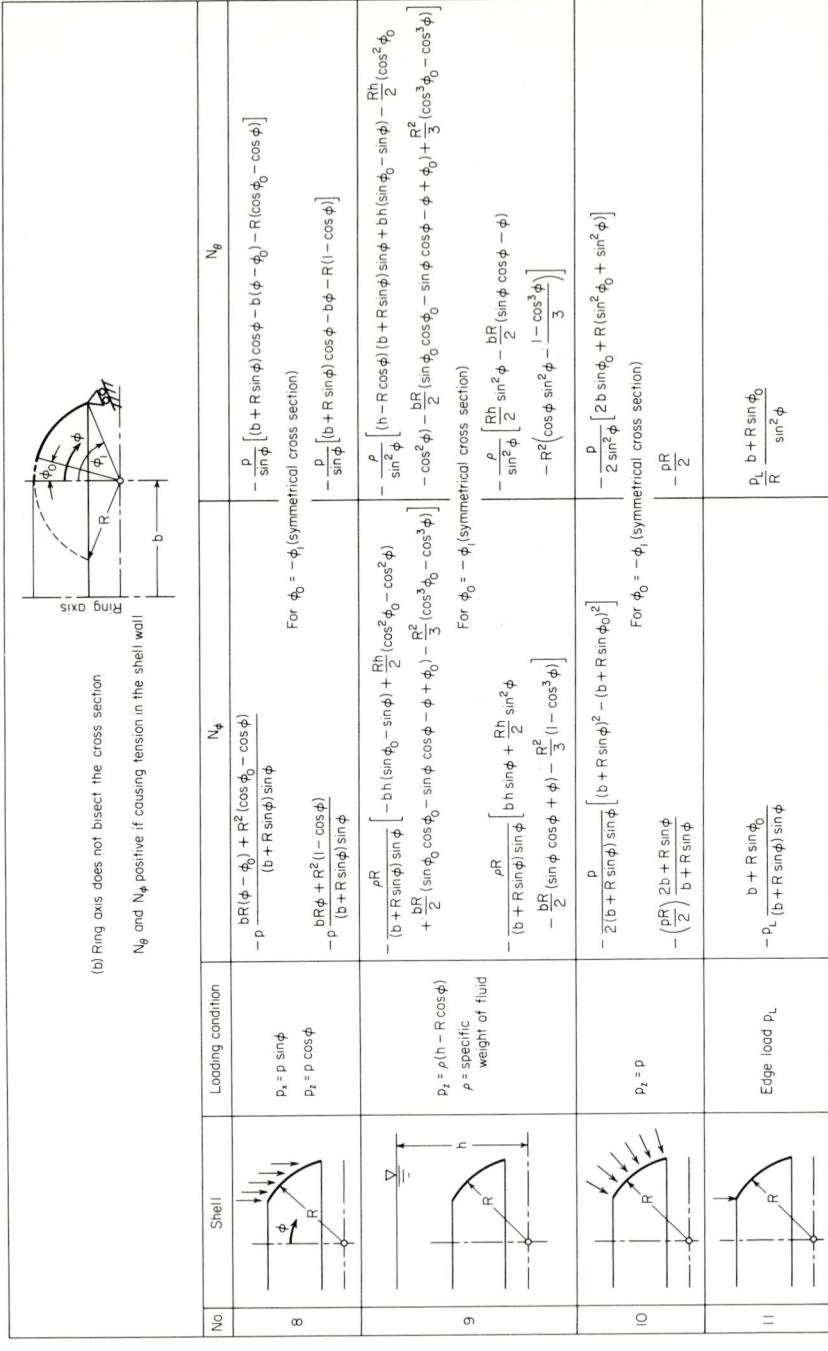

(b) Ring axis does not bisect the cross section

N_θ and N_ϕ positive if causing tension in the shell wall

No.	Shell	Loading condition	N_ϕ	N_θ
8		$p_x = p \sin\phi$ $p_z = p \cos\phi$	$-\dfrac{bR(\phi - \phi_0) + R^2(\cos\phi_0 - \cos\phi)}{(b + R\sin\phi)\sin\phi}$ $-p\dfrac{bR\phi + R^2(1 - \cos\phi)}{(b + R\sin\phi)\sin\phi}$	$-\dfrac{p}{\sin\phi}\left[(b + R\sin\phi)\cos\phi - b(\phi - \phi_0) - R(\cos\phi_0 - \cos\phi)\right]$ $-\dfrac{p}{\sin\phi}\left[(b + R\sin\phi)\cos\phi - b\phi - R(1 - \cos\phi)\right]$
			For $\phi_0 = -\phi_1$ (symmetrical cross section)	For $\phi_0 = -\phi_1$ (symmetrical cross section)
9		$p_z = \rho(h - R\cos\phi)$ ρ = specific weight of fluid	$-\dfrac{\rho R}{(b + R\sin\phi)\sin\phi}\left[-bh(\sin\phi_0 - \sin\phi) + \dfrac{Rh}{2}(\cos^2\phi_0 - \cos^2\phi)\right.$ $\left. + \dfrac{bR}{2}(\sin\phi_0 \cos\phi_0 - \sin\phi \cos\phi - \phi + \phi_0) - \dfrac{R^2}{3}(\cos^3\phi_0 - \cos^3\phi)\right]$	$-\dfrac{\rho}{\sin^2\phi}\left[(h - R\cos\phi)(b + R\sin\phi)\sin\phi + bh(\sin\phi_0 - \sin\phi) - \dfrac{Rh}{2}(\cos^2\phi_0 \right.$ $\left. - \cos^2\phi) - \dfrac{bR}{2}(\sin\phi_0 \cos\phi_0 - \sin\phi \cos\phi - \phi + \phi_0) + \dfrac{R^2}{3}(\cos^3\phi_0 - \cos^3\phi)\right]$
			$-\dfrac{\rho R}{(b + R\sin\phi)\sin\phi}\left[bh\sin\phi + \dfrac{Rh}{2}\sin^2\phi \right.$ $\left. -\dfrac{bR}{2}(\sin\phi \cos\phi + \phi) - \dfrac{R^2}{3}(1 - \cos^3\phi)\right]$	$-\dfrac{\rho}{\sin^2\phi}\left[\dfrac{Rh}{2}\sin^2\phi - \dfrac{bR}{2}(\sin\phi \cos\phi - \phi)\right.$ $\left. -R^2\left(\cos\phi \sin^2\phi - \dfrac{1 - \cos^3\phi}{3}\right)\right]$
			For $\phi_0 = -\phi_1$ (symmetrical cross section)	For $\phi_0 = -\phi_1$ (symmetrical cross section)
10		$p_z = p$	$-\dfrac{p}{2(b + R\sin\phi)\sin\phi}\left[(b + R\sin\phi)^2 - (b + R\sin\phi_0)^2\right]$ $-\left(\dfrac{pR}{2}\right)\dfrac{2b + R\sin\phi}{b + R\sin\phi}$	$-\dfrac{p}{2\sin^2\phi}\left[2b\sin\phi_0 + R(\sin^2\phi_0 + \sin^2\phi)\right]$ $-\dfrac{pR}{2}$
11		Edge load p_L	$-p_L\dfrac{b + R\sin\phi_0}{(b + R\sin\phi)\sin\phi}$	$\dfrac{p_L}{R}\dfrac{b + R\sin\phi_0}{\sin^2\phi}$

3-11 Other Geometries of Shells

Some solutions for shells of other geometries—modified elliptical, pointed, parabolical, and cycloidal shells—are presented here. The shells considered are loaded by internal and external pressure, or by uniformly distributed loading over the shell surface or shell base. For parabolical shells, hydrostatic pressure is also considered. All loadings are axisymmetrical.

The formulas for the internal forces are obtained by linear membrane theory; consequently, the boundaries must be free to rotate and deflect normal to the shell middle surface. Table 3-12 presents the stresses in various shells. For determining displacements, the procedure described in Sec. 3-3 should be used.

The membrane forces are positive if causing tension in the shell.

70 *Structural Analysis of Shells*

TABLE 3-12 Other Shells of Revolution with Curved Meridian (Ref. 6)

r_0 — Radius of curvature at vertex
X, Z — Surface load components
N_ϕ, N_θ — Membrane forces (positive if carrying tension in wall)

Case	Shell geometry	Loading	N_ϕ	N_θ
1	Parabola	$X = p \sin\phi$ $Z = p \cos\phi$	$-p \dfrac{r_0}{3} \dfrac{1 - \cos^3\phi}{\sin^2\phi \cos^2\phi}$	$-p \dfrac{r_0}{3} \dfrac{2 - 3\cos^2\phi + \cos^3\phi}{\sin^2\phi}$
2	Parabola	$X = p \sin\phi \cos\phi$ $Z = p \cos^2\phi$	$-p \dfrac{r_0}{2} \dfrac{1}{\cos\phi}$	$-p \dfrac{r_0}{2} \cos\phi$
3	Parabola	$Z = \rho\left(h + \dfrac{r_0}{2}\tan^2\phi\right)$	$-\rho\dfrac{r_0}{2}\left(h + \dfrac{r_0}{4}\tan^2\phi\right)\dfrac{1}{\cos\phi}$	$-\rho \dfrac{r_0}{2}\left[h(2\tan^2\phi + 1) + r_0 \tan^2\phi\left(\tan^2\phi + \dfrac{3}{4}\right)\right]\cos\phi$
4	Parabola	$Z = p$	$-p \dfrac{r_0}{2} \dfrac{1}{\cos\phi}$	$-p \dfrac{r_0}{2} \dfrac{1 + \sin^2\phi}{\cos\phi}$
5	Cycloid	$X = p \sin\phi$ $Z = p \cos\phi$	$-p\,2 r_0 \dfrac{\phi \sin\phi + \cos\phi - \frac{1}{3}\cos^3\phi - \frac{2}{3}}{(2\phi + \sin 2\phi)\sin\phi}$	$-p\, r_0 \left(\dfrac{1}{3}\dfrac{1 - \cos^3\phi}{\sin^2\phi \cos\phi} - \dfrac{\phi}{2}\tan\phi - \dfrac{1}{2}\sin^2\phi\right)$
6	Cycloid	$X = p \sin\phi \cos\phi$ $Z = p \cos^2\phi$	$-p \dfrac{r_0}{8} \dfrac{2\phi + \sin 2\phi}{\sin\phi}$	$-p \dfrac{r_0}{16} \dfrac{2\phi + \sin 2\phi}{\sin\phi} \cdot \left(4\cos^2\phi - \dfrac{2\phi}{\sin 2\phi} - 1\right)$

Case	Geometry	Loading	Stresses N_ϕ	Stresses N_θ	Remarks
7	Modified elliptical shell equation of meridian $y = -\int \dfrac{x^3 dx}{\sqrt{(1-x^2)(x^2 - a_1)(x^2 - a_2)}}$; $x_0 \rightarrow$ Opening; width a	Pressurization	$\dfrac{a\, p\, n^{\frac{1}{2}} m^{\frac{1}{2}}}{2 x^2}$ For the bottom edge: $\dfrac{pa}{2}$	$N_\phi\left(2 - \dfrac{x}{n}\right)$ $N_\phi\left(2 - \dfrac{x}{n}\right)$	$a_1 = \dfrac{1}{2}\left(\sqrt{1 - \dfrac{4 x_0^2}{1 - x_0^2}} - 1\right)$ $a_2 = -\dfrac{1}{2}\left(\sqrt{1 + \dfrac{4 x_0^2}{1 - x_0^2}} + 1\right)$ $m = (1 - x^2)(x^2 - a_1)(x^2 - a_2)$ $n = 1 + \dfrac{x^6}{m}$
8	Pointed shell — Apex; Circle of radius R, angle ϕ_0; base $R\sin\phi_0$	Pressurization	$\dfrac{pRm}{2}$ where $m = 1 - \dfrac{\sin\phi_0}{\sin\phi}$	$pm\left(1 - \dfrac{m}{2}\right)R$	Solution is not valid at apex Ref. 3-4

3-12 Irregular Shell

Regardless of the shape of the meridian and type of loading, the determination of membrane forces appears to be relatively simple because it is a statically determinate problem. Table 3-1 may be used for this purpose.

Determination of the displacements, however, is a complicated and possibly time-consuming problem. A simplified method for obtaining the approximate solution is presented by Salvadori (Ref. 3-5).

Consider a shell of revolution generated by the rotation of a meridian curve $y(x)$ around the y axis, as is shown in Fig. 3-18. The following conditions must be satisfied:

1. The shell is vertical at the edge ($y = 0$).
2. The meridian curve is symmetrical about the equator.
3. The shell thickness is practically constant near the edge.

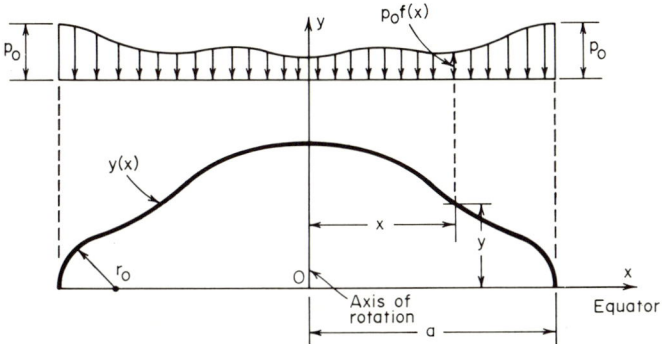

figure 3-18 *Shell of revolution under vertical load.*

The shell is loaded by a distributed load p per unit of horizontal projection, whose intensity varies with the law

$$p(x) = p_0 f(x)$$

It is shown in Ref. (3-5) that, under conditions 1, 2, and 3 and assuming membrane conditions, the following is correct:

1. The edge of the shell does not rotate.
2. The radial displacement of the edge of the shell is independent of the meridian shape.
3. The displacement (positive outward) is given by

$$\Delta r = \frac{p_0 r^2}{E t_s} \frac{r}{r_0} S_p$$

72 Structural Analysis of Shells

where r = equator radius = a
r_0 = radius of curvature of meridian at equator
t_s = shell thickness at equator
E = modulus of elasticity
$S_p = \dfrac{2}{r^2} \int_0^r f(x)\, x\, dx$ is the static moment of the load about the axis of rotation in nondimensional form

The Poisson ratio μ is assumed to be zero. For a load distribution not represented by a simple formula, S_p may be evaluated numerically by means of the approximate summation formula

$$S_p = \frac{2}{p_0 r^2} \sum_{i=1}^{n} p(x_i)\, x_i\, \Delta x$$

as shown in Fig. 3-19.

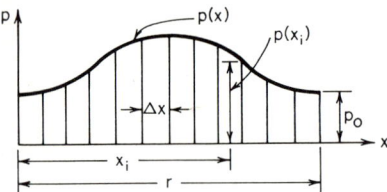

figure 3-19 *Numerical evaluation of S_p.*

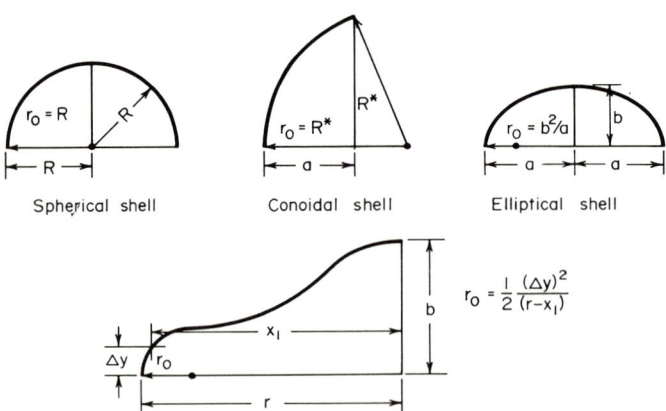

figure 3-20 *Radius of curvature r_0 for various types of shells.*

Primary Solutions 73

The value of the radius of curvature r_0 in connection with Fig. 3-20 is

$$r_0 = R \quad \text{for spherical shells}$$
$$r_0 = R^* \quad \text{for conoidal shells}$$
$$r_0 = \frac{b^2}{a} \quad \text{for elliptical shells of rise } b$$

Finally, for common loading distributions $p(x)$, Table 3-13 gives the corresponding S_p factors. This approximate method is very handy if time is pressing.

TABLE 3-13 S_p **Factors**

Distribution of loading	Equation of loading intensity	S_p factor
(uniform rectangle)	$p = p_0$	1
(linear increasing)	$p = p_0 \frac{x}{a}$	$\frac{2}{3}$
(linear decreasing)	$p = p_0(1 - \frac{x}{a})$	$\frac{1}{3}$
(parabolic increasing)	$p = p_0(\frac{x}{a})^2$	$\frac{1}{2}$
(parabolic decreasing)	$p = p_0(1 - \frac{x}{a})^2$	$\frac{1}{6}$
(parabolic complement)	$p = p_0\left[1 - (\frac{x}{a})^2\right]$	$\frac{1}{2}$
(cubic increasing)	$p = p_0(\frac{x}{a})^3$	$\frac{2}{5}$
(cubic decreasing)	$p = p_0(1 - \frac{x}{a})^3$	$\frac{1}{10}$
(cubic complement)	$p = p_0\left[1 - (\frac{x}{a})^3\right]$	$\frac{3}{5}$

A similar "short cut" for determining deformations due to unit edge loadings is given in Ref. 3-5 and will be presented later.

REFERENCES

3-1. Hampe, E.: *Statik Rotationssymmetrischer Flächentragwerke* (in German), vols. 1 to 4, VEB Verlag für Bauwesen, Berlin, 1964.
3-2. Föppl, L., and G. Sonntag: *Tafeln und Tabellen zur Festigkeitlehre* (in German), Munich, 1951.
3-3. Pflüger, Alf.: *Elementary Statics of Shells*, 2d ed., Mc Graw-Mill Book Company, New York, 1961.
3-4. Flügge, W.: *Stresses in Shells*, Springer-Verlag OHG, Berlin, 1960.
3-5. Salvadori, M. G.: Live Load and Temperature Moments in Shell of Rotation Built into Cylinders, *J. Am. Concrete Inst.*, October, 1955.
3-6. Timoshenko, S.: *Theory of Plates and Shells*, McGraw-Hill Book Company, New York, 1940.

Chapter 4

SECONDARY SOLUTIONS

4-1 Introduction

Unit loadings are defined as unit moments or unit shears acting on the edges of a shell, that is, $M = 1$ in.-lb/in., $H = 1$ lb/in. Secondary solutions are the deformations and forces in a shell as a result of the unit loadings. The disturbances due to the unit edge loadings are of local extent and do not progress very far into the shell from the disturbed edge if the shell is not shallow. Moments and shears not restricted to unit values are defined as edge loadings.

Bending theory is used to obtain the solutions for the unit loadings. The solutions for unit edge loadings can be used with the primary solutions, to obtain discontinuity stresses (Sec. 2-8).

An infinitesimal element of the shell of revolution (which is loaded with axisymmetrical edge loadings) is shown in Fig. 4-1a. The element is in equilibrium under the indicated loadings. The nomenclature is the same as before, except that in addition we have

M_ϕ = moment in the meridional direction, in.-lb/in.
M_θ = moment in the circumferential direction, in.-lb/in.
H = transverse shear in the meridional direction, lb/in.

76 *Structural Analysis of Shells*

The twisting moments, membrane inplane shears, and transverse shear in the circumferential direction for a shell of revolution loaded by axisymmetrical loading are zero.

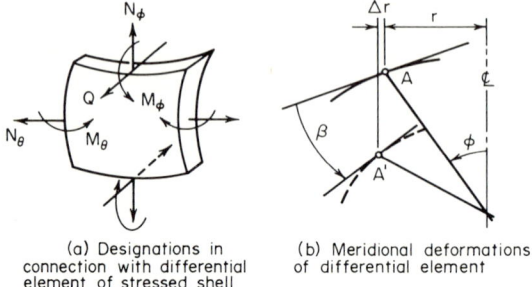

(a) Designations in connection with differential element of stressed shell

(b) Meridional deformations of differential element

figure 4-1 *Designations, positive as shown.*

Figure 4-1b shows the meridional deformations of the element from Fig. 4-1a. Because of the deformation, point A is displaced to the position A'.

Δr = horizontal displacement (usually positive if r is increased)
β = angle of rotation of element (usually positive if observed in the $+\theta$ meridional direction, rotates clockwise)

The circumferential deformations are obtained directly from the meridional deformations and can be obtained from Δr in the tables.

4-2 Spherical Shells (Open, Closed)

This section presents the solutions for nonshallow spherical shells. In a nonshallow shell the edge disturbances die out before reaching the apex of the shell and the relation $\cot \phi \approx 1/\phi$ is not satisfied as in the case of a shallow shell (Ref. 4-1).

For the cases shown, the boundaries of the shells are assumed to be free to rotate and deflect vertically and horizontally because of the action of edge loadings. Abrupt discontinuities in the shell thickness must not be present. The thickness of the shell must be uniform in the range in which the stresses are present.

Formulas are listed for closed and open spherical shells. Open spherical shells are shells that have an axisymmetrical circular opening at the apex. The spherical segment must have a meridional length such that the disturbances due to the edge loading will die out or become insignificant before the opposite edge ($\alpha_{min} \approx 20°$) is reached. Edge loadings may act at the lower or upper edge of the open shell. Linear bending theory is used for the derivation of the formulas.

The following designations are used:

$$k = \sqrt[4]{\left(\frac{R}{t}\right)^2 3(1-\mu^2)} \qquad \alpha = \phi_1 - \phi$$

Table 4-1 presents the formulas for closed spherical shells. The table can also be used for open shells if the segment is large enough so that the disturbances due to the unit loadings will die out before reaching the edge of the opening (see Sec. 4-6).

Usually, the central opening is in an "unstressed" area of the shell. Therefore, for the analysis, all formulas as presented in Table 4-1 may be used for open shells provided that $\alpha_0 \geqslant 20°$. (For α_0 see Fig. 4-6.)

TABLE 4-1 Spherical Shell   (Ref. 4-3)

	H (load case)	M (load case)
Q_ϕ	$\left[-\sqrt{2}\sin\phi_1\, e^{-k\alpha}\cos(k\alpha+\tfrac{\pi}{4})\right]H$	$\left(+\tfrac{2k}{R}e^{-k\alpha}\sin k\alpha\right)M$
N_ϕ	$-Q_\phi\cot\phi$	$-Q_\phi\cot\phi$
N_θ	$\left(2k\sin\phi_1\, e^{-k\alpha}\cos k\alpha\right)H$	$\left[+2\sqrt{2}\,\tfrac{k^2}{R}e^{-k\alpha}\cos(k\alpha+\tfrac{\pi}{4})\right]M$
M_ϕ	$\left(\tfrac{R}{k}\sin\phi_1\, e^{-k\alpha}\sin k\alpha\right)H$	$\left[\sqrt{2}\,e^{-k\alpha}\sin(k\alpha+\tfrac{\pi}{4})\right]M$
M_θ	$\left[-\tfrac{R}{k^2\sqrt{2}}\sin\phi_1\cot\phi\, e^{-k\alpha}\sin(k\alpha+\tfrac{\pi}{4})\right]H+\mu M_\phi$	$\left(\tfrac{1}{k}\cot\phi\, e^{-k\alpha}\cos k\alpha\right)M+\mu M_\phi$
Deformations		
$Et\beta$	$\left[-2\sqrt{2}\,k^2\sin\phi_1\, e^{-k\alpha}\sin(k\alpha+\tfrac{\pi}{4})\right]H$	$\left(-\tfrac{4k^3}{R}e^{-k\alpha}\cos k\alpha\right)M$
$Et(\Delta r)$	$\left\{R\sin\phi_1\, e^{-k\alpha}\left[2k\sin\phi\cos k\alpha-\sqrt{2}\,\mu\cos\phi\cos(k\alpha+\tfrac{\pi}{4})\right]\right\}H$	$+\left[R\sin\phi(N_\theta-\mu N_\phi)\right]M=+2ke^{-k\alpha}\left[\sqrt{2}\,k\sin\phi\cdot\cos(k\alpha+\tfrac{\pi}{4})+\mu\cos\phi\sin k\alpha\right]M$
For $\alpha=0$ and $\phi=\phi_1$		
$Et\beta$	$(-2k^2\sin\phi_1)H$	$\left(-\tfrac{4k^3}{R}\right)M$
$Et(\Delta r)$	$\left[R\sin\phi_1(2k\sin\phi_1-\mu\cos\phi_1)\right]H$	$(+2k^2\sin\phi_1)M$
For $\phi_1=90°$		
$Et\beta$	$-2k^2 H$	$-\tfrac{4k^3}{R}M$
$Et(\Delta r)$	$2RkH$	$+2k^2 M$

For k factors, see Sec. 4-2
For sign convention see Fig. 4-1

Open Spherical Shell

Figure 4-2 represents the case in which the M and H loading acts on the upper edge of an open shell. These cases can be reduced easily to the previous case, as shown in Fig. 4-3, and the same formulas can be

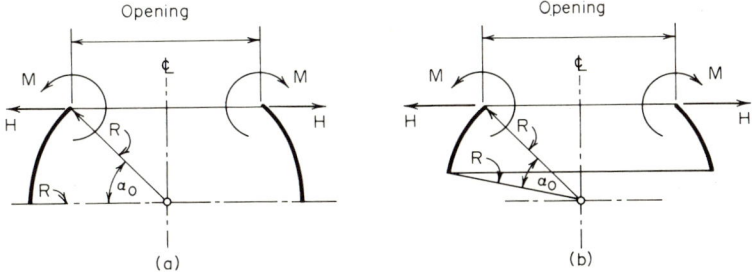

figure 4-2 *Loadings at upper edge of open spherical shell.*

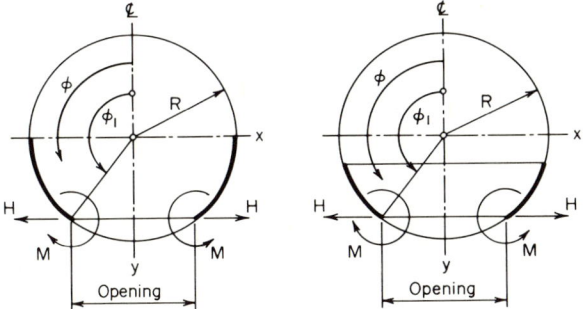

figure 4-3 *Spherical shell, $\phi_1 > 90°$.*

used, except that $\phi_1 > 90°$. The actual shell (Fig. 4-2) is imagined to be turned 180° (Fig. 4-3), and the solution is calculated in the usual way.

An additional set of formulas for spherical open shells is presented in Tables 4-2 and 4-3. These formulas are expressed in terms of the functions F_i and $F_i(\alpha)$. These functions are given in Table 5-10 or Figs. 5-2, 5-3, and 5-4. In the case of a spherical shell, the parameter α should be used in the tables instead of ξ. In all these cases α_0 may be less than 20°.

The following tables (4-2 and 4-3) are adapted from Ref. 4-1, which uses the sign convention indicated in Fig. 4-4.

TABLE 4-2 Open Spherical Shell, Edge Loading at Lower Boundary (Ref. 4-1)

a_0 arbitrary

	① H_{ik}	⑩ M_{ik}	① M_{ki}
Boundary conditions	$a = 0(\phi = \phi_1)$: $M_\phi = 0$, $H_{ik} = Q_{ik}\sin\phi_1 + N_{ik}\cos\phi_1$	$a = 0$: $M_\phi = -M_{ik}$, $Q_\phi = 0$	$a = 0(\phi = \phi_1)$: $M_\phi = 0$, $Q_\phi = 0$
	$a = a_0(\phi = \phi_2)$: $M_\phi = 0$, $Q = 0$	$a = a_0$: $M_\phi = 0$, $Q_\phi = 0$	$a = a_0(\phi = \phi_2)$: $M_\phi = M_{ki}$, $Q_\phi = 0$

Internal forces and deformations

N_ϕ	$H_{ik}\sin\phi_1\cot\phi\left[F_7(a) - \dfrac{F_4}{F_1}F_{10}(a) + \dfrac{F_2}{F_1}F_8(a)\right]$	$M_{ik}\cot\phi\,\dfrac{2k}{R}\left[\dfrac{F_6}{F_1}F_{15}(a) + \dfrac{F_5}{F_1}F_{16}(a) - \dfrac{F_3}{F_1}F_8(a)\right]$	$M_{ki}\dfrac{2k}{R}\cot\phi\left[\dfrac{F_8}{F_1}F_{10}(a) - \dfrac{F_{10}}{F_1}F_8(a)\right]$
N_θ	$-H_{ik}k\sin\phi_1\left[-F_9(a) - 2\dfrac{F_4}{F_1}F_7(a) + \dfrac{F_5}{F_1}F_{10}(a)\right]$	$-M_{ik}\dfrac{2k^2}{R}\left[\dfrac{F_6}{F_1}F_{14}(a) + \dfrac{F_5}{F_1}F_{13}(a) - \dfrac{F_3}{F_1}F_{10}(a)\right]$	$M_{ki}\dfrac{2k^2}{R}\left[\dfrac{F_8}{F_1}F_7(a) + \dfrac{F_{10}}{F_1}F_{10}(a)\right]$
Q_ϕ	$H_{ik}\sin\phi_1\left[F_7(a) - \dfrac{F_4}{F_1}F_{10}(a) + \dfrac{F_2}{F_1}F_8(a)\right]$	$M_{ik}\dfrac{2k}{R}\left[\dfrac{F_6}{F_1}F_{15}(a) + \dfrac{F_5}{F_1}F_{16}(a) - \dfrac{F_3}{F_1}F_8(a)\right]$	$-M_{ki}\dfrac{2k}{R}\left[\dfrac{F_8}{F_1}F_{10}(a) + \dfrac{F_{10}}{F_1}F_8(a)\right]$
M_ϕ	$-H_{ik}\dfrac{R}{2k}\sin\phi_1\left[-F_{10}(a) + 2\dfrac{F_4}{F_1}F_8(a) - \dfrac{F_2}{F_1}F_9(a)\right]$	$-M_{ik}\left[\dfrac{F_6}{F_1}F_{13}(a) - \dfrac{F_5}{F_1}F_{14}(a) + \dfrac{F_3}{F_1}F_9(a)\right]$	$M_{ki}\left[2\dfrac{F_8}{F_1}F_8(a) - \dfrac{F_{10}}{F_1}F_9(a)\right]$
M_θ	$H_{ik}\dfrac{R}{2k}\sin\phi_1\left\{-\left[\dfrac{\cot\phi}{k}F_8(a) - \mu F_{10}(a)\right] + \dfrac{F_4}{F_1}\left[\dfrac{\cot\phi}{k}F_9(a)\right.\right.$ $\left.\left.+2\mu F_8(a)\right] + \dfrac{F_2}{F_1}\left[\dfrac{\cot\phi}{k}F_7(a) + \mu F_9(a)\right]\right\}$	$M_{ik}\left\{\left[\dfrac{F_6}{F_1}\dfrac{\cot\phi}{k}F_{16}(a) - \mu F_{13}(a)\right] - \dfrac{F_5}{F_1}\left[\dfrac{\cot\phi}{k}F_{15}(a)\right.\right.$ $\left.-\mu F_{14}(a)\right] + \dfrac{F_3}{F_1}\left[\dfrac{\cot\phi}{k}F_7(a) + \mu F_9(a)\right]\right\}$	$-M_{ki}\left\{\dfrac{F_8}{F_1}\left[\dfrac{\cot\phi}{k}F_9(a) - \mu 2F_8(a)\right] + \dfrac{F_{10}}{F_1}\left[\dfrac{\cot\phi}{k}F_7(a)\right.\right.$ $\left.\left.+\mu F_9(a)\right]\right\}$
Δr	$-H_{ik}\dfrac{Rk}{Et}\sin\phi\sin\phi_1\left[-F_9(a) - 2\dfrac{F_4}{F_1}F_7(a) + \dfrac{F_2}{F_1}F_{10}(a)\right]$	$-M_{ik}\dfrac{2k^2}{Et}\sin\phi\left[\dfrac{F_6}{F_1}F_{14}(a) + \dfrac{F_5}{F_1}F_{13}(a) - \dfrac{F_3}{F_1}F_{10}(a)\right]$	$M_{ki}\dfrac{2k^2}{Et}\sin\phi\left[-2\dfrac{F_8}{F_1}F_7(a) + \dfrac{F_{10}}{F_1}F_{10}(a)\right]$
β	$-H_{ik}\dfrac{2k^2}{Et}\sin\phi_1\left[-F_8(a) - \dfrac{F_4}{F_1}F_9(a) + \dfrac{F_2}{F_1}F_7(a)\right]$	$-M_{ik}\dfrac{4k^3}{EtR}\sin\phi_1\left[\dfrac{F_6}{F_1}F_{16}(a) - \dfrac{F_5}{F_1}F_{15}(a) - \dfrac{F_3}{F_1}F_7(a)\right]$	$M_{ki}\dfrac{4k^3}{EtR}\left[\dfrac{F_8}{F_1}F_9(a) + \dfrac{F_{10}}{F_1}F_7(a)\right]$

For factors F, $F(a)$, and k see under "Definition of F Factors" in Sec. 5-2.

TABLE 4-3 Open Spherical Shell, Edge Loadings at Upper Boundary (Ref. 4-1)

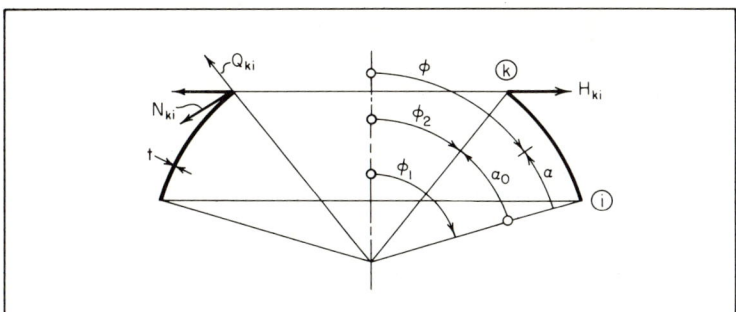

Boundary conditions
$a = 0 \; (\phi = \phi_1) \quad M_\phi = 0$
$\hspace{3.5cm} Q_\phi = 0$
$a = a_0 \; (\phi = \phi_2) \quad M_\phi = 0$
$\hspace{3.5cm} H_{ki} = -Q_{ki}\sin\phi_2 - N_{ki}\cos\phi_2$

Internal forces and deformations	
N_ϕ	$H_{ki}\cot\phi\sin\phi_2 \left[\dfrac{F_9}{F_1}F_{10}(a) - 2\dfrac{F_8}{F_1}F_8(a) \right]$
N_θ	$H_{ki}\, 2k\sin\phi_2 \left[-\dfrac{F_9}{F_1}F_7(a) + \dfrac{F_8}{F_1}F_{10}(a) \right]$
Q_ϕ	$H_{ki}\sin\phi_2 \left[\dfrac{F_9}{F_1}F_{10}(a) - \dfrac{2F_8}{F_1}F_8(a) \right]$
M_ϕ	$H_{ki}\dfrac{R}{k}\sin\phi_2 \left[\dfrac{F_9}{F_1}F_8(a) - \dfrac{F_8}{F_1}F_9(a) \right]$
M_θ	$H_{ki}\dfrac{R}{2k}\sin\phi_2 \left\{ \dfrac{F_9}{F_1}\left[-\dfrac{\cot\phi}{k}F_9(a) + \mu 2 F_8(a) \right] - \dfrac{2F_8}{F_1}\left[\dfrac{\cot\phi}{k}F_7(a) + \mu F_9(a) \right] \right\}$
Δr	$-H_{ki}\sin\phi_2 \dfrac{Rk}{Et} 2\sin\phi \left[\dfrac{F_9}{F_1}F_7(a) - \dfrac{F_8}{F_1}F_{10}(a) \right]$
β	$H_{ki}\sin\phi_2 \dfrac{2k^2}{Et}\left[\dfrac{F_9}{F_1}F_9(a) + \dfrac{2F_8}{F_1}F_7(a) \right]$

For factors F, $F(a)$, and k see under "Definition of F Factors" in Sec. 5-2.

82 Structural Analysis of Shells

Distortions

In connection with some problems, it may be of interest to know the stresses and displacements of a spherical shell (closed or open), if displacements at the edges are prescribed instead of M and H:

At lower edge i:

$$\Delta r_{ik} = \text{displacement in } r \text{ direction}$$
$$\Delta V_{ik} = \text{displacement in vertical direction}$$
$$\beta_{ik} = \text{rotation}$$

At upper edge k:

$$\Delta r_{ki} = \text{displacement in } r \text{ direction}$$
$$\Delta V_{ki} = \text{displacement in vertical direction}$$
$$\beta_{ki} = \text{rotation}$$

Tables 4-4, 4-5, and 4-6 contain the answer to this problem.

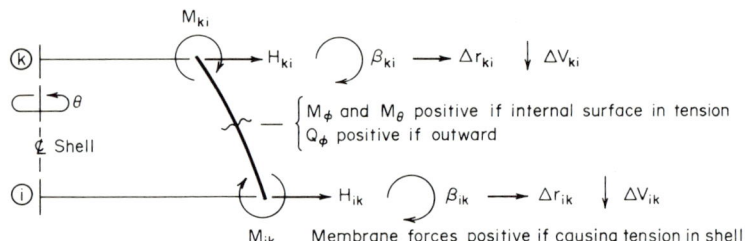

figure 4-4 *Sign convention for spherical shell.*

TABLE 4-4 Closed Spherical Shell Solutions Due to Edge Deformations (Ref. 4-1)

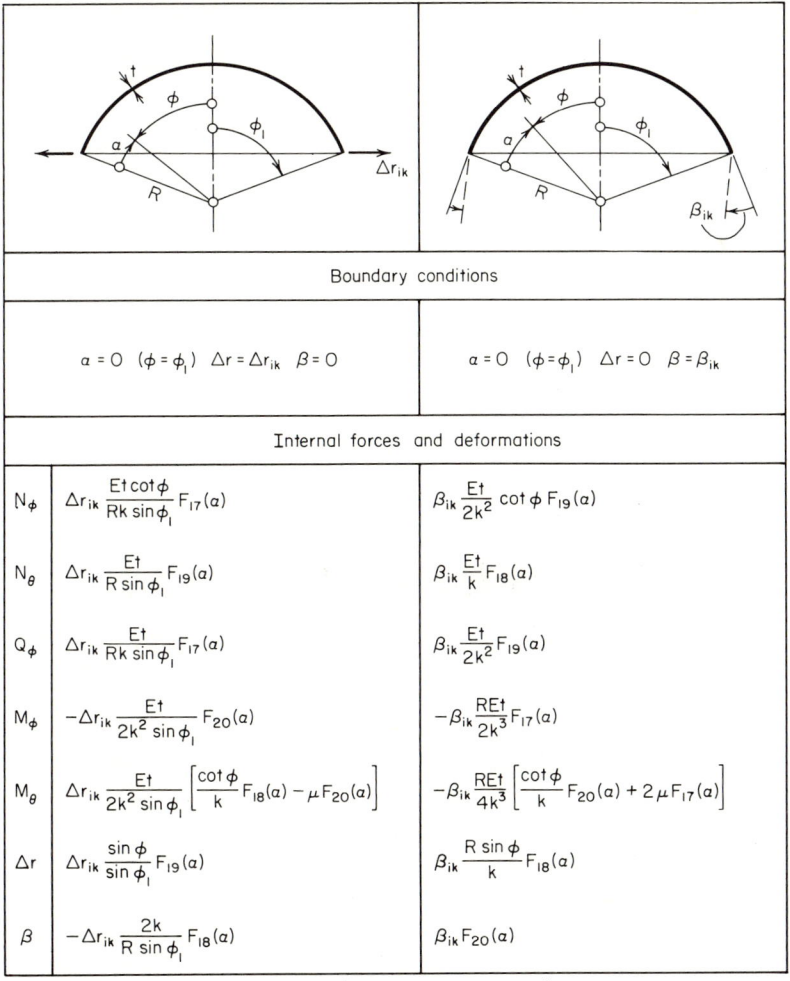

	Boundary conditions	
	$a=0$ ($\phi=\phi_1$) $\Delta r = \Delta r_{ik}$ $\beta=0$	$a=0$ ($\phi=\phi_1$) $\Delta r = 0$ $\beta = \beta_{ik}$
	Internal forces and deformations	
N_ϕ	$\Delta r_{ik} \dfrac{Et \cot\phi}{Rk \sin\phi_1} F_{17}(a)$	$\beta_{ik} \dfrac{Et}{2k^2} \cot\phi \, F_{19}(a)$
N_θ	$\Delta r_{ik} \dfrac{Et}{R \sin\phi_1} F_{19}(a)$	$\beta_{ik} \dfrac{Et}{k} F_{18}(a)$
Q_ϕ	$\Delta r_{ik} \dfrac{Et}{Rk \sin\phi_1} F_{17}(a)$	$\beta_{ik} \dfrac{Et}{2k^2} F_{19}(a)$
M_ϕ	$-\Delta r_{ik} \dfrac{Et}{2k^2 \sin\phi_1} F_{20}(a)$	$-\beta_{ik} \dfrac{REt}{2k^3} F_{17}(a)$
M_θ	$\Delta r_{ik} \dfrac{Et}{2k^2 \sin\phi_1} \left[\dfrac{\cot\phi}{k} F_{18}(a) - \mu F_{20}(a) \right]$	$-\beta_{ik} \dfrac{REt}{4k^3} \left[\dfrac{\cot\phi}{k} F_{20}(a) + 2\mu F_{17}(a) \right]$
Δr	$\Delta r_{ik} \dfrac{\sin\phi}{\sin\phi_1} F_{19}(a)$	$\beta_{ik} \dfrac{R \sin\phi}{k} F_{18}(a)$
β	$-\Delta r_{ik} \dfrac{2k}{R \sin\phi_1} F_{18}(a)$	$\beta_{ik} F_{20}(a)$

For factors F, F(a), and k see under "Definition of F Factors" in Sec. 5-2.

TABLE 4-5 Open (or Closed) Spherical Shell Exposed to Distortions at Lower Edge (Ref. 4-1)

	Rotation β_{ik}	Displacement Δv_{ik}	Displacement Δr_{ik}
Boundary conditions	$a=0(\phi=\phi_1)$ $\quad \beta=\beta_{ik}$ $\quad \Delta r=0$ $a=a_0(\phi=\phi_2)$ $\quad \beta=0$ $\quad \Delta r=0$	$a=0(\phi=\phi_1)$ $\quad \beta=0$ $\quad \Delta v=\Delta v_{ik}$ $a=a_0(\phi=\phi_2)$ $\quad \beta=0$ $\quad \Delta v=0$	$a=0(\phi=\phi_1)$ $\quad \Delta r=\Delta r_{ik}$ $\quad \beta=0$ $a=a_0(\phi=\phi_2)$ $\quad \Delta r=0$ $\quad \beta=0$
Internal forces and deformations			
N_ϕ	$\beta_{ik}\cot\phi + \dfrac{Et}{2k^2}\left[\dfrac{F_2}{F_1}F_7(a) + \dfrac{F_4}{F_1}F_9(a) - F_8(a)\right]$	$\dfrac{\Delta v_{ik} Et \cot\phi}{R[(1+\mu)\sin\phi_1 F_3 + k\cos\phi_1 F_1]}\left[-F_3 F_7(a) - F_5 F_{15}(a) + F_6 F_{16}(a)\right]$	$\Delta r_{ik}\cot\phi \dfrac{Et}{Rk\sin\phi_1}\left[\dfrac{F_3}{F_1}F_7(a) + \dfrac{F_5}{F_1}F_{15}(a) - \dfrac{F_6}{F_1}F_{16}(a)\right]$
N_θ	$-\beta_{ik}\dfrac{Et}{2k}\left[-\dfrac{F_2}{F_1}F_9(a) + 2\dfrac{F_4}{F_1}F_8(a) - F_{10}(a)\right]$	$\dfrac{-\Delta v_{ik} Et k}{R[(1+\mu)\sin\phi_1 F_3 + k\cos\phi_1 F_1]}\left[F_3 F_9(a) - F_5 F_{14}(a) + F_6 F_{13}(a)\right]$	$\Delta r_{ik}\dfrac{Et}{R\sin\phi_1}\left[\dfrac{F_3}{F_1}F_9(a) - \dfrac{F_5}{F_1}F_{14}(a) + \dfrac{F_6}{F_1}F_{13}(a)\right]$
Q_ϕ	$\beta_{ik}\dfrac{REt}{2k^2}\left[\dfrac{F_2}{F_1}F_7(a) + \dfrac{F_4}{F_1}F_9(a) - F_8(a)\right]$	$\dfrac{\Delta v_{ik} Et}{R[(1+\mu)\sin\phi_1 F_3 + k\cos\phi_1 F_1]}\left[-F_3 F_7(a) - F_5 F_{15}(a) + F_6 F_{16}(a)\right]$	$\Delta r_{ik}\dfrac{Et}{Rk\sin\phi_1}\left[\dfrac{F_3}{F_1}F_7(a) - \dfrac{F_5}{F_1}F_{15}(a) - \dfrac{F_6}{F_1}F_{16}(a)\right]$
M_ϕ	$-\beta_{ik}\dfrac{REt}{4k^3}\left[-\dfrac{F_2}{F_1}F_{10}(a) + 2\dfrac{F_4}{F_1}F_7(a) + F_9(a)\right]$	$\dfrac{-\Delta v_{ik} Et}{2k[(1+\mu)\sin\phi_1 F_3 + k\cos\phi_1 F_1]}\left[F_3 F_{10}(a) - F_5 F_{13}(a) - F_6 F_{14}(a)\right]$	$\Delta r_{ik}\dfrac{Et}{2k^2\sin\phi_1}\left[\dfrac{F_3}{F_1}F_{10}(a) - \dfrac{F_5}{F_1}F_{13}(a) - \dfrac{F_6}{F_1}F_{14}(a)\right]$
M_θ	$\beta_{ik}\dfrac{REt}{4k^3}\left\{-\dfrac{F_2}{F_1}\left[\dfrac{\cot\phi}{k}F_8(a) - \mu F_{10}(a)\right] + \dfrac{F_4}{F_1}\left[\dfrac{\cot\phi}{k}F_{10}(a)\right]\right\}$	$\dfrac{\Delta v_{ik} Et}{2k[(1+\mu)\sin\phi_1 F_3 + k\cos\phi_1 F_1]}\left\{F_3\left[\dfrac{\cot\phi}{k}F_8(a)-\mu F_{10}(a)\right]-F_6\left[\dfrac{\cot\phi}{k}F_{15}(a)-\mu F_{14}(a)\right]\right\}$	$\Delta r_{ik}\dfrac{Et}{2k^2\sin\phi_1}\left\{-\dfrac{F_3}{k}\left[\dfrac{\cot\phi}{k}F_8(a)-\mu F_{10}(a)\right]+\dfrac{F_5}{k}\left[\dfrac{\cot\phi}{k}F_{15}(a)-\mu F_{14}(a)\right]\right\}$
Δr	$-2\mu F_7(a) - \left[\dfrac{\cot\phi}{k}F_7(a) + \mu F_9(a)\right]$	$\dfrac{-\Delta v_{ik}\sin\phi}{[(1+\mu)\sin\phi_1 F_3 + k\cos\phi_1 F_1]}\left[F_3 F_9(a) - F_5 F_{14}(a) + F_6 F_{13}(a)\right]$	$-\mu F_{13}(a) + \dfrac{\sin\phi}{\sin\phi_1}\left[\dfrac{F_3}{F_1}F_9(a) - \dfrac{F_5}{F_1}F_{14}(a) + \dfrac{F_6}{F_1}F_{13}(a)\right]$
β	$-\beta_{ik}\left[-\dfrac{F_2}{F_1}F_8(a) + \dfrac{F_4}{F_1}F_{10}(a) - F_7(a)\right]$	$\dfrac{2k^2 \Delta v_{ik}}{R[(1+\mu)\sin\phi_1 F_3 + k\cos\phi_1 F_1]}\left[F_3 F_8(a) - F_5 F_{16}(a) - F_6 F_{15}(a)\right]$	$\Delta r_{ik}\dfrac{2k}{R\sin\phi_1}\left[\dfrac{F_3}{F_1}F_8(a) - \dfrac{F_5}{F_1}F_{16}(a) - \dfrac{F_6}{F_1}F_{15}(a)\right]$

For factors F, F(a), and k see under "Definition of F Factors" in Sec. 5-2.

TABLE 4-6 Open Spherical Shell, Edge Deformations at Upper Boundary Ref. 4-1)

	$\alpha = 0 \ (\phi = \phi_1)$, $\alpha = \alpha_0 \ (\phi = \phi_2)$	$\alpha = 0 \ (\phi = \phi_1)$, $\alpha = \alpha_0 \ (\phi = \phi_2)$	$\alpha = 0 \ (\phi = \phi_1)$, $\alpha = \alpha_0 \ (\phi = \phi_2)$
Boundary conditions	$\Delta r = 0$, $\beta = 0$ $\Delta r = \Delta r_{ki}$, $\beta = 0$	$\Delta r = 0$, $\beta = 0$ $\Delta V = \Delta V_{ki}$, $\beta = 0$	$\Delta r = 0$, $\beta = 0$ $\Delta r = 0$, $\beta = \beta_{ki}$

Internal forces and deformations

N_ϕ	$-\Delta r_{ki} \dfrac{Et \cot\phi}{R k \sin\phi_2} \left[\dfrac{F_{10}}{F_1} F_7(\alpha) + \dfrac{F_8}{F_1} F_9(\alpha) \right]$	$\cot\phi \dfrac{\Delta V_{ki} Et}{R} \dfrac{[-F_{10} F_7(\alpha) - F_8 F_9(\alpha)]}{(1+\mu)\sin\phi_2 F_3 - k \cos\phi_2 F_1}$	$\beta_{ki} \dfrac{Et \cot\phi}{2k^2} \left[\dfrac{2F_8}{F_1} F_7(\alpha) + \dfrac{F_9}{F_1} F_9(\alpha) \right]$
N_θ	$\Delta r_{ki} \dfrac{Et}{R k \sin\phi_2} \left[-\dfrac{F_{10}}{F_1} F_9(\alpha) + 2\dfrac{F_8}{F_1} F_8(\alpha) \right]$	$-k \dfrac{\Delta V_{ki} Et}{R} \dfrac{[-F_{10} F_9(\alpha) + 2 F_8 F_8(\alpha)]}{(1+\mu)\sin\phi_2 F_3 - k \cos\phi_2 F_1}$	$-\beta_{ki} \dfrac{Et}{k} \left[-\dfrac{F_8}{F_1} F_9(\alpha) + \dfrac{F_9}{F_1} F_8(\alpha) \right]$
Q_ϕ	$-\Delta r_{ki} \dfrac{Et}{R \sin\phi_2} \left[\dfrac{F_{10}}{F_1} F_7(\alpha) + \dfrac{F_8}{F_1} F_9(\alpha) \right]$	$\dfrac{\Delta V_{ki} Et}{R} \dfrac{[-F_{10} F_7(\alpha) - F_8 F_9(\alpha)]}{(1+\mu)\sin\phi_2 F_3 - k \cos\phi_2 F_1}$	$\beta_{ki} \dfrac{Et}{2k^2} \left[\dfrac{2F_8}{F_1} F_7(\alpha) + \dfrac{F_9}{F_1} F_9(\alpha) \right]$
M_ϕ	$\Delta r_{ki} \dfrac{Et}{2k^2 \sin\phi_2} \left[-\dfrac{F_{10}}{F_1} F_{10}(\alpha) + 2\dfrac{F_8}{F_1} F_7(\alpha) \right]$	$-\dfrac{R}{2k} \dfrac{\Delta V_{ki} Et}{R} \dfrac{[F_{10} F_{10}(\alpha) - 2 F_8 F_7(\alpha)]}{(1+\mu)\sin\phi_2 F_3 - k \cos\phi_2 F_1}$	$-\beta_{ki} \dfrac{REt}{2k^3} \left[-\dfrac{F_8}{F_1} F_{10}(\alpha) + \dfrac{F_9}{F_1} F_7(\alpha) \right]$
M_θ	$\Delta r_{ki} \dfrac{Et}{2k^2 \sin\phi_2} \left\{ \dfrac{F_{10}}{F_1} \left[\dfrac{\cot\phi}{k} F_8(\alpha) - \mu F_{10}(\alpha) \right] - \dfrac{F_8}{F_1} \left[\dfrac{\cot\phi}{k} F_{10}(\alpha) - \mu 2 F_7(\alpha) \right] \right\}$	$\mu \dfrac{\Delta V_{ki} Et}{2k} \dfrac{[F_{10} F_{10}(\alpha) - 2 F_8 F_7(\alpha)]}{(1+\mu)\sin\phi_2 F_3 - k \cos\phi_2 F_1}$	$\beta_{ki} \dfrac{REt}{4k^3} \left\{ -\dfrac{2F_8}{F_1} \left[\dfrac{\cot\phi}{k} F_8(\alpha) - \mu F_{10}(\alpha) \right] + \dfrac{F_9}{F_1} \left[\dfrac{\cot\phi}{k} F_{10}(\alpha) - \mu 2 F_7(\alpha) \right] \right\}$
Δr	$\Delta r_{ki} \dfrac{\sin\phi}{R \sin\phi_2} \left[-\dfrac{F_{10}}{F_1} F_9(\alpha) + 2\dfrac{F_8}{F_1} F_8(\alpha) \right]$	$-\sin\phi k \dfrac{\Delta V_{ki}}{R} \dfrac{[F_{10} F_9(\alpha) - 2 F_8 F_8(\alpha)]}{(1+\mu)\sin\phi_2 F_3 - k \cos\phi_2 F_1}$	$-\beta_{ki} \dfrac{R \sin\phi}{k} \left[-\dfrac{F_8}{F_1} F_9(\alpha) + \dfrac{F_9}{F_1} F_8(\alpha) \right]$
β	$\Delta r_{ki} \dfrac{2k}{R \sin\phi_2} \left[-\dfrac{F_{10}}{F_1} F_8(\alpha) + \dfrac{F_8}{F_1} F_{10}(\alpha) \right]$	$2k^2 \dfrac{\Delta V_{ki}}{R} \dfrac{[-F_{10} F_8(\alpha) + F_8 F_{10}(\alpha)]}{(1+\mu)\sin\phi_2 F_3 + k \cos\phi_2 F_1}$	$-\beta_{ki} \left[-2 \dfrac{F_8}{F_1} F_8(\alpha) + \dfrac{F_9}{F_1} F_{10}(\alpha) \right]$

For factors F, F(α), and k see under "Definition of F Factors" in Sec. 5-2.

4-3 Conical Shells

This section presents the solutions for nonshallow open or closed conical shells in which α_0 is not small (Fig. 4-5). There is no exact information about the limiting angle α_0. It is recommended that the solutions be limited to the range of $\alpha_0 \geqslant 45°$. If $\alpha_0 = 90°$, the cone degenerates into a cylinder.

Another limitation must be applied to the height of the cone. As in the case of the sphere, the disturbances due to edge loadings will die at a short distance from the disturbed edge (for practical purposes, approximately at a distance $\sqrt{R_0 t}$) where $R_0 = \max R$. Consequently, a "high" cone is characterized by an undisturbed edge (or apex) due to loading influences on the opposite edge.

The boundaries must be free to rotate and deflect vertically and horizontally because of the action of the edge loadings. Abrupt discontinuities in the shell thickness must not be present. The thickness of shell must be uniform in the region in which stresses are present.

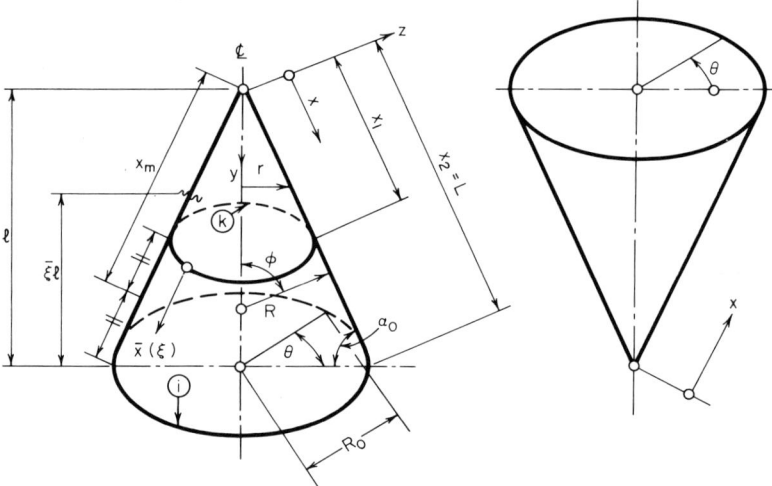

figure 4-5 *Cone nomenclature.*

Formulas are assembled for closed and open conical shells. Open conical shells are characterized by removal of the part above some circumference in a plane parallel to the base.

Linear bending theory was used to derive the following formulas. If the height of the segment is less than $\sqrt{R_0 t}$, the segment is essentially a circular ring instead of a shell.

The following terms are defined:

$$k = \frac{l}{\sqrt{R_0 t} \sin \phi} \sqrt[4]{3(1-\mu^2)} \qquad R_0 = \max R$$

$$D = \frac{Et^3}{12(1-\mu^2)}$$

Additional designations are shown in Fig. 4-5.

R is variable and perpendicular to the meridian. The angle ϕ is constant. Table 4-7 presents the formulas for a closed conical shell.

Open Conical Shell—Loading at Lower Edge

Since the disturbances due to edge loadings do not progress very far from the edge into the cone, the formulas presented in Table 4-7 can be used for a cone with an opening at the vertex (Fig. 4-6).

TABLE 4-7 Conical Shell, Edge-loading Solutions (Ref. 4-4)

	Horizontal load	Moment loading
	(conical shell diagram with ϕ, R, r, ℓ, $a\ell$, α_0, H)	(conical shell diagram with ϕ, θ, R, r, ℓ, $a\ell$, M)
N_ϕ	$\left[-\sqrt{2}\cos\phi\, e^{-k\alpha}\cos(k\alpha+\tfrac{\pi}{4})\right]H$	$\left(-\dfrac{2k\cos\phi}{\ell}e^{-k\alpha}\sin k\alpha\right)M$
N_θ	$\left(-\dfrac{2Rk\sin^2\phi}{\ell}e^{-k\alpha}\cos k\alpha\right)H$	$\left[\dfrac{2\sqrt{2}\,Rk^2\sin^2\phi}{\ell^2}e^{-k\alpha}\cos(k\alpha+\tfrac{\pi}{4})\right]M$
M_ϕ	$H\dfrac{\ell}{k}e^{-k\alpha}\sin k\alpha$	$\left[-\sqrt{2}\,e^{-k\alpha}\sin(k\alpha+\tfrac{\pi}{4})\right]M$
M_θ	$H\dfrac{\ell^2}{\sqrt{2}\,Rk^2}\dfrac{\cot\phi}{\sin\phi}e^{-k\alpha}\sin(k\alpha+\tfrac{\pi}{4})+\mu M_\phi$	$\left(-\dfrac{\ell\cot\phi\, e^{-k\alpha}\cos k\alpha}{Rk\sin\alpha}+\mu M_\phi\right)M$
Q	$H\left[-\sqrt{2}\sin\phi\, e^{-k\alpha}\cos(k\alpha+\tfrac{\pi}{4})\right]$	$\left(-\dfrac{2k\sin\phi}{\ell}e^{-k\alpha}\sin k\alpha\right)M$
	Deformations	
Δr	$H\dfrac{\ell^3 e^{-k\alpha}}{2Dk^3\sin\phi}\left[\cos k\alpha-\mu\dfrac{\ell}{\sqrt{2}\,Rk}\dfrac{\cot\phi}{\sin\phi}\cos(k\alpha+\tfrac{\pi}{4})\right]$	$-\dfrac{\ell^2 e^{-k\alpha}}{2Dk^2\sin\phi}\left[\sqrt{2}\cos(k\alpha+\tfrac{\pi}{4})+\mu\dfrac{1}{R}\dfrac{\cos\phi\sin k\alpha}{k\sin^2\phi}\right]M$
β	$-\dfrac{\ell^2 e^{-k\alpha}\sin(k\alpha+\pi/4)}{\sqrt{2}\,Dk^2\sin\phi}H$	$\dfrac{\ell e^{-k\alpha}\cos k\alpha}{Dk\sin\phi}M$
	For $\alpha=0$	
Δr	$\dfrac{\ell^3}{2Dk^3\sin\phi}\left(1-\dfrac{\mu\ell\cot\phi}{2Rk\sin\phi}\right)H$	$-\dfrac{\ell^2}{2Dk^2\sin\phi}M$
β	$-\dfrac{\ell^2}{2Dk^2\sin\phi}H$	$\dfrac{\ell}{Dk\sin\phi}M$

For k factors, see Sec. 4-3.
NOTE: In this table a is a coefficient which locates the section, (not to be mistaken for angle α_0).
For sign convention see Fig. 4-1.

Open Conical Shell—Loading at Upper Edge

If it is imagined that the shell, loaded as per Fig. 4-7a, is replaced by a shell as per Fig. 4-7b, the result is a conical shell loaded with loading at the lower edge. The same formulas are used for determining edge disturbances with $\phi > 90°$.

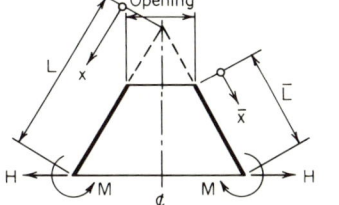

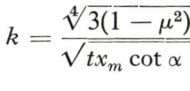

figure 4-6 *Open conical shell loading at lower edge.*

An additional set of formulas for open conical shells is presented in Table 4-8. These formulas may also be used for closed cones. The formulas are expressed in terms of the functions F_i and $F_i(\xi)$, which are tabulated in Chap. 5. The factor k is defined as follows:

$$k = \frac{\sqrt[4]{3(1-\mu^2)}}{\sqrt{tx_m \cot \alpha}}$$

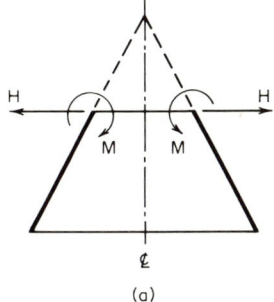
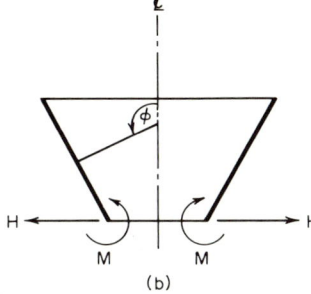

(a)   (b)

figure 4-7 *Open conical shell loading at upper edge.*

The sign convention used in Ref. 4-1 is as shown in Fig. 4-8. Non-dimensional parameter

$$\xi = \frac{\bar{x}}{L}$$

In the case of a cone in which the edge displacements are known, solutions in terms of the stresses and displacements throughout the cone are given in Table 4-9. Displacements and rotations are defined as follows:

90   Structural Analysis of Shells

At the lower boundary i:

Δr_{ik} = displacement in horizontal direction

β_{ik} = rotation

At the upper boundary k:

Δr_{ki} = displacement in horizontal direction

β_{ki} = rotation

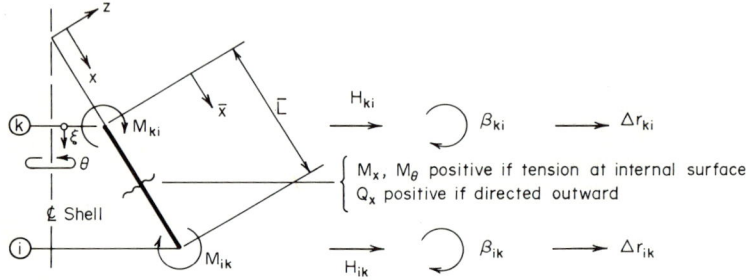

figure 4-8 *Sign convention for open conical shell.*

TABLE 4-8 Open Conical Shell—Edge-loading Solutions (Ref. 4-1)

N_x	$M_{ki} 2k \cot \alpha_0 \left[\dfrac{F_6}{F_1} F_{15}(\xi) + \dfrac{F_5}{F_1} F_{16}(\xi) - \dfrac{F_3}{F_1} F_8(\xi) \right]$	$M_{ik} 2k \cot \alpha_0 \left[\dfrac{F_8}{F_2} F_{10}(\xi) - \dfrac{F_{10}}{F_1} F_8(\xi) \right]$
N_θ	$M_{ki} 2k^2 x_m \cot \alpha_0 \left[\dfrac{F_6}{F_1} F_{14}(\xi) + \dfrac{F_5}{F_1} F_{13}(\xi) - \dfrac{F_3}{F_1} F_{10}(\xi) \right]$	$M_{ik} 2k^2 x_m \cot \alpha_0 \left[\dfrac{2F_8}{F_1} F_7(\xi) - \dfrac{F_{10}}{F_1} F_{10}(\xi) \right]$
M_x	$M_{ki} \left[\dfrac{F_6}{F_1} F_{13}(\xi) - \dfrac{F_5}{F_1} F_{14}(\xi) + \dfrac{F_3}{F_1} F_9(\xi) \right]$	$-M_{ik} \left[\dfrac{2F_8}{F_1} F_8(\xi) - \dfrac{F_{10}}{F_1} F_9(\xi) \right]$
Q_x	$M_{ki} 2k \left[\dfrac{F_6}{F_1} F_{15}(\xi) + \dfrac{F_5}{F_1} F_{16}(\xi) - \dfrac{F_3}{F_1} F_8(\xi) \right]$	$M_{ik} 2k \left[\dfrac{F_8}{F_1} F_{10}(\xi) - \dfrac{F_{10}}{F_1} F_8(\xi) \right]$
Δr	$M_{ki} \dfrac{\sin \alpha_0}{2Dk^2} \left[\dfrac{F_6}{F_1} F_{14}(\xi) + \dfrac{F_5}{F_1} F_{13}(\xi) - \dfrac{F_3}{F_1} F_{10}(\xi) \right]$	$M_{ik} \dfrac{\sin \alpha_0}{2Dk^2} \left[\dfrac{2F_8}{F_1} F_7(\xi) - \dfrac{F_{10}}{F_1} F_{10}(\xi) \right]$
β	$-\dfrac{M_{ki}}{Dk} \left[\dfrac{F_6}{F_1} F_{16}(\xi) - \dfrac{F_5}{F_1} F_{15}(\xi) - \dfrac{F_3}{F_1} F_7(\xi) \right]$	$M_{ik} \dfrac{1}{Dk} \left[\dfrac{F_8}{F_1} F_9(\xi) + \dfrac{F_{10}}{F_1} F_7(\xi) \right]$

For factors F, $F(\xi)$, and k see under "Definition of F Factors" in Sec. 5-2.

TABLE 4-8 Continued

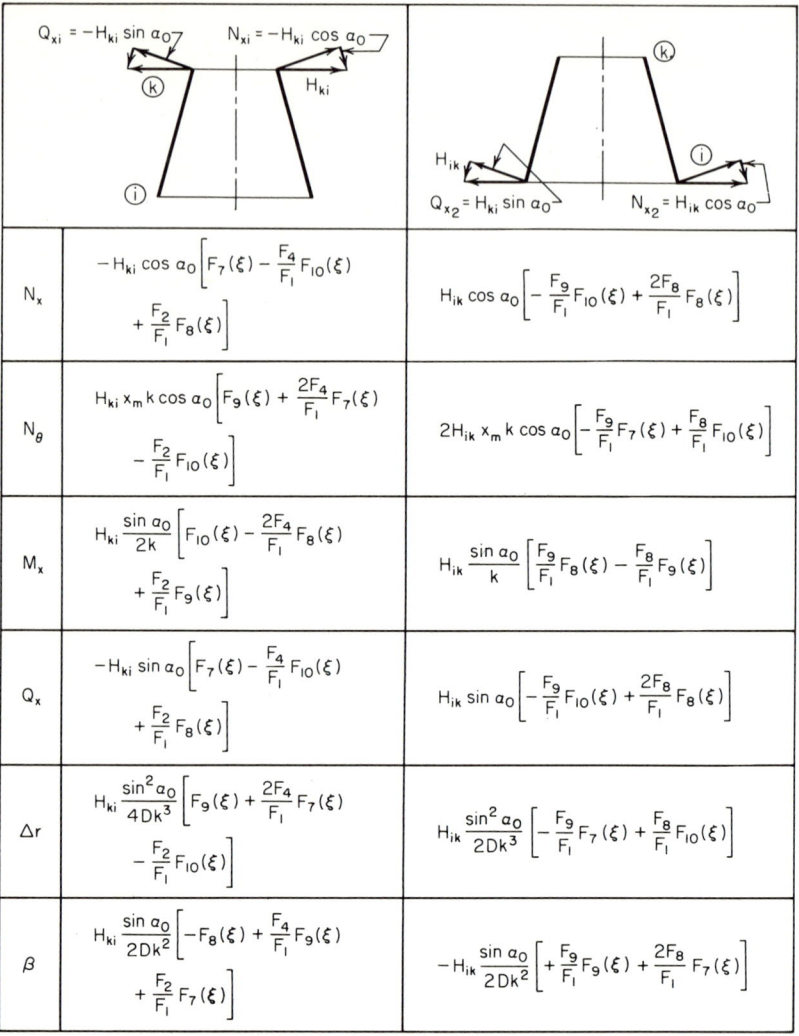

N_x	$-H_{ki}\cos\alpha_0 \left[F_7(\xi) - \dfrac{F_4}{F_1} F_{10}(\xi) \right]$ $+ \dfrac{F_2}{F_1} F_8(\xi)$	$H_{ik}\cos\alpha_0 \left[-\dfrac{F_9}{F_1} F_{10}(\xi) + \dfrac{2F_8}{F_1} F_8(\xi) \right]$
N_θ	$H_{ki} x_m k \cos\alpha_0 \left[F_9(\xi) + \dfrac{2F_4}{F_1} F_7(\xi) \right]$ $- \dfrac{F_2}{F_1} F_{10}(\xi)$	$2H_{ik} x_m k \cos\alpha_0 \left[-\dfrac{F_9}{F_1} F_7(\xi) + \dfrac{F_8}{F_1} F_{10}(\xi) \right]$
M_x	$H_{ki} \dfrac{\sin\alpha_0}{2k} \left[F_{10}(\xi) - \dfrac{2F_4}{F_1} F_8(\xi) \right]$ $+ \dfrac{F_2}{F_1} F_9(\xi)$	$H_{ik} \dfrac{\sin\alpha_0}{k} \left[\dfrac{F_9}{F_1} F_8(\xi) - \dfrac{F_8}{F_1} F_9(\xi) \right]$
Q_x	$-H_{ki}\sin\alpha_0 \left[F_7(\xi) - \dfrac{F_4}{F_1} F_{10}(\xi) \right]$ $+ \dfrac{F_2}{F_1} F_8(\xi)$	$H_{ik}\sin\alpha_0 \left[-\dfrac{F_9}{F_1} F_{10}(\xi) + \dfrac{2F_8}{F_1} F_8(\xi) \right]$
Δr	$H_{ki} \dfrac{\sin^2\alpha_0}{4Dk^3} \left[F_9(\xi) + \dfrac{2F_4}{F_1} F_7(\xi) \right]$ $- \dfrac{F_2}{F_1} F_{10}(\xi)$	$H_{ik} \dfrac{\sin^2\alpha_0}{2Dk^3} \left[-\dfrac{F_9}{F_1} F_7(\xi) + \dfrac{F_8}{F_1} F_{10}(\xi) \right]$
β	$H_{ki} \dfrac{\sin\alpha_0}{2Dk^2} \left[-F_8(\xi) + \dfrac{F_4}{F_1} F_9(\xi) \right]$ $+ \dfrac{F_2}{F_1} F_7(\xi)$	$-H_{ik} \dfrac{\sin\alpha_0}{2Dk^2} \left[+\dfrac{F_9}{F_1} F_9(\xi) + \dfrac{2F_8}{F_1} F_7(\xi) \right]$

For factors F, F(ξ), and k see under "Definition of F Factors" in Sec. 5-2.

TABLE 4-9 Open Conical Shell—Edge-loading Solutions (Ref. 4-1)

	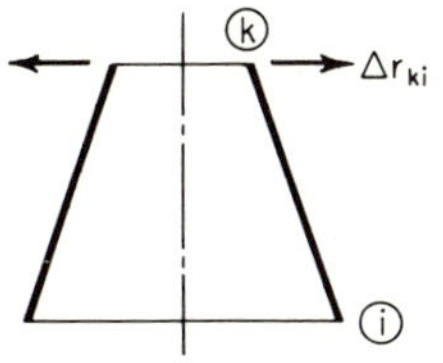	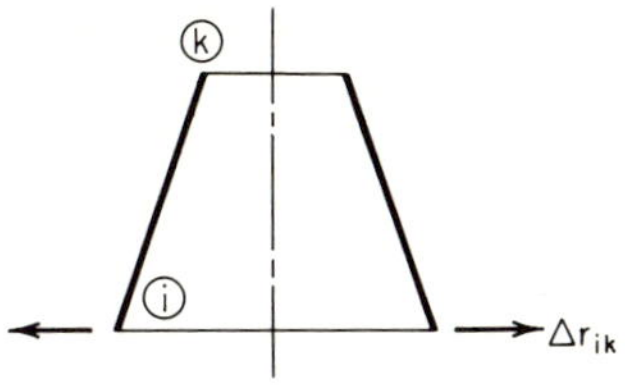
	(left cone, Δr_{ki} at top)	(right cone, Δr_{ik} at top)
N_x	$\dfrac{4Dk^3 \cot \alpha_0\, \Delta r_{ki}}{\sin \alpha_0} \left[-\dfrac{F_3}{F_1} F_7(\xi) - \dfrac{F_5}{F_1} F_{15}(\xi) + \dfrac{F_6}{F_1} F_{16}(\xi) \right]$	$\dfrac{4Dk^3 \cot \alpha_0\, \Delta r_{ik}}{\sin \alpha_0} \left[\dfrac{F_{10}}{F_1} F_7(\xi) + \dfrac{F_8}{F_1} F_9(\xi) \right]$
N_θ	$\dfrac{Et\, \Delta r_{ki}}{x_m \cos \alpha_0} \left[\dfrac{F_3}{F_1} F_9(\xi) - \dfrac{F_5}{F_1} F_{14}(\xi) + \dfrac{F_6}{F_1} F_{13}(\xi) \right]$	$\dfrac{Et\, \Delta r_{ik}}{x_m \cos \alpha_0} \left[-\dfrac{F_{10}}{F_1} F_9(\xi) + \dfrac{2F_8}{F_1} F_8(\xi) \right]$
M_x	$\dfrac{2Dk^2\, \Delta r_{ki}}{\sin \alpha_0} \left[\dfrac{F_3}{F_1} F_{10}(\xi) - \dfrac{F_5}{F_1} F_{13}(\xi) - \dfrac{F_6}{F_1} F_{14}(\xi) \right] \quad M_\phi = \mu M_x$	$\dfrac{2Dk^2\, \Delta r_{ik}}{\sin \alpha_0} \left[-\dfrac{F_{10}}{F_1} F_{10}(\xi) + \dfrac{2F_8}{F_1} F_7(\xi) \right]$
Q_x	$\dfrac{4Dk^3\, \Delta r_{ki}}{\sin \alpha_0} \left[-\dfrac{F_3}{F_1} F_7(\xi) - \dfrac{F_5}{F_1} F_{15}(\xi) + \dfrac{F_6}{F_1} F_{16}(\xi) \right]$	$\dfrac{4Dk^3\, \Delta r_{ik}}{\sin \alpha_0} \left[\dfrac{F_{10}}{F_1} F_7(\xi) + \dfrac{F_8}{F_1} F_9(\xi) \right]$
Δr	$\Delta r_{ki} \left[\dfrac{F_3}{F_1} F_9(\xi) - \dfrac{F_5}{F_1} F_{14}(\xi) + \dfrac{F_6}{F_1} F_{13}(\xi) \right]$	$\Delta r_{ik} \left[-\dfrac{F_{10}}{F_1} F_9(\xi) + \dfrac{2F_8}{F_1} F_8(\xi) \right]$
β	$-\dfrac{2k}{\sin \alpha_0} \Delta r_{ki} \left[\dfrac{F_3}{F_1} F_8(\xi) - \dfrac{F_5}{F_1} F_{16}(\xi) - \dfrac{F_6}{F_1} F_{15}(\xi) \right]$	$\dfrac{2k\, \Delta r_{ik}}{\sin \alpha_0} \left[\dfrac{F_{10}}{F_1} F_8(\xi) - \dfrac{F_8}{F_1} F_{16}(\xi) - \dfrac{F_8}{F_1} F_{15}(\xi) \right]$

For factors F, $F(\xi)$, and k see under "Definition of F Factors" in Sec. 5-2.

94 *Structural Analysis of Shells*

TABLE 4-9 Continued

	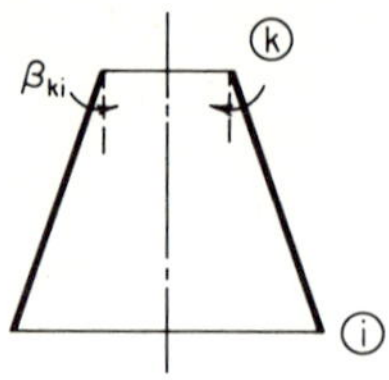 β_{ki} (trapezoid with k at top right, i at bottom right)	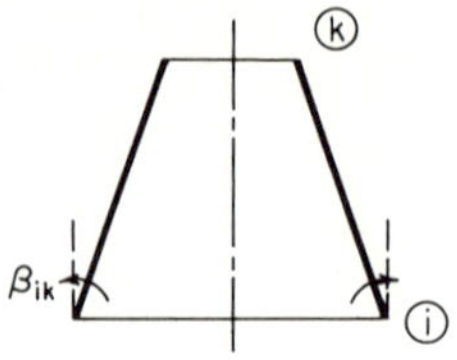 β_{ik} (trapezoid with k at top right, i at bottom right)
N_x	$2Dk^2 \cot a_0 \, \beta_{ki} \left[\dfrac{F_2}{F_1} F_7(\xi) + \dfrac{F_4}{F_1} F_9(\xi) - F_8(\xi) \right]$	$2Dk^2 \cot a_0 \, \beta_{ik} \left[\dfrac{2F_8}{F_1} F_7(\xi) + \dfrac{F_9}{F_1} F_9(\xi) \right]$
N_θ	$2Dk^3 x_m \cot a_0 \, \beta_{ki} \left[-\dfrac{F_2}{F_1} F_9(\xi) + \dfrac{2F_4}{F_1} F_8(\xi) - F_{10}(\xi) \right]$	$4Dk^3 x_m \cot a_0 \, \beta_{ik} \left[-\dfrac{F_8}{F_1} F_9(\xi) + \dfrac{F_9}{F_1} F_8(\xi) \right]$
M_x	$Dk \, \beta_{ki} \left[-\dfrac{F_2}{F_1} F_{10}(\xi) + \dfrac{2F_4}{F_1} F_7(\xi) + F_9(\xi) \right]$	$2Dk \, \beta_{ik} \left[-\dfrac{F_8}{F_1} F_{10}(\xi) + \dfrac{F_9}{F_1} F_7(\xi) \right]$
Q_x	$2Dk^2 \beta_{ki} \left[\dfrac{F_2}{F_1} F_7(\xi) + \dfrac{F_4}{F_1} F_9(\xi) - F_8(\xi) \right]$	$2Dk^2 \beta_{ik} \left[\dfrac{2F_8}{F_1} F_7(\xi) + \dfrac{F_9}{F_1} F_9(\xi) \right]$
Δr	$\dfrac{\sin a_0}{2k} \beta_{ki} \left[-\dfrac{F_2}{F_1} F_9(\xi) + \dfrac{2F_4}{F_1} F_8(\xi) - F_{10}(\xi) \right]$	$\dfrac{\sin a_0}{k} \beta_{ik} \left[-\dfrac{F_8}{F_1} F_9(\xi) + \dfrac{F_9}{F_1} F_8(\xi) \right]$
β	$\beta_{ki} \left[\dfrac{F_2}{F_1} F_8(\xi) - \dfrac{F_4}{F_1} F_{10}(\xi) + F_7(\xi) \right]$	$\beta_{ik} \left[\dfrac{2F_8}{F_1} F_8(\xi) - \dfrac{F_9}{F_1} F_{10}(\xi) \right]$

For factors F, $F(\xi)$, and k see under "Definition of F Factors" in Sec. 5-2.

4-4 Cylindrical Shells

This section presents the solutions for long and short cylinders, loaded along a boundary by edge loadings (moment, shear, forced horizontal displacement, forced rotation at the boundary). All disturbances in the cylindrical wall caused by edge loading will become, for practical purposes, negligible at distance $\sqrt{Rt}$ from the disturbed edge. If the length of the cylinder is less than $\sqrt{Rt}$, the cylinder is essentially a circular ring instead of a shell. Furthermore, to be on the safe side, the following precautions should be observed:

1. If $\beta L \geqslant 5$, the simplified formulas are used, and this is a special case of the more general case 2.
2. If $\beta L < 5$, the more exact theory is used, and such cylinders are designated as short cylinders.

The factor β is defined as follows:

$$\beta^4 = \frac{3(1-\mu^2)}{R^2 t^2}$$

and should be distinguished from the similar designation β which is an angle of rotation due to the edge loadings.

The primary solutions (membrane theory) will not be affected by the length of the cylinder.

The boundaries must be free to rotate and deflect because of the assumed action of the edge loadings. The shell thickness must be uniform in the rengion where the stresses are present.

Long Cylinders

The following constants are defined:

$$k = \frac{L}{\sqrt{Rt}} \sqrt[4]{3(1-\mu^2)}$$

$$D = \frac{Et^3}{12(1-\mu^2)}$$

The formulas for the disturbances caused by the edge loadings are presented in Table 4-10.

96 *Structural Analysis of Shells*

TABLE 4-10 Long Cylindrical Shells, Edge-loading Solutions (Ref. 4-4)

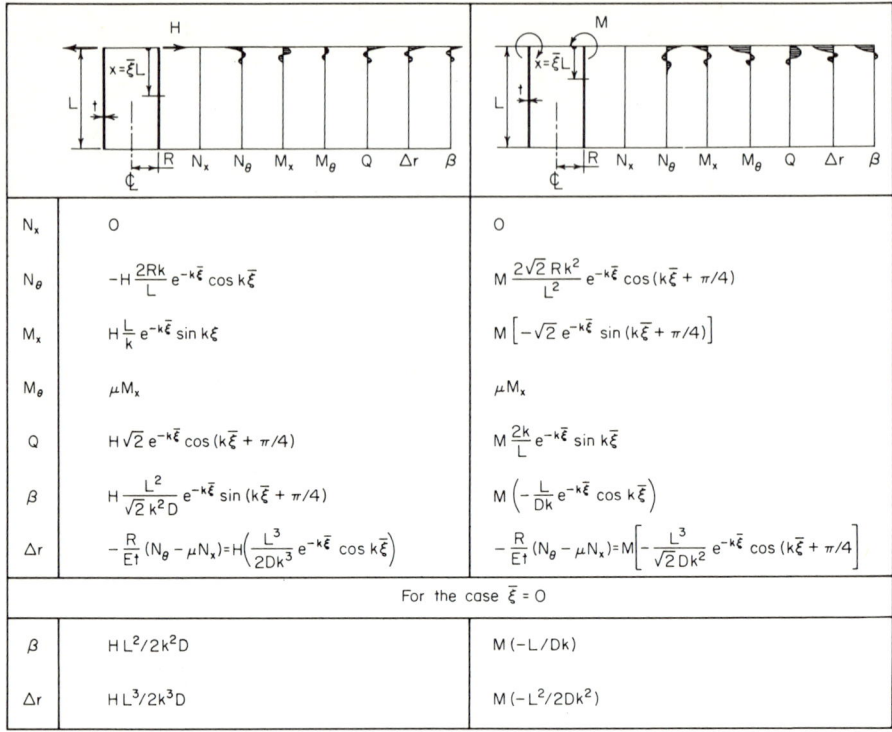

N_x	0	0
N_θ	$-H \dfrac{2Rk}{L} e^{-k\bar{\xi}} \cos k\bar{\xi}$	$M \dfrac{2\sqrt{2}\, Rk^2}{L^2} e^{-k\bar{\xi}} \cos(k\bar{\xi} + \pi/4)$
M_x	$H \dfrac{L}{k} e^{-k\bar{\xi}} \sin k\bar{\xi}$	$M\left[-\sqrt{2}\, e^{-k\bar{\xi}} \sin(k\bar{\xi} + \pi/4)\right]$
M_θ	μM_x	μM_x
Q	$H\sqrt{2}\, e^{-k\bar{\xi}} \cos(k\bar{\xi} + \pi/4)$	$M \dfrac{2k}{L} e^{-k\bar{\xi}} \sin k\bar{\xi}$
β	$H \dfrac{L^2}{\sqrt{2}\, k^2 D} e^{-k\bar{\xi}} \sin(k\bar{\xi} + \pi/4)$	$M\left(-\dfrac{L}{Dk} e^{-k\bar{\xi}} \cos k\bar{\xi}\right)$
Δr	$-\dfrac{R}{Et}(N_\theta - \mu N_x) = H\left(\dfrac{L^3}{2Dk^3} e^{-k\bar{\xi}} \cos k\bar{\xi}\right)$	$-\dfrac{R}{Et}(N_\theta - \mu N_x) = M\left[-\dfrac{L^3}{\sqrt{2}\, Dk^2} e^{-k\bar{\xi}} \cos(k\bar{\xi} + \pi/4)\right]$
	For the case $\bar{\xi} = 0$	
β	$HL^2/2k^2 D$	$M(-L/Dk)$
Δr	$HL^3/2k^3 D$	$M(-L^2/2Dk^2)$

For k factors see Sec. 4-4.
For sign convention see Fig. 4-1.

Short Cylinders with Uniform Wall Thickness without Abrupt Discontinuity

Hampe (Ref. 4-1) gives exact solutions for cylinders, based on linear bending theory. Tables 4-11 and 4-12 present the solutions for the edge loadings M_{ik} and H_{ik}, forced rotation of the edge β_{ik}, and forced displacement of the edge Δr_{ik}. All formulas are valid for both long and short cylinders.

If the length of cylinder is such that $L \geqslant 3.1\sqrt{Rt}$, the formulas can be simplified. The simplification is also considered in Tables 4-13 and 4-14. The functions F_i and $F_i(\xi)$ and factor k are from Sec. 5-3. These formulas are for long cylinders and are slightly more accurate than in Table 4-10.

The sign convention in Fig. 4-9 is used by Hampe (Ref. 4-1).
In Table 4-15 ξ is measured from the unloaded boundary.

TABLE 4-11 Cylindrical Shells—Exact Formulas for Edge-loading Solutions (Ref. 4-1)

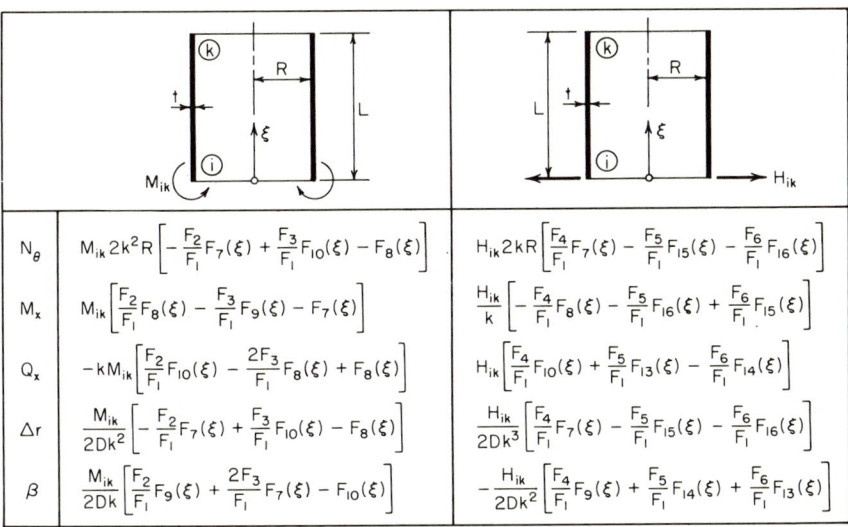

N_θ	$M_{ik} 2k^2 R \left[-\dfrac{F_2}{F_1} F_7(\xi) + \dfrac{F_3}{F_1} F_{10}(\xi) - F_8(\xi) \right]$	$H_{ik} 2kR \left[\dfrac{F_4}{F_1} F_7(\xi) - \dfrac{F_5}{F_1} F_{15}(\xi) - \dfrac{F_6}{F_1} F_{16}(\xi) \right]$
M_x	$M_{ik} \left[\dfrac{F_2}{F_1} F_8(\xi) - \dfrac{F_3}{F_1} F_9(\xi) - F_7(\xi) \right]$	$\dfrac{H_{ik}}{k} \left[-\dfrac{F_4}{F_1} F_8(\xi) - \dfrac{F_5}{F_1} F_{16}(\xi) + \dfrac{F_6}{F_1} F_{15}(\xi) \right]$
Q_x	$-kM_{ik} \left[\dfrac{F_2}{F_1} F_{10}(\xi) - \dfrac{2F_3}{F_1} F_8(\xi) + F_8(\xi) \right]$	$H_{ik} \left[\dfrac{F_4}{F_1} F_{10}(\xi) + \dfrac{F_5}{F_1} F_{13}(\xi) - \dfrac{F_6}{F_1} F_{14}(\xi) \right]$
Δr	$\dfrac{M_{ik}}{2Dk^2} \left[-\dfrac{F_2}{F_1} F_7(\xi) + \dfrac{F_3}{F_1} F_{10}(\xi) - F_8(\xi) \right]$	$\dfrac{H_{ik}}{2Dk^3} \left[\dfrac{F_4}{F_1} F_7(\xi) - \dfrac{F_5}{F_1} F_{15}(\xi) - \dfrac{F_6}{F_1} F_{16}(\xi) \right]$
β	$\dfrac{M_{ik}}{2Dk} \left[\dfrac{F_2}{F_1} F_9(\xi) + \dfrac{2F_3}{F_1} F_7(\xi) - F_{10}(\xi) \right]$	$-\dfrac{H_{ik}}{2Dk^2} \left[\dfrac{F_4}{F_1} F_9(\xi) + \dfrac{F_5}{F_1} F_{14}(\xi) + \dfrac{F_6}{F_1} F_{13}(\xi) \right]$

For factors F, $F(\xi)$, and k see under "Definition of F Factors" in Sec. 5-2.

TABLE 4-12 Cylindrical Shells—Exact Formulas for Edge Deformations (Ref. 4-1)

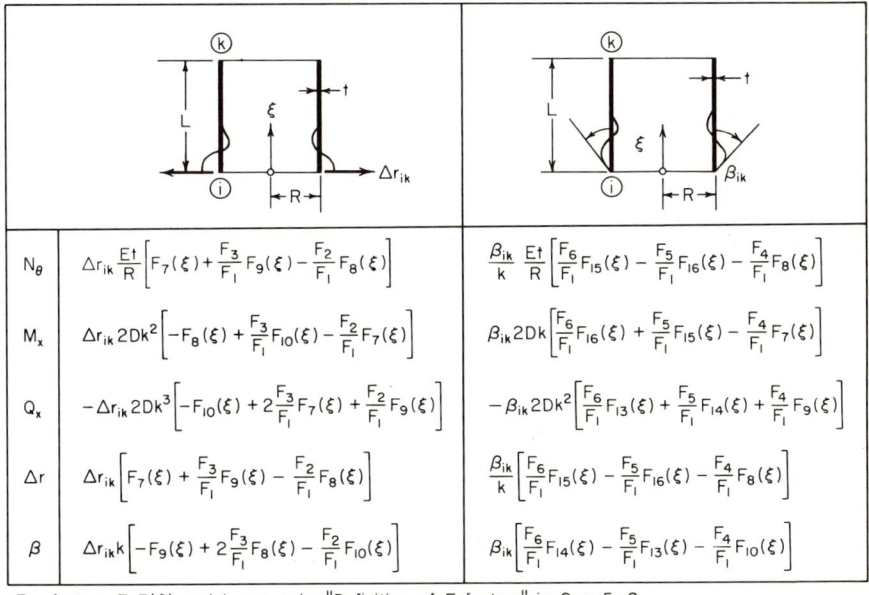

N_θ	$\Delta r_{ik} \dfrac{Et}{R} \left[F_7(\xi) + \dfrac{F_3}{F_1} F_9(\xi) - \dfrac{F_2}{F_1} F_8(\xi) \right]$	$\dfrac{\beta_{ik}}{k} \dfrac{Et}{R} \left[\dfrac{F_6}{F_1} F_{15}(\xi) - \dfrac{F_5}{F_1} F_{16}(\xi) - \dfrac{F_4}{F_1} F_8(\xi) \right]$
M_x	$\Delta r_{ik} 2Dk^2 \left[-F_8(\xi) + \dfrac{F_3}{F_1} F_{10}(\xi) - \dfrac{F_2}{F_1} F_7(\xi) \right]$	$\beta_{ik} 2Dk \left[\dfrac{F_6}{F_1} F_{16}(\xi) + \dfrac{F_5}{F_1} F_{15}(\xi) - \dfrac{F_4}{F_1} F_7(\xi) \right]$
Q_x	$-\Delta r_{ik} 2Dk^3 \left[-F_{10}(\xi) + 2\dfrac{F_3}{F_1} F_7(\xi) + \dfrac{F_2}{F_1} F_9(\xi) \right]$	$-\beta_{ik} 2Dk^2 \left[\dfrac{F_6}{F_1} F_{13}(\xi) + \dfrac{F_5}{F_1} F_{14}(\xi) + \dfrac{F_4}{F_1} F_9(\xi) \right]$
Δr	$\Delta r_{ik} \left[F_7(\xi) + \dfrac{F_3}{F_1} F_9(\xi) - \dfrac{F_2}{F_1} F_8(\xi) \right]$	$\dfrac{\beta_{ik}}{k} \left[\dfrac{F_6}{F_1} F_{15}(\xi) - \dfrac{F_5}{F_1} F_{16}(\xi) - \dfrac{F_4}{F_1} F_8(\xi) \right]$
β	$\Delta r_{ik} k \left[-F_9(\xi) + 2\dfrac{F_3}{F_1} F_8(\xi) - \dfrac{F_2}{F_1} F_{10}(\xi) \right]$	$\beta_{ik} \left[\dfrac{F_6}{F_1} F_{14}(\xi) - \dfrac{F_5}{F_1} F_{13}(\xi) - \dfrac{F_4}{F_1} F_{10}(\xi) \right]$

For factors F, $F(\xi)$, and k see under "Definition of F factors" in Sec. 5-2.

98 *Structural Analysis of Shells*

TABLE 4-13 Long Cylindrical Shells—Approximate Solutions for Edge Loading (Ref. 4-1)

External forces and deformations	M_{ik} applied	H_{ik} applied	Remarks
N_θ	$-M_{ik}2k^2RF_{20}(\xi)$	$H_{ik}2kRF_{17}(\xi)$	
M_x	$-M_{ik}F_{19}(\xi)$	$\dfrac{H_{ik}}{k}F_{18}(\xi)$	
Q_x	$-M_{ik}2kF_{18}(\xi)$	$-H_{ik}F_{20}(\xi)$	$kL \geq 4$ or $L \geq 3.1\sqrt{Rt}$ must be satisfied
Δr	$-\dfrac{M_{ik}}{2Dk^2}F_{20}(\xi)$	$\dfrac{H_{ik}}{2Dk^3}F_{17}(\xi)$	
β	$\dfrac{M_{ik}}{Dk}F_{17}(\xi)$	$-\dfrac{H_{ik}}{2Dk^2}F_{19}(\xi)$	
Approximation used	$\dfrac{F_3}{F_1} \approx \dfrac{F_2}{F_1} \approx 1$	$\dfrac{F_4}{F_1} \approx \dfrac{F_6}{F_1} \approx 1 \quad \dfrac{F_5}{F_1} \approx 0$	

For factors F, F(ξ), and k see under "Definition of F Factors" in Sec. 5-2.

TABLE 4-14 Cylindrical Shells—Approximate Solutions for Edge Deformations (Ref. 4-1)

Internal forces and deformations	Δr_{ik} applied	β_{ik} applied	Remarks
N_θ	$\Delta r_{ik}\dfrac{Et}{R}F_{19}(\xi)$	$\dfrac{\beta_{ik}}{k}\dfrac{Et}{R}F_{18}(\xi)$	
M_x	$-\Delta r_{ik}2Dk^2F_{20}(\xi)$	$-\beta_{ik}2DkF_{17}(\xi)$	
Q_x	$-\Delta r_{ik}4Dk^3F_{17}(\xi)$	$-\beta_{ik}2Dk^2F_{19}(\xi)$	$kL \geq 4$ or $L \geq 3.1\sqrt{Rt}$ must be satisfied
Δr	$\Delta r_{ik}F_{19}(\xi)$	$\dfrac{\beta_{ik}}{k}F_{18}(\xi)$	
β_i	$-\Delta r_{ik}2kF_{18}(\xi)$	$\beta_{ik}F_{20}(\xi)$	
Approximation used	$\dfrac{F_3}{F_1} \approx \dfrac{F_2}{F_1} \approx 1$	$\dfrac{F_6}{F_1} = \dfrac{F_4}{F_1} \approx 1 \quad \dfrac{F_5}{F_1} \approx 0$	

For factors F, F(ξ), and k see under "Definition of F Factors" in Sec. 5-2.

TABLE 4-15   Cylindrical Shell—Edge Loading at Upper Edge   (Ref. 4-1)

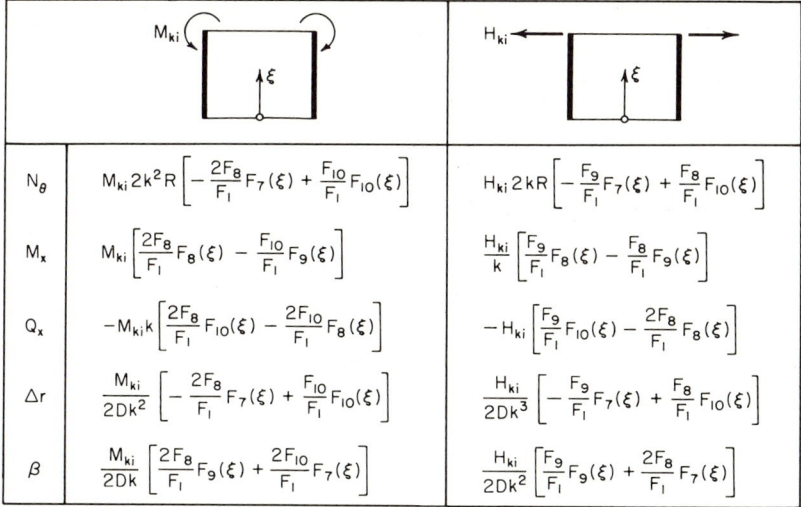

	M_{ki}	H_{ki}
N_θ	$M_{ki} 2k^2 R \left[-\dfrac{2F_8}{F_1} F_7(\xi) + \dfrac{F_{10}}{F_1} F_{10}(\xi) \right]$	$H_{ki} 2kR \left[-\dfrac{F_9}{F_1} F_7(\xi) + \dfrac{F_8}{F_1} F_{10}(\xi) \right]$
M_x	$M_{ki} \left[\dfrac{2F_8}{F_1} F_8(\xi) - \dfrac{F_{10}}{F_1} F_9(\xi) \right]$	$\dfrac{H_{ki}}{k} \left[\dfrac{F_9}{F_1} F_8(\xi) - \dfrac{F_8}{F_1} F_9(\xi) \right]$
Q_x	$-M_{ki} k \left[\dfrac{2F_8}{F_1} F_{10}(\xi) - \dfrac{2F_{10}}{F_1} F_8(\xi) \right]$	$-H_{ki} \left[\dfrac{F_9}{F_1} F_{10}(\xi) - \dfrac{2F_8}{F_1} F_8(\xi) \right]$
Δr	$\dfrac{M_{ki}}{2Dk^2} \left[-\dfrac{2F_8}{F_1} F_7(\xi) + \dfrac{F_{10}}{F_1} F_{10}(\xi) \right]$	$\dfrac{H_{ki}}{2Dk^3} \left[-\dfrac{F_9}{F_1} F_7(\xi) + \dfrac{F_8}{F_1} F_{10}(\xi) \right]$
β	$\dfrac{M_{ki}}{2Dk} \left[\dfrac{2F_8}{F_1} F_9(\xi) + \dfrac{2F_{10}}{F_1} F_7(\xi) \right]$	$\dfrac{H_{ki}}{2Dk^2} \left[\dfrac{F_9}{F_1} F_9(\xi) + \dfrac{2F_8}{F_1} F_7(\xi) \right]$

For factors F, F(ξ), and k see under "Definition of F Factors" in Sec. 5-2.

4-5 Approximate Method

Consider a shell of revolution of any meridional shape loaded by an axisymmetrical loading with restricting assumptions as stated in Sec. 3-12.

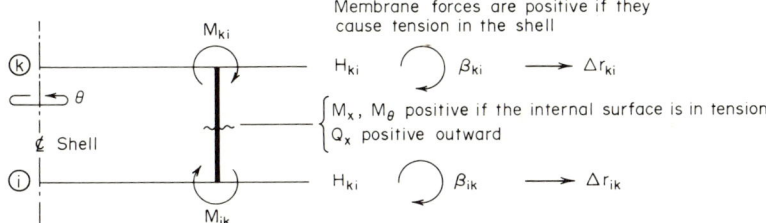

figure 4-9 *Sign convention for cylindrical shell.*

Structural Analysis of Shells

Displacements

The simplified formulas for displacements presented in Table 4-16 are from Ref. 4-5.

The ratio t_s/a is assumed to be small enough for the displacements, rotations, and stresses in the shell to be approximated by the same quantities as in a cylinder of constant thickness t_s tangent to the shell

TABLE 4-16 Irregular Shells—Edge-loading Solutions

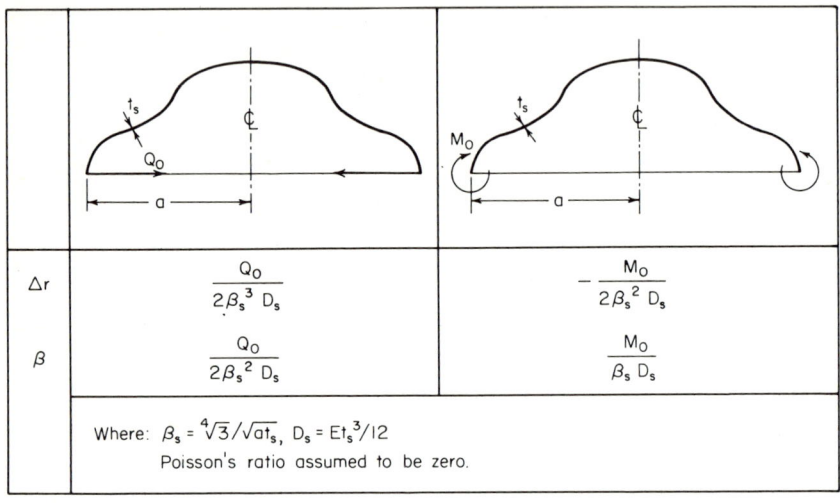

	Q_o	M_o
Δr	$\dfrac{Q_o}{2\beta_s^3 D_s}$	$-\dfrac{M_o}{2\beta_s^2 D_s}$
β	$\dfrac{Q_o}{2\beta_s^2 D_s}$	$\dfrac{M_o}{\beta_s D_s}$

Where: $\beta_s = \sqrt[4]{3}/\sqrt{at_s}$, $D_s = Et_s^3/12$
Poisson's ratio assumed to be zero.

at the equator. This approximation is satisfactory for shells whose ratio t_s/a is less than 1/50 and whose thickness near the edge does not change abruptly.

Interaction

Considering displacements of the primary and secondary solutions, the interaction analysis leads to the following formulas for discontinuity forces:

$$M_0 = m_0 p_0 a t_s \frac{a}{r_0} S_p$$

$$Q_0 = q_0 p_0 a \sqrt{\frac{t_s}{a}} \frac{a}{r_0} S_p$$

where

$$m_0 = \frac{1}{4\sqrt{3}} \frac{1-\chi^2}{(1+\chi^2)^2 + 2\chi^{3/2}(1+\chi)}$$

$$q_0 = \frac{1}{2\sqrt[4]{3}} \frac{1+\chi^{5/2}}{(1+\chi^2)^2 + 2\chi^{3/2}(1+\chi)}$$

$$\chi = \frac{t_s}{t_c}$$

where t_c = thickness of cylinder
S_p = factor, dependent on the loading (see Table 3-13)

The Maximum Values

The maximum value of M may be larger than M_0; consequently, it has to be determined. It was shown in Ref. 4-5 that a maximum value of M appears at a distance $\bar{x}_s$, in the bulkhead, defined as follows:

$$\tan \beta_s \bar{x}_s = \frac{q_0}{q_0 - 2\sqrt[4]{3}\, m_0}$$

This distance is measured along the meridian, starting from the discontinuity section. In a cylinder, the corresponding distance is $\bar{x}_c$, defined as follows:

$$\tan \beta_c \bar{x}_c = \frac{q_0}{q_0 + (2\sqrt[4]{3}\sqrt{\chi})\, m_0}$$

Then, for both bulkhead and cylinder, special values M_s (bulkhead) and M_c (cylinder) have to be determined. The larger value, M_0 or M, as defined above represents the actual maximum moment.

This analysis is performed using Fig. 4-10 and the formulas that are given on this graph for M and H. To obtain M_s or M_c, use the formula in Fig. 4-10, entering m_s, m_c, or m, as shown in the formula.

The simplicity of the results obtained allows checking of the shell design for boundary moments (the most critical loads in most practical cases) without any difficulty and in a routine manner.

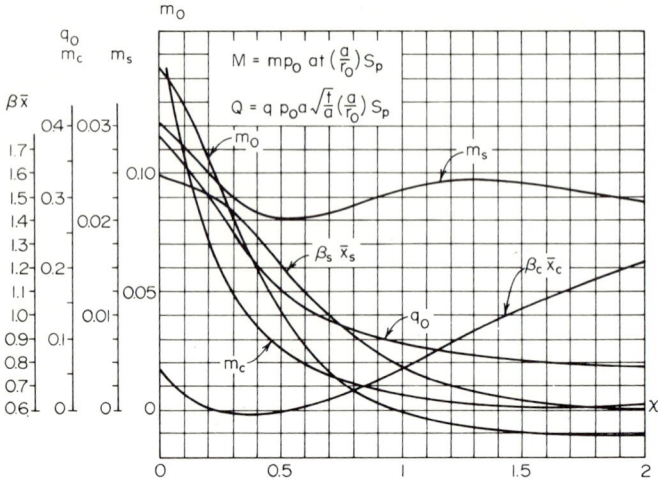

figure 4-10 *Graphical form of data.*

4-6 Some Practical Considerations

Table 4-17 illustrates an important fact for nonshallow shells. Deflections and internal loads due to edge loadings are of importance only locally near the edge. Observe that all stresses and deflections due to unit edge loadings disappear for $\alpha > 20°$ and are negligible for $\alpha > 10°$, as is shown in Table 4-17 for a spherical shell.

In the edge-loading cases the portion of the shell $\alpha > 20°$ is not needed for satisfying equilibrium. We can ignore in the analysis all material above $\alpha = 20°$, because this material does not contribute to the stresses

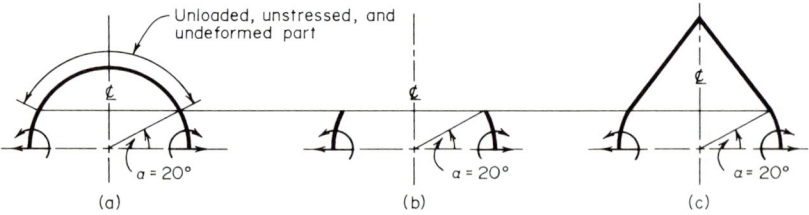

figure 4-11 *Statically analogous shells.*

or strains, which are computed for the zone defined by $0 < \alpha < 20°$. Nothing is changed in the regime of stresses or deformations in the zone $0 < \alpha < 20°$ if we replace the original shell with any shape of

TABLE 4-17 Edge-loading and Deformation Solutions

	N_θ	N_ϕ	M_θ	M_ϕ	H	Δr	θ

shell, as shown in Fig. 4-11. Consequently, cases A, B, and C in Fig. 4-11 are statically equivalent for the indicated loadings. This leads to the following conclusions:

1. The spherical shell of revolution, loaded by edge loadings M and H, acts as the lower segment acts under the same loading (segment defined by $\alpha = 20°$). Consequently, the shape of the rest of the shell does not affect the solution (see Fig. 4-12).

2. If the lower portion of a shell can be approximated by a spherical shell to a satisfactory degree, the solution obtained for the spherical shell, which is loaded by M and H all around the edges (Figs. 4-12 and 4-13), can be used for the actual shell.

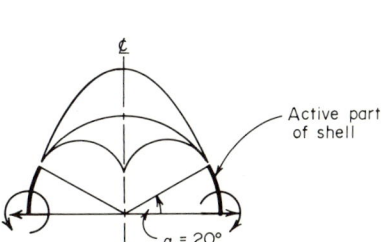

figure 4-12 *Different variants for unstressed portion.*

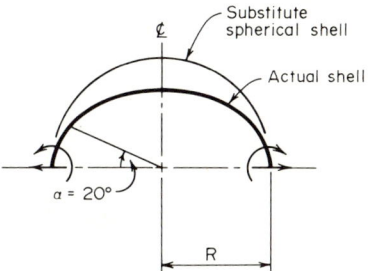

figure 4-13 *Approximation with a sphere.*

104 *Structural Analysis of Shells*

3. When extreme accuracy is required, $\alpha = 10°$ may be considered in place of $\alpha = 20°$.

Another approximation, known as Geckeler's assumption (Ref. 4-2), may be useful, i.e.:

If the thickness of the shell t is small in comparison with the equatorial radius and is limited by the relation $R/t > 50$, the bending stresses at the edge may be determined by cylindrical shell theory. This means that the bulkhead shell can be approximated by a cylinder to find solutions for the edge loadings.

Sometimes, there are bulkheads that cannot be easily approximated at the boundary zone with the spherical or cylindrical shell to obtain unit solutions. Figure 4-14 is an example of such shells. It is clear that

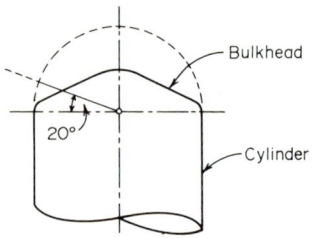

figure 4-14 *Bulkhead, which cannot be approximated with a sphere alone.*

within the range of 20°, the shell at the junction with the cylinder cannot be approximated by a sphere. In such a case, however, two spherical shells or a toroidal and a spherical shell will be needed for the approximation, as shown in Fig. 4-15. The determination of discontinuity

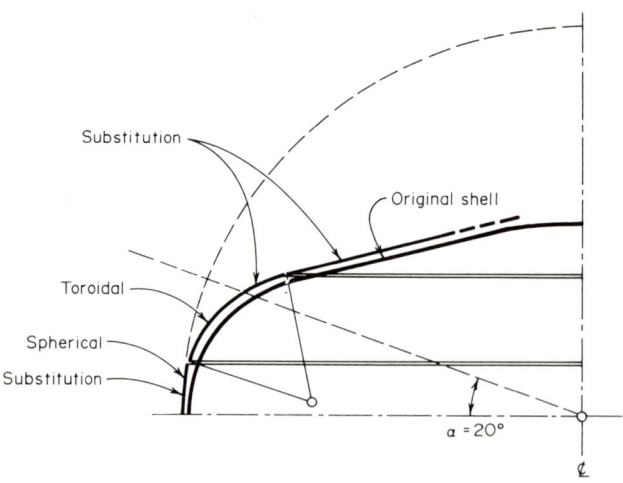

figure 4-15 *Combined bulkhead.*

stresses will be performed with the more complicated process, as will be shown in Chap. 6.

Up to this point, solutions have been discussed and presented for homogeneous and isotropic shells of constant thickness t, as far as deformation is concerned.

REFERENCES

4-1. Hampe, E.: *Statik Rotationssymmetrischer Flächentragwerke*, VEB Verlag für Bauwesen, Berlin, 1963.
4-2. Pflüger, Alf.: *Elementary Statics of Shells*, 2d ed., McGraw-Hill Book Company, Inc., New York, 1961.
4-3. Worch, G.: *Elastische Schalen* (in German), *Beton-Kalendar*, Wilhelm Ernst & Sohn KG, Berlin, 1958.
4-4. Worch, G.: Elastiche Schalen (in German), *Beton-Kalendar*, Wilhelm Ernst & Sohn KG, Berlin, 1941.
4-5. Salvadori, M. G.: Live Load and Temperature Moments in Shell of Rotation Built into Cylinders, *J. Am. Concrete Inst.*, October, 1955.

Chapter 5

SPECIAL SOLUTIONS

5-1 Introduction

Special solutions for specific cases of shells and plates are presented in this chapter. Solutions are presented for cylindrical shells under various loadings. The boundary conditions include fixed and pinned edges, which are defined as follows:

A free edge is unrestrained and permits displacements and rotations along the boundary.

A fixed or built-in edge does not permit any displacements or rotations along the boundary.

A pinned edge permits rotation of the support, but not displacement.

A special case of pinned edge, the simply supported edge, permits rotation of the edge and movement in one prescribed direction.

Many special cases have been solved; in particular, Hampe's solutions (Ref. 5-1) are presented. These solutions cover many cases of circular cylinders and spheres with various support conditions along the boundaries. Similarly, discontinuities in the wall thickness may be covered by the method.

Special Solutions   107

In addition, formulas are presented for circular plates (with and without a circular hole at the center) under various loading conditions. Solutions are also given for circular rings.

5-2 Hampe's Solutions

Hampe (Ref. 5-1) derived a set of general formulas for the stresses and deformations of cylindrical and spherical shells with various boundary conditions and loadings. The results of his derivations (based on classical linear bending theory) are presented in tabular form in this section.

The w factors listed below are for particular solutions of the differential equations:

w_p = ordinates of deflection line (deflection normal to shell surface)
w_p' = inclination of deflection line (rotation of meridian)
w_p'' = second derivative of deflection line
w_p''' = third derivative of deflection line

These factors, for different loadings, are tabulated in Table 5-1. The following constants are used:

F and $F(\xi)$: These functions of geometry are presented under "Definition of F Factors" (Sec. 5-2) and in Table 5-10.
S_i : These factors are dependent on the type of boundary and loading (Tables 5-2 through 5-7).

The scope of this book does not permit a breakdown of the derivations or a more rigorous explanation of the method. For further information refer to Hampe (Ref. 5-1).

Cylinders with Uniform Thickness

The stresses and deformations of cylinders with any fixity at the boundary can be described by the formulas given below. The factors w_p, w_p', w_p'', and w_p''' are functions of the loading and are presented in Table 5-1.

Stresses: GENERAL: The circumferential load is

$$N_\theta = \frac{Et}{R}[w_p(\xi) + S_1 F_7(\xi) + S_2 F_{15}(\xi) + S_3 F_{16}(\xi) + S_4 F_8(\xi)]$$

The location ξ_N of the max N_θ is obtained from the relation $(N_\theta)' = 0$:

$$S_1 F_9(\xi_N) - S_2 F_{14}(\xi_N) - S_3 F_{13}(\xi_N) + S_4 F_{10}(\xi_N) - \frac{w_p'(\xi_N)}{k} = 0$$

TABLE 5-1 w_p, w_p', w_p'', w_p''' **Functions** (Ref. 5-1)

Load condition	$w_p(\xi)$	$w_p'(\xi)$	$w_p''(\xi)$	$w_p'''(\xi)$
$p(\xi) = p_0 = $ const	$+\dfrac{p_0 R^2}{Et}$	0	0	0
$p(\xi) = p_v(1-\xi)$	$+\dfrac{p_v R^2}{Et}(1-\xi)$	$-\dfrac{p_v R^2}{EtL}$	0	0
$p(\xi) = p_0 \sin \alpha\xi$ $\quad a < kL$	$+\dfrac{p_0 R^2}{Et}\dfrac{4(kL)^4}{a^4+4(kL)^4}\sin\alpha\xi$	$+\dfrac{p_0 R^2}{Et}\left(\dfrac{a}{L}\right)\dfrac{4(kL)^4}{a^4+4(kL)^4}\cos\alpha\xi$	$-\dfrac{p_0 R^2}{Et}\left(\dfrac{a}{L}\right)^2\dfrac{4(kL)^4}{a^4+4(kL)^4}\sin\alpha\xi$	$-\dfrac{p_0 R^2}{Et}\left(\dfrac{a}{L}\right)^3\dfrac{4(kL)^4}{a^4+4(kL)^4}\cos\alpha\xi$
$p(\xi) = p_0 \cos \alpha\xi$ $\quad a < kL$	$+\dfrac{p_0 R^2}{Et}\dfrac{4(kL)^4}{a^4+4(kL)^4}\cos\alpha\xi$	$+\dfrac{p_0 R^2}{Et}\left(\dfrac{a}{L}\right)\sin\alpha\xi$	$-\dfrac{p_0 R^2}{Et}\left(\dfrac{a}{L}\right)^2\dfrac{4(kL)^4}{a^4+4(kL)^4}\cos\alpha\xi$	$-\dfrac{p_0 R^2}{Et}\left(\dfrac{a}{L}\right)^3\cos\alpha\xi$
	$+\dfrac{p_0 R^2}{Et}\cos\alpha\xi$	$-\dfrac{p_0 R^2}{Et}\dfrac{a}{L}\sin\alpha\xi$	$-\dfrac{p_0 R^2}{Et}\left(\dfrac{a}{L}\right)^2\cos\alpha\xi$	$+\dfrac{p_0 R^2}{Et}\left(\dfrac{a}{L}\right)^3\sin\alpha\xi$
$p(\xi) = p_0 \exp(-\alpha\xi)$ $\quad a < kL$	$+\dfrac{p_0 R^2}{Et}\dfrac{4(kL)^4}{a^4+4(kL)^4}\exp(-\alpha\xi)$	$-\dfrac{p_0 R^2}{Et}\dfrac{a}{L}\dfrac{4(kL)^4}{a^4+4(kL)^4}\exp(-\alpha\xi)$	$+\dfrac{p_0 R^2}{Et}\left(\dfrac{a}{L}\right)^2\dfrac{4(kL)^4}{a^4+4(kL)^4}\exp(-\alpha\xi)$	$-\dfrac{p_0 R^2}{Et}\left(\dfrac{a}{L}\right)^3\dfrac{4(kL)^4}{a^4+4(kL)^4}\exp(-\alpha\xi)$
	$+\dfrac{p_0 R^2}{Et}\exp(-\alpha\xi)$	$-\dfrac{p_0 R^2}{Et}\exp(-\alpha\xi)$	$+\dfrac{p_0 R^2}{Et}\exp(-\alpha\xi)$	$-\dfrac{p_0 R^2}{Et}\exp(-\alpha\xi)$

For k factors see under "Definition of F Factors" in Sec. 5-2.

The moment M_x (defined in Chap. 4) is

$$M_x = D\{w_p''(\xi) + 2k^2[-S_1F_8(\xi) + S_2F_{16}(\xi) - S_3F_{15}(\xi) + S_4F_7(\xi)]\}$$

The location ξ_M of the max M_x is obtained from the relation $M_x' = 0$:

$$S_1F_{10}(\xi_M) - S_2F_{13}(\xi_M) + S_3F_{14}(\xi_M) + S_4F_9(\xi_M) - \frac{w_p'''(\xi_M)}{2k^3} = 0$$

The shear (defined in Chap. 4) is

$$Q_x = -D\{w_p'''(\xi) + 2k^3[-S_1F_{10}(\xi) + S_2F_{13}(\xi) - S_3F_{14}(\xi) - S_4F_9(\xi)]\}$$

SPECIAL CASE: $kL \geqslant 4$ (long cylinders). With the same notation as before, the circumferential force is

$$N_\theta = \frac{Et}{R}[w_p(\xi) + S_1F_{17}(\xi) + S_2F_{18}(\xi)]$$

The location ξ_N of the maximum N_θ is determined from the relation

$$S_1F_{19}(\xi_N) - S_2F_{20}(\xi_N) - \frac{w_p'(\xi_N)}{k} = 0$$

The moment is

$$M_x = D\{w_p''(\xi) + 2k^2[S_1F_{18}(\xi) - S_2F_{17}(\xi)]\}$$

The location ξ_M of max M is determined by the relation

$$S_1F_{20}(\xi_M) + S_2F_{19}(\xi_M) + \frac{w_p'''(\xi_M)}{2k^3} = 0$$

The shear is

$$Q_x = -D\{w_p'''(\xi) + 2k^3[S_1F_{20}(\xi_M) + S_2F_{19}(\xi_M)]\}$$

Deformations: GENERAL: The deflection line is represented by the equation

$$w = w_p(\xi) + S_1F_7(\xi) + S_2F_{15}(\xi) + S_3F_{16}(\xi) + S_4F_8(\xi)$$

The inclination of the tangent to the deflection line is

$$w' = w_p'(\xi) + k[-S_1F_9(\xi) + S_2F_{14}(\xi) + S_3F_{13}(\xi) + S_4F_{10}(\xi)]$$

SPECIAL CASE: $kL \geqslant 4$ (long cylinders). The deflection line is

$$w = w_p(\xi) + S_1F_{17}(\xi) + S_2F_{18}(\xi)$$

The inclination of the tangent to the deflection line is

$$w' = w_p'(\xi) + k[S_1 F_{19}(\xi) + S_2 F_{20}(\xi)]$$

Coefficients and Tables: The w_p to $w_p^{(n)}$ coefficients for the different loadings and geometries are presented in Table 5-1. S_i coefficients are presented similarly in Tables 5-2 through 5-7.

These tables are prepared for different kinds of loadings and boundary conditions:

Tables 5-2 and 5-3 are for cylinders with one edge free and the other edge fixed.

Tables 5-4 and 5-5 are for cylinders with one edge pinned and the other edge free.

Table 5-6 is for a cylinder with both edges fixed.

Table 5-7 is for a cylinder with one edge pinned and the other edge fixed.

TABLE 5-2 S_i Coefficients, $kL < 4.0$ (Ref. 5-1)

Loading conditions	S_1	S_2	S_3	S_4
$p = p(\xi)$	$-\dfrac{p_0 R^2}{Et}$	$-\dfrac{p_0 R^2}{Et}\dfrac{F_3}{F_1+2}$	$+\dfrac{p_0 R^2}{Et}\dfrac{F_3}{F_1+2}$	$+\dfrac{p_0 R^2}{Et}\dfrac{F_2}{F_1+2}$
$p(\xi) = p_0 = \text{const}$	$-\dfrac{p_v R^2}{Et}$	$-\dfrac{p_v R^2}{Et}\left(\dfrac{F_3}{F_1+2} - \dfrac{1}{kL}\dfrac{F_6+1}{F_1+2}\right)$	$+\dfrac{p_v R^2}{Et}\left(\dfrac{F_3}{F_1+2} - \dfrac{1}{kL}\dfrac{F_5-1}{F_1+2}\right)$	$+\dfrac{p_v R^2}{Et}\left(\dfrac{F_2}{F_1+2} - \dfrac{1}{kL}\dfrac{F_4}{F_v+2}\right)$
$p(\xi) = p_v(1-\xi)$	0	$-\dfrac{p_0 R^2}{Et}\dfrac{4(kL)^4}{a^4+4(kL)^4}\left[\dfrac{a}{kL}\dfrac{F_6+1}{F_1+2} - \dfrac{a^2}{2(kL)^2}\right]$ $\cdot\left(\dfrac{F_9}{F_1+2}\sin\alpha + \dfrac{a}{kL}\dfrac{F_7}{F_1+2}\cos\alpha\right)$	$+\dfrac{p_0 R^2}{Et}\dfrac{4(kL)^4}{a^4+4(kL)^4}\left[\dfrac{a}{kL}\dfrac{F_5-1}{F_1+2} - \dfrac{a^2}{2(kL)^2}\right]$ $\cdot\left(\dfrac{F_9}{F_1+2}\sin\alpha + \dfrac{a}{kL}\dfrac{F_7}{F_1+2}\cos\alpha\right)$	$+\dfrac{p_0 R^2}{Et}\dfrac{4(kL)^4}{a^4+4(kL)^4}\left[\dfrac{a}{kL}\dfrac{F_4}{F_1+2} + \dfrac{a^2}{2(kL)^2}\right]$ $\cdot\left(\dfrac{2F_7}{F_1+2}\sin\alpha - \dfrac{a}{kL}\dfrac{F_{10}}{F_1+2}\cos\alpha\right)$
$p(\xi) = p_0 \sin\alpha\xi$, $a < kL$	0	$-\dfrac{p_0 R^2}{Et}\left[\dfrac{a}{kL}\dfrac{F_6+1}{F_1+2} - \dfrac{a^2}{2(kL)^2}\right]$ $\cdot\left(\dfrac{F_9}{F_1+2}\sin\alpha + \dfrac{a}{kL}\dfrac{F_7}{F_1+2}\cos\alpha\right)$	$+\dfrac{p_0 R^2}{Et}\left[\dfrac{a}{kL}\dfrac{F_5-1}{F_1+2} - \dfrac{a^2}{2(kL)^2}\right]$ $\cdot\left(\dfrac{F_9}{F_1+2}\sin\alpha + \dfrac{a}{kL}\dfrac{F_7}{F_1+2}\cos\alpha\right)$	$+\dfrac{p_0 R^2}{Et}\left[\dfrac{a}{kL}\dfrac{F_4}{F_1+2} + \dfrac{a^2}{2(kL)^2}\right]$ $\cdot\left(\dfrac{2F_7}{F_1+2}\sin\alpha - \dfrac{a}{kL}\dfrac{F_{10}}{F_1+2}\cos\alpha\right)$

TABLE 5-2 Continued

Loading				
$p(\xi) = p_0 \cos \alpha \xi$	$-\dfrac{p_0 R^2}{Et} \dfrac{4(kL)^4}{\alpha^4 + 4(kL)^4}$	$-\dfrac{p_0 R^2}{Et} \dfrac{4(kL)^4}{\alpha^4 + 4(kL)^4} \left[\dfrac{F_3}{F_1 + 2} - \dfrac{\alpha^2}{2(kL)^2} \right]$ $\cdot \left(\dfrac{F_9}{F_1 + 2} \cos\alpha - \dfrac{\alpha}{kL} \dfrac{F_7}{F_1 + 2} \sin\alpha \right)$	$+\dfrac{p_0 R^2}{Et} \dfrac{4(kL)^4}{\alpha^4 + 4(kL)^4} \left[\dfrac{F_3}{F_1 + 2} - \dfrac{\alpha^2}{2(kL)^2} \right]$ $\cdot \left(\dfrac{F_9}{F_1 + 2} \cos\alpha - \dfrac{\alpha}{kL} \dfrac{F_7}{F_1 + 2} \sin\alpha \right)$	$+\dfrac{p_0 R^2}{Et} \dfrac{4(kL)^4}{\alpha^4 + 4(kL)^4} \left[\dfrac{F_2}{F_1 + 2} - \dfrac{\alpha^2}{2(kL)^2} \right]$ $\cdot \left(\dfrac{2F_7}{F_1 + 2} \cos\alpha + \dfrac{\alpha}{kL} \dfrac{F_{10}}{F_1 + 2} \sin\alpha \right)$
$\alpha < kL$	$-\dfrac{p_0 R^2}{Et}$			$-\dfrac{p_0 R^2}{Et} \left[\dfrac{F_2}{F_1 + 2} - \dfrac{\alpha^2}{2(kL)^2} \right]$ $\cdot \left(\dfrac{2F_7}{F_1 + 2} \cos\alpha + \dfrac{\alpha}{kL} \dfrac{F_{10}}{F_1 + 2} \sin\alpha \right)$
$p(\xi) = p_0 \exp(-\alpha \xi)$	$-\dfrac{p_0 R^2}{Et} \dfrac{4(kL)^4}{\alpha^4 + 4(kL)^4}$	$-\dfrac{p_0 R^2}{Et} \dfrac{4(kL)^4}{\alpha^4 + 4(kL)^4} \left[\dfrac{F_3}{F_1 + 2} - \dfrac{F_6 + 1}{kL} \dfrac{F_7}{F_1 + 2} \right]$ $+ \left(\dfrac{F_9}{F_1 + 2} - \dfrac{\alpha}{kL} \dfrac{F_7}{(F_1 + 2)} \right) \dfrac{\alpha^2}{2(kL)^2} \exp(-\alpha)$	$+\dfrac{p_0 R^2}{Et} \dfrac{4(kL)^4}{\alpha^4 + 4(kL)^4} \left[\dfrac{F_3}{F_1 + 2} - \dfrac{\alpha}{kL} \dfrac{F_5 - 1}{F_1 + 2} \right]$ $+ \left(\dfrac{F_9}{F_1 + 2} - \dfrac{\alpha}{kL} \dfrac{F_7}{F_1 + 2} \right) \dfrac{\alpha^2}{2(kL)^2} \exp(-\alpha)$	$+\dfrac{p_0 R^2}{Et} \dfrac{4(kL)^4}{\alpha^4 + 4(kL)^4} \left[\dfrac{F_2}{F_1 + 2} - \dfrac{\alpha}{kL} \dfrac{F_4}{F_1 + 2} \right]$ $- \left(\dfrac{2F_7}{F_1 + 2} - \dfrac{\alpha}{kL} \dfrac{F_{10}}{F_1 + 2} \right) \dfrac{\alpha^2}{2(kL)^2} \exp(-\alpha)$
$\alpha < kL$	$-\dfrac{p_0 R^2}{Et}$			$-\dfrac{p_0 R^2}{Et} \left[\dfrac{F_2}{F_1 + 2} - \dfrac{\alpha}{kL} \dfrac{F_4}{F_1 + 2} \right]$ $- \left(\dfrac{2F_7}{F_1 + 2} - \dfrac{\alpha}{kL} \dfrac{F_{10}}{F_1 + 2} \right) \dfrac{\alpha^2}{2(kL)^2} \exp(-\alpha)$
M_i (moment)	0	$\dfrac{M_i R^2}{Et} 2k^2 \dfrac{F_9}{F_1 + 2}$	$-\dfrac{M_i R^2}{Et} 2k^2 \dfrac{F_9}{F_1 + 2}$	$+\dfrac{M_i R^2}{Et} 2k^2 \dfrac{2F_7}{F_1 + 2}$
H_i (force)	0	$-\dfrac{H_i R^2}{Et} 2k \dfrac{F_7}{F_1 + 2}$	$\dfrac{H_i R^2}{Et} 2k \dfrac{F_7}{F_1 + 2}$	$+\dfrac{H_i R^2}{Et} 2k \dfrac{F_{10}}{F_1 + 2}$

For F and k factors see under "Definition of F Factors" in Sec. 5-2.

TABLE 5-3 S_1 and S_2 **Coefficients,** $kL \geq 4.0$ (Ref. 5-1)

Loading condition	S_1		S_2
$p = p(\xi)$			
$p(\xi) = p_0 = \text{const}$	$-\dfrac{p_0 R^2}{Et}$		$-\dfrac{p_0 R^2}{Et}$
$p(\xi) = p_v(1-\xi)$	$-\dfrac{p_v R^2}{Et}$		$-\dfrac{p_v R^2}{Et}\left(1-\dfrac{1}{kL}\right)$
$p(\xi) = p_0 \sin a\xi$	0		$-\dfrac{p_0 R^2}{Et}\dfrac{a}{kL}\dfrac{4(kL)^4}{a^4+4(kL)^4}$
	$a < kL$	0	$-\dfrac{p_0 R^2}{Et}\dfrac{a}{kL}$
$p(\xi) = p_0 \cos a\xi$	$-\dfrac{p_0 R^2}{Et}\dfrac{4(kL)^4}{a^4+4(kL)^4}$		$-\dfrac{p_0 R^2}{Et}\dfrac{4(kL)^4}{a^4+4(kL)^4}$
	$a < kL$	$-\dfrac{p_0 R^2}{Et}$	$-\dfrac{p_0 R^2}{Et}$
$p(\xi) = p_0 \exp(-a\xi)$	$-\dfrac{p_0 R^2}{Et}\dfrac{4(kL)^4}{a^4+4(kL)^4}$		$-\dfrac{p_0 R^2}{Et}\dfrac{4(kL)^4}{a^4+4(kL)^4}\left(1-\dfrac{a}{kL}\right)$
	$a < kL$	$-\dfrac{p_0 R^2}{Et}$	$-\dfrac{p_0 R^2}{Et}\left(1-\dfrac{a}{kL}\right)$
H_i, M_i	0		0

For k factors see under "Definition of F Factors" in Sec. 5-2.

TABLE 5-4 S_i Values, $kL < 4.0$ (Ref. 5-1)

Load condition	S_1	S_2	S_3	S_4
$p = p(\xi)$				
$p(\xi) = p_0 = \text{const}$	$-\dfrac{p_0 R^2}{Et}$	$+\dfrac{p_0 R^2}{Et}\dfrac{F_5}{F_4}$	$+\dfrac{p_0 R^2}{Et}\dfrac{F_6}{F_4}$	0
$p(\xi) = p_v(1-\xi)$	$-\dfrac{p_v R^2}{Et}$	$+\dfrac{p_v R^2}{Et}\dfrac{F_5}{F_4}$	$+\dfrac{p_v R^2}{Et}\dfrac{F_6}{F_4}$	0
p_0 (triangular)	0	$+\dfrac{p_0 R^2}{Et}\dfrac{4(kL)^4}{a^4+4(kL)^4}\dfrac{a^2}{2(kL)^2}$ $\cdot\left(\dfrac{F_{14}\sin a}{F_4} - \dfrac{a}{kL}\dfrac{F_{15}\cos a}{F_4}\right)$	$+\dfrac{p_0 R^2}{Et}\dfrac{4(kL)^4}{a^4+4(kL)^4}\dfrac{a^2}{2(kL)^2}$ $\cdot\left(\dfrac{F_{13}\sin a}{F_4} - \dfrac{a}{kL}\dfrac{F_{16}\cos a}{F_4}\right)$	0
$p(\xi) = p_0 \sin a\xi$	$a < kL$: 0	$+\dfrac{p_0 R^2}{Et}\dfrac{a^2}{2(kL)^2}$ $\cdot\left(\dfrac{F_{14}\sin a}{F_4} - \dfrac{a}{kL}\dfrac{F_{15}\cos a}{F_4}\right)$	$+\dfrac{p_0 R^2}{Et}\dfrac{a^2}{2(kL)^2}$ $\cdot\left(\dfrac{F_{13}\sin a}{F_4} - \dfrac{a}{kL}\dfrac{F_{16}\cos a}{F_4}\right)$	0

TABLE 5-4 Continued

Loading				
$p(\xi) = p_0 \cos \alpha \xi$	$-\frac{p_0 R^2}{Et} \frac{4(kL)^4}{\alpha^4 + 4(kL)^4}$	$+\frac{p_0 R^2}{Et} \frac{4(kL)^4}{\alpha^4 + 4(kL)^4} \left[\frac{F_5}{F_4} - \frac{\alpha^2}{2(kL)^2} \frac{F_6 + 1}{F_4}\right]$ $+\frac{\alpha^2}{2(kL)^2}\left(\frac{F_{14}}{F_4} \cos\alpha + \frac{\alpha}{kL} \frac{F_{15}}{F_4} \sin\alpha\right)$	$+\frac{p_0 R^2}{Et} \frac{4(kL)^4}{\alpha^4 + 4(kL)^4} \left[\frac{F_6}{F_4} + \frac{\alpha^2}{2(kL)^2} \frac{F_5 - 1}{F_4}\right]$ $-\frac{\alpha^2}{2(kL)^2}\left(\frac{F_{13}}{F_4} \cos\alpha + \frac{\alpha}{kL} \frac{F_{16}}{F_4} \sin\alpha\right)$	$+\frac{p_0 R^2}{Et} \frac{\alpha^2}{2(kL)^2} \frac{4(kL)^4}{\alpha^4 + 4(kL)^4}$
	$-\frac{p_0 R^2}{Et}$	$+\frac{p_0 R^2}{Et}\left[\frac{F_5}{F_4} - \frac{\alpha^2}{2(kL)^2}\frac{F_6+1}{F_4}\right.$ $\left. +\frac{\alpha^2}{2(kL)^2}\left(\frac{F_{14}}{F_4}\cos\alpha + \frac{\alpha}{kL}\frac{F_{15}}{F_4}\sin\alpha\right)\right]$	$+\frac{p_0 R^2}{Et}\left[\frac{F_6}{F_4} + \frac{\alpha^2}{2(kL)^2}\frac{F_5-1}{F_4}\right.$ $\left. -\frac{\alpha^2}{2(kL)^2}\left(\frac{F_{13}}{F_4}\cos\alpha + \frac{\alpha}{kL}\frac{F_{16}}{F_4}\sin\alpha\right)\right]$	$+\frac{p_0 R^2}{Et} \frac{\alpha^2}{2(kL)^2}$
$p(\xi) = p_0 \exp(-\alpha\xi)$	$-\frac{p_0 R^2}{Et}\frac{4(kL)^4}{\alpha^4+4(kL)^4}$	$+\frac{p_0 R^2}{Et}\frac{4(kL)^4}{\alpha^4+4(kL)^4}\left[\frac{F_5}{F_4} + \frac{\alpha^2}{2(kL)^2}\frac{F_6+1}{F_4}\right.$ $\left.-\frac{\alpha^2}{2(kL)^2}\left(\frac{F_{14}}{F_4} + \frac{\alpha}{kL}\frac{F_{15}}{F_4}\right)\exp(-\alpha)\right]$	$+\frac{p_0 R^2}{Et}\frac{4(kL)^4}{\alpha^4+4(kL)^4}\left[\frac{F_6}{F_4} - \frac{\alpha^2}{2(kL)^2}\frac{F_5-1}{F_4}\right.$ $\left. -\frac{\alpha^2}{2(kL)^2}\left(\frac{F_{13}}{F_4} + \frac{\alpha}{kL}\frac{F_{16}}{F_4}\right)\exp(-\alpha)\right]$	$-\frac{p_0 R^2}{Et}\frac{\alpha^2}{2(kL)^2}\frac{4(kL)^4}{\alpha^4+4(kL)^4}$
	$-\frac{p_0 R^2}{Et}$, $\alpha < kL$	$+\frac{p_0 R^2}{Et}\left[\frac{F_5}{F_4} + \frac{\alpha^2}{2(kL)^2}\frac{F_6+1}{F_4}\right.$ $\left.-\frac{\alpha^2}{2(kL)^2}\left(\frac{F_{14}}{F_4} + \frac{\alpha}{kL}\frac{F_{15}}{F_4}\right)\exp(-\alpha)\right]$	$+\frac{p_0 R^2}{Et}\left[\frac{F_6}{F_4} - \frac{\alpha^2}{2(kL)^2}\frac{F_5-1}{F_4}\right.$ $\left.-\frac{\alpha^2}{2(kL)^2}\left(\frac{F_{13}}{F_4} + \frac{\alpha}{kL}\frac{F_{16}}{F_4}\right)\exp(-\alpha)\right]$	$-\frac{p_0 R^2}{Et}\frac{\alpha^2}{2(kL)^2}$
M_i	0	$+\frac{M_i R^2}{Et} 2k^2 \frac{F_{14}}{F_4}$	$+\frac{M_i R^2}{Et} 2k^2 \frac{F_{13}}{F_4}$	0
H_i	0	$+\frac{H_i R^2}{Et} 2k \frac{F_{15}}{F_4}$	$+\frac{H_i R^2}{Et} 2k \frac{F_{16}}{F_4}$	0

For F factors see Table 5-10. For k factors see under "Definition of F Factors" in Sec. 5-2.

TABLE 5-5 S_1 and S_2 **Coefficients,** $kL > 4.0$ (Ref. 5-1)

Load condition	S_1		S_2
$p = p(\xi)$			
$p(\xi) = p_0 = $ const	$-\dfrac{p_0 R^2}{Et}$		0
$p(\xi) = p_v(1-\xi)$	$-\dfrac{p_v R^2}{Et}$		0
$p(\xi) = p_0 \sin a\xi$	0		0
$p(\xi) = p_0 \cos a\xi$	$-\dfrac{p_0 R^2}{Et} \dfrac{4(kL)^4}{a^4+4(kL)^4}$		$-\dfrac{p_0 R^2}{Et} \dfrac{4(kL)^4}{a^4+4(kL)^4} \dfrac{a^2}{2(kL)^2}$
	$a < kL$	$-\dfrac{p_0 R^2}{Et}$	$-\dfrac{p_0 R^2}{Et} \dfrac{a^2}{2(kL)^2}$
$p(\xi) = p_0 \exp(-a\xi)$	$-\dfrac{p_0 R^2}{Et} \dfrac{4(kL)^4}{a^4+4(kL)^4}$		$+\dfrac{p_0 R^2}{Et} \dfrac{4(kL)^4}{a^4+4(kL)^4} \dfrac{a^2}{2(kL)^2}$
	$a < kL$	$-\dfrac{p_0 R^2}{Et}$	$+\dfrac{p_0 R^2}{Et} \dfrac{a^2}{2(kL)^2}$
H_i, M_i	0		0

For k factors see under "Definition of F Factors" in Sec. 5-2.

TABLE 5-6 S_i Coefficients (Ref. 5-1)

Load condition	S_1	S_2	S_3	S_4
$p = p(\xi)$				
$p(\xi) = p_0 = \text{const}$	$-\dfrac{p_0 R^2}{Et}$	$-\dfrac{p_0 R^2}{Et}\left[\dfrac{F_3}{F_1} - \dfrac{F_{10}}{F_1}\right]$	$+\dfrac{p_0 R^2}{Et}\left[\dfrac{F_3}{F_1} - \dfrac{F_{10}}{F_1}\right]$	$+\dfrac{p_0 R^2}{Et}\left[\dfrac{F_2}{F_1} - \dfrac{2F_8}{F_1}\right]$
$p(\xi) = p_v(1-\xi)$	$-\dfrac{p_v R^2}{Et}$	$-\dfrac{p_v R^2}{Et}\left[\dfrac{F_3}{F_1} - \dfrac{1}{kL}\left(\dfrac{F_6}{F_1} + \dfrac{F_8}{F_1}\right)\right]$	$+\dfrac{p_v R^2}{Et}\left[\dfrac{F_3}{F_1} - \dfrac{1}{kL}\left(\dfrac{F_5}{F_1} + \dfrac{F_8}{F_1}\right)\right]$	$+\dfrac{p_v R^2}{Et}\left[\dfrac{F_2}{F_1} - \dfrac{1}{kL}\left(\dfrac{F_4}{F_1} + \dfrac{F_9}{F_1}\right)\right]$
$p(\xi) = p_0 \sin \alpha \xi$, $\alpha < kL$	0	$+\dfrac{p_0 R^2}{Et}\dfrac{4(kL)^4}{a^4 + 4(kL)^4}\left[\dfrac{F_{10}}{F_1}\sin\alpha - \dfrac{a}{kL}\left(\dfrac{F_6}{F_1} + \dfrac{F_8}{F_1}\cos\alpha\right)\right]$	$-\dfrac{p_0 R^2}{Et}\dfrac{4(kL)^4}{a^4 + 4(kL)^4}\left[\dfrac{F_{10}}{F_1}\sin\alpha - \dfrac{a}{kL}\left(\dfrac{F_5}{F_1} + \dfrac{F_8}{F_1}\cos\alpha\right)\right]$	$-\dfrac{p_0 R^2}{Et}\dfrac{4(kL)^4}{a^4 + 4(kL)^4}\left[\dfrac{2F_8}{F_1}\sin\alpha - \dfrac{a}{kL}\left(\dfrac{F_4}{F_1} + \dfrac{F_9}{F_1}\cos\alpha\right)\right]$
	0	$+\dfrac{p_0 R^2}{Et}\left[\dfrac{F_{10}}{F_1}\sin\alpha - \dfrac{a}{kL}\left(\dfrac{F_6}{F_1} + \dfrac{F_8}{F_1}\cos\alpha\right)\right]$	$-\dfrac{p_0 R^2}{Et}\left[\dfrac{F_{10}}{F_1}\sin\alpha - \dfrac{a}{kL}\left(\dfrac{F_5}{F_1} + \dfrac{F_8}{F_1}\cos\alpha\right)\right]$	$-\dfrac{p_0 R^2}{Et}\left[\dfrac{2F_8}{F_1}\sin\alpha - \dfrac{a}{kL}\left(\dfrac{F_4}{F_1} + \dfrac{F_9}{F_1}\cos\alpha\right)\right]$

TABLE 5-6 Continued

Loading				
$p(\xi) = p_0 \cos \alpha\, \xi$	$-\dfrac{p_0 R^2}{Et}\dfrac{4(kL)^4}{\alpha^4 + 4(kL)^4}$	$-\dfrac{p_0 R^2}{Et}\left(\dfrac{F_3}{F_1} - \dfrac{F_{10}}{F_1}\cos\alpha\right)\dfrac{4(kL)^4}{\alpha^4 + 4(kL)^4} - \dfrac{\alpha}{kL}\dfrac{F_8}{F_1}\sin\alpha$	$-\dfrac{p_0 R^2}{Et}\dfrac{4(kL)^4}{\alpha^4 + 4(kL)^4}\left(\dfrac{F_3}{F_1} - \dfrac{F_{10}}{F_1}\cos\alpha\right) - \dfrac{\alpha}{kL}\dfrac{F_8}{F_1}\sin\alpha$	$+\dfrac{p_0 R^2}{Et}\dfrac{4(kL)^4}{\alpha^4 + 4(kL)^4}\left(\dfrac{F_2}{F_1} - \dfrac{2F_8}{F_1}\cos\alpha\right) - \dfrac{\alpha}{kL}\dfrac{F_9}{F_1}\sin\alpha$
	$-\dfrac{p_0 R^2}{Et}\quad \alpha < kL$	$-\dfrac{p_0 R^2}{Et}\left(\dfrac{F_3}{F_1} - \dfrac{F_{10}}{F_1}\cos\alpha - \dfrac{\alpha}{kL}\dfrac{F_8}{F_1}\sin\alpha\right)$	$+\dfrac{p_0 R^2}{Et}\left(\dfrac{F_3}{F_1} - \dfrac{F_{10}}{F_1}\cos\alpha - \dfrac{\alpha}{kL}\dfrac{F_8}{F_1}\sin\alpha\right)$	$+\dfrac{p_0 R^2}{Et}\left(\dfrac{F_2}{F_1} - \dfrac{2F_8}{F_1}\cos\alpha - \dfrac{\alpha}{kL}\dfrac{F_9}{F_1}\sin\alpha\right)$
$p(\xi) = p_0 \exp(-\alpha\xi)$	$-\dfrac{p_0 R^2}{Et}\dfrac{4(kL)^4}{\alpha^4 + 4(kL)^4}$	$-\dfrac{p_0 R^2}{Et}\dfrac{4(kL)^4}{\alpha^4 + 4(kL)^4}\left[\dfrac{F_3}{F_1} - \dfrac{F_{10}}{F_1}\exp(-\alpha)\right] - \dfrac{\alpha}{kL}\left(\dfrac{F_6}{F_1} + \dfrac{F_8}{F_1}\exp(-\alpha)\right)$	$+\dfrac{p_0 R^2}{Et}\dfrac{4(kL)^4}{\alpha^4 + 4(kL)^4}\left[\dfrac{F_3}{F_1} - \dfrac{F_{10}}{F_1}\exp(-\alpha)\right] - \dfrac{\alpha}{kL}\left(\dfrac{F_5}{F_1} + \dfrac{F_8}{F_1}\exp(-\alpha)\right)$	$+\dfrac{p_0 R^2}{Et}\dfrac{4(kL)^4}{\alpha^4 + 4(kL)^4}\left[\dfrac{F_2}{F_1} - \dfrac{2F_8}{F_1}\exp(-\alpha)\right] - \dfrac{\alpha}{kL}\left(\dfrac{F_4}{F_1} + \dfrac{F_9}{F_1}\exp(-\alpha)\right)$
	$-\dfrac{p_0 R^2}{Et}\quad \alpha < kL$	$-\dfrac{p_0 R^2}{Et}\left[\dfrac{F_3}{F_1} - \dfrac{F_{10}}{F_1}\exp(-\alpha)\right] - \dfrac{\alpha}{kL}\left(\dfrac{F_6}{F_1} + \dfrac{F_8}{F_1}\exp(-\alpha)\right)$	$+\dfrac{p_0 R^2}{Et}\left[\dfrac{F_3}{F_1} - \dfrac{F_{10}}{F_1}\exp(-\alpha)\right] - \dfrac{\alpha}{kL}\left(\dfrac{F_5}{F_1} + \dfrac{F_8}{F_1}\exp(-\alpha)\right)$	$+\dfrac{p_0 R^2}{Et}\left[\dfrac{F_2}{F_1} - \dfrac{2F_8}{F_1}\exp(-\alpha)\right] - \dfrac{\alpha}{kL}\left(\dfrac{F_4}{F_1} + \dfrac{F_9}{F_1}\exp(-\alpha)\right)$

For F factors see Table 5-10.
For k factors see under "Definition of F Factors" in Sec. 5-2.

TABLE 5-7 S_i Coefficients, $kL \geqslant 4.0$ (Ref. 5-1)

Load conditions	S_1	S_2	S_3	S_4
$p = p(\xi)$				
$p(\xi) = p_0 = $ const	$-\dfrac{p_0 R^2}{Et}$	$+\dfrac{p_0 R^2}{Et}\left(\dfrac{F_5-1}{F_4}+\dfrac{F_{13}}{F_4}\right)$	$+\dfrac{p_0 R^2}{Et}\left(\dfrac{F_6+1}{F_4}-\dfrac{F_{14}}{F_4}\right)$	0
$p(\xi) = p_v(1-\xi)$	$-\dfrac{p_v R^2}{Et}$	$+\dfrac{p_v R^2}{Et}\left(\dfrac{F_5-1}{F_4}+\dfrac{1}{kL}\dfrac{F_{16}}{F_4}\right)$	$+\dfrac{p_v R^2}{Et}\left(\dfrac{F_6+1}{F_4}-\dfrac{1}{kL}\dfrac{F_{15}}{F_4}\right)$	0
$p(\xi) = p_0 \sin\alpha\xi$	0	$+\dfrac{p_0 R^2}{Et}\dfrac{4(kL)^4}{\alpha^4+4(kL)^4}\left(\dfrac{F_{13}}{F_4}\sin\alpha-\dfrac{\alpha}{kL}\dfrac{F_{16}}{F_4}\cos\alpha\right)$	$-\dfrac{p_0 R^2}{Et}\dfrac{4(kL)^4}{\alpha^4+4(kL)^4}\left(\dfrac{F_{14}}{F_4}\sin\alpha-\dfrac{\alpha}{kL}\dfrac{F_{15}}{F_4}\cos\alpha\right)$	0
$\alpha < kL$	0	$+\dfrac{p_0 R^2}{Et}\left(\dfrac{F_{13}}{F_4}\sin\alpha-\dfrac{\alpha}{kL}\dfrac{F_{16}}{F_4}\cos\alpha\right)$	$-\dfrac{p_0 R^2}{Et}\left(\dfrac{F_{14}}{F_4}\sin\alpha-\dfrac{\alpha}{kL}\dfrac{F_{15}}{F_4}\cos\alpha\right)$	0

TABLE 5-7 Continued

Loading			
$p(\xi) = p_0 \cos \alpha \xi$	$-\dfrac{p_0 R^2}{Et}\dfrac{4(kL)^4}{\alpha^4+4(kL)^4}$	$+\dfrac{p_0 R^2}{Et}\dfrac{4(kL)^4}{\alpha^4+4(kL)^4}\left[\dfrac{F_5-1}{F_4}+\dfrac{F_{13}}{F_4}\cos\alpha\right] + \dfrac{\alpha}{kL}\left(\dfrac{F_{16}}{F_4}\sin\alpha - \dfrac{\alpha}{2kL}\dfrac{F_6}{F_4}\cos\alpha\right)$	$+\dfrac{p_0 R^2}{Et}\dfrac{4(kL)^4}{\alpha^4+4(kL)^4}\left[\dfrac{F_6+1}{F_4}-\dfrac{F_{14}}{F_4}\cos\alpha\right] - \dfrac{\alpha}{kL}\left(\dfrac{F_{15}}{F_4}\sin\alpha - \dfrac{\alpha}{2kL}\dfrac{F_5}{F_4}\cos\alpha\right)$
	$\alpha < kL$: $\;-\dfrac{p_0 R^2}{Et}$	$+\dfrac{p_0 R^2}{Et}\left[\dfrac{F_5-1}{F_4}+\dfrac{F_{13}}{F_4}\cos\alpha\right] + \dfrac{\alpha}{kL}\left(\dfrac{F_{16}}{F_4}\sin\alpha - \dfrac{\alpha}{2kL}\dfrac{F_6}{F_4}\cos\alpha\right)$	$+\dfrac{p_0 R^2}{Et}\dfrac{\alpha^2}{2(kL)^2}\cos\alpha$
$p(\xi) = p_0 \exp(-\alpha\xi)$	$-\dfrac{p_0 R^2}{Et}\dfrac{4(kL)^4}{\alpha^4+4(kL)^4}$	$+\dfrac{p_0 R^2}{Et}\dfrac{4(kL)^4}{\alpha^4+4(kL)^4}\left[\dfrac{F_5-1}{F_4}+\dfrac{F_{13}}{F_4}\exp(-\alpha)\right] + \dfrac{\alpha}{kL}\left(\dfrac{F_{16}}{F_4}\exp(-\alpha) + \dfrac{\alpha}{2kL}\dfrac{F_6}{F_4}\right)$	$+\dfrac{p_0 R^2}{Et}\dfrac{4(kL)^4}{\alpha^4+4(kL)^4}\left[\dfrac{F_6+1}{F_4}-\dfrac{F_{14}}{F_4}\exp(-\alpha)\right] - \dfrac{\alpha}{kL}\left(\dfrac{F_{15}}{F_4}\exp(-\alpha) + \dfrac{\alpha}{2kL}\dfrac{F_5}{F_4}\right)$
	$\alpha < kL$: $\;-\dfrac{p_0 R^2}{Et}$	$+\dfrac{p_0 R^2}{Et}\left[\dfrac{F_5-1}{F_4}+\dfrac{F_{13}}{F_4}\exp(-\alpha)\right] + \dfrac{\alpha}{kL}\left(\dfrac{F_{16}}{F_4}\exp(-\alpha) + \dfrac{\alpha}{2kL}\dfrac{F_6}{F_4}\right)$	$-\dfrac{p_0 R^2}{Et}\dfrac{\alpha^2}{2(kL)^2}$

For F factors see Table 5-10.
For k factors see under "Definition of F Factors" in Sec. 5-2.

Cylinders with Rotationally Symmetric Discontinuities in Geometry or Loading

Rotationally symmetric discontinuities are of special importance, especially if they are located not far from the edges of the cylinder. As before, the tables are prepared for many possible discontinuity types (Ref. 5-1).

Assume that ⓘ represents the place where discontinuities will occur. Figure 5-1 shows the designations and the system of coordinates which will be used.

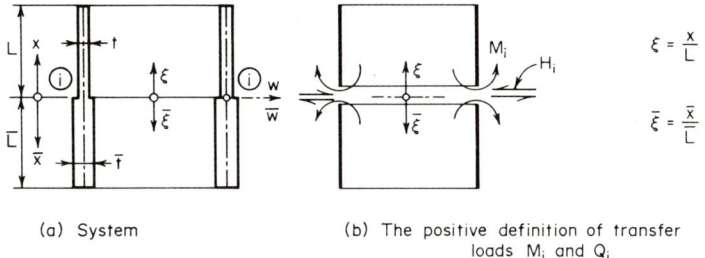

(a) System

(b) The positive definition of transfer loads M_i and Q_i

figure 5-1 *Cylinders with discontinuities.*

Parameters with bar over them refer to the lower cylinder (such as $\bar{F}, \bar{\xi}, \bar{w}, \bar{L}$, etc.).

The imaginary cut at ⓘ is introduced, and an attempt is made to determine discontinuity loads M_i and H_i. This is a usual problem of interaction; and it is solved as such. The following formulas are obtained:

$$M_i = \frac{2Dk}{\phi}(k\phi_{M_2}\Delta w_p + \phi_{M_1}\Delta w_p')$$

$$H_i = \frac{2Dk^2}{\phi}(k\phi_{H_2}\Delta w_p + \phi_{M_2}\Delta w_p')$$

where

$$\Delta w_p = \bar{w}_{pi} - w_{pi}$$

$$\Delta w_p' = \bar{w}'_{pi} - w_{pi}$$

$$\phi = \phi_{M_2}^2 - \phi_{M_1}\phi_{H_2}$$

The values ϕ_M and ϕ_H are given in Table 5-8. Some special cases of discontinuity loads M_1 and H_1 are presented in Table 5-9.

122 Structural Analysis of Shells

TABLE 5-8 ϕ **Functions** (Ref. 5-1)

Shell geometry	ϕ_{M_1}	$\phi_{H_1} = \phi_{M_2}$	ϕ_{H_2}
(fixed–fixed)	$\dfrac{F_4}{F_1} + \left(\dfrac{t}{\bar{t}}\right)^{3/2} \dfrac{\bar{F}_4}{\bar{F}_1}$	$\dfrac{F_2}{F_1} - \left(\dfrac{t}{\bar{t}}\right)^2 \dfrac{\bar{F}_2}{\bar{F}_1}$	$2\left[\dfrac{F_3}{F_1} + \left(\dfrac{t}{\bar{t}}\right)^{5/2} \dfrac{\bar{F}_3}{\bar{F}_1}\right]$
(fixed–pinned)	$\dfrac{F_4}{F_1} + \left(\dfrac{t}{\bar{t}}\right)^{3/2} \dfrac{\bar{F}_2}{\bar{F}_4}$	$\dfrac{F_2}{F_1} - \left(\dfrac{t}{\bar{t}}\right)^2 \dfrac{\bar{F}_3}{\bar{F}_4}$	$2\left[\dfrac{F_3}{F_1} + \left(\dfrac{t}{\bar{t}}\right)^{5/2} \dfrac{\bar{F}_1 + 1}{\bar{F}_4}\right]$
(fixed–free)	$\dfrac{F_4}{F_1} + \left(\dfrac{t}{\bar{t}}\right)^{3/2} \dfrac{\bar{F}_4}{\bar{F}_1 + 2}$	$\dfrac{F_2}{F_1} - \left(\dfrac{t}{\bar{t}}\right)^2 \dfrac{\bar{F}_2}{\bar{F}_1 + 2}$	$2\left[\dfrac{F_3}{F_1} + \left(\dfrac{t}{\bar{t}}\right)^{5/2} \dfrac{\bar{F}_3}{\bar{F}_1 + 2}\right]$
(pinned–fixed)	$\dfrac{F_2}{F_4} + \left(\dfrac{t}{\bar{t}}\right)^{3/2} \dfrac{\bar{F}_4}{\bar{F}_1}$	$\dfrac{F_3}{F_4} - \left(\dfrac{t}{\bar{t}}\right)^2 \dfrac{\bar{F}_2}{\bar{F}_1}$	$2\left[\dfrac{F_1 + 1}{F_4} + \left(\dfrac{t}{\bar{t}}\right)^{5/2} \dfrac{\bar{F}_3}{\bar{F}_1}\right]$
(pinned–pinned)	$\dfrac{F_2}{F_4} + \left(\dfrac{t}{\bar{t}}\right)^{3/2} \dfrac{\bar{F}_2}{\bar{F}_4}$	$\dfrac{F_3}{F_4} - \left(\dfrac{t}{\bar{t}}\right)^2 \dfrac{\bar{F}_3}{\bar{F}_4}$	$2\left[\dfrac{F_1 + 1}{F_4} + \left(\dfrac{t}{\bar{t}}\right)^{5/2} \dfrac{\bar{F}_1 + 1}{\bar{F}_4}\right]$
(pinned–free)	$\dfrac{F_2}{F_4} + \left(\dfrac{t}{\bar{t}}\right)^{3/2} \dfrac{\bar{F}_4}{\bar{F}_1 + 2}$	$\dfrac{F_3}{F_4} - \left(\dfrac{t}{\bar{t}}\right)^2 \dfrac{\bar{F}_2}{\bar{F}_1 + 2}$	$2\left[\dfrac{F_1 + 1}{F_4} + \left(\dfrac{t}{\bar{t}}\right)^{5/2} \dfrac{\bar{F}_3}{\bar{F}_1 + 2}\right]$
(free–fixed)	$\dfrac{F_4}{F_1 + 2} + \left(\dfrac{t}{\bar{t}}\right)^{3/2} \dfrac{\bar{F}_4}{\bar{F}_1}$	$\dfrac{F_2}{F_1 + 2} - \left(\dfrac{t}{\bar{t}}\right)^2 \dfrac{\bar{F}_3}{\bar{F}_4}$	$2\left[\dfrac{F_3}{F_1 + 2} + \left(\dfrac{t}{\bar{t}}\right)^{5/2} \dfrac{\bar{F}_3}{\bar{F}_1}\right]$
(free–pinned)	$\dfrac{F_4}{F_1 + 2} + \left(\dfrac{t}{\bar{t}}\right)^{3/2} \dfrac{\bar{F}_2}{\bar{F}_4}$	$\dfrac{F_2}{F_1 + 2} - \left(\dfrac{t}{\bar{t}}\right)^2 \dfrac{\bar{F}_3}{\bar{F}_4}$	$2\left[\dfrac{F_3}{F_1 + 2} + \left(\dfrac{t}{\bar{t}}\right)^{5/2} \dfrac{\bar{F}_1 + 1}{\bar{F}_4}\right]$
(free–free)	$\dfrac{F_4}{F_1 + 2} + \left(\dfrac{t}{\bar{t}}\right)^{3/2} \dfrac{\bar{F}_4}{\bar{F}_1 + 2}$	$\dfrac{F_2}{F_1 + 2} - \left(\dfrac{t}{\bar{t}}\right)^2 \dfrac{\bar{F}_2}{\bar{F}_1 + 2}$	$2\left[\dfrac{F_3}{F_1 + 2} + \left(\dfrac{t}{\bar{t}}\right)^{5/2} \dfrac{\bar{F}_3}{\bar{F}_1 + 2}\right]$

For F factors see Table 5-10.

TABLE 5-9 Determination of M_i and H_i (Ref. 5-1)

Load condition	M_i	H_i
$P \leftarrow\!\!-\!\!-\!\!-\!\!\rightarrow P$ (hinged)	$+\dfrac{P}{k\phi}\left(\phi_{M2}\dfrac{\bar{F}_4}{\bar{F}_1+2}+\phi_{M1}\dfrac{\bar{F}_2}{\bar{F}_1+2}\right)$	$+\dfrac{P}{\phi}\left(\phi_{Q2}\dfrac{\bar{F}_4}{\bar{F}_1+2}+\phi_{M2}\dfrac{\bar{F}_2}{\bar{F}_1+2}\right)$
$M \curvearrowleft \cdots \curvearrowright M$ (hinged)	$+\dfrac{M}{\phi}\left(\phi_{M2}\dfrac{\bar{F}_2}{\bar{F}_1+2}+\phi_{M1}\dfrac{2\bar{F}_3}{\bar{F}_1+2}\right)$	$+\dfrac{M\bar{k}}{\phi}\left(\phi_{Q2}\dfrac{\bar{F}_2}{\bar{F}_1+2}+\phi_{M1}\dfrac{2\bar{F}_3}{\bar{F}_1+2}\right)$
$p_v\lambda_p$, $p_v(1+\lambda_p)$ (hinged)	$+\dfrac{p_v}{\phi 2\bar{k}^2}\left\{\phi_{M2}\left[\lambda_p-\dfrac{2\bar{F}_7}{\bar{F}_1+2}(\lambda_p+1)+\dfrac{1}{kL}\dfrac{\bar{F}_{10}}{\bar{F}_1+2}\right]\right.$ $\left.+\phi_{M1}\left[\dfrac{2\bar{F}_9}{\bar{F}_1+2}(\lambda_p+1)-\dfrac{1}{kL}\left(1-\dfrac{2\bar{F}_7}{\bar{F}_1+2}\right)\right]\right\}$	$+\dfrac{p_v}{\phi 2k}\left\{\phi_{Q2}\left[\lambda_p-\dfrac{2\bar{F}_7}{\bar{F}_1+2}(\lambda_p+1)+\dfrac{1}{kL}\dfrac{\bar{F}_{10}}{\bar{F}_1+2}\right]\right.$ $\left.+\phi_{M2}\left[\dfrac{2\bar{F}_9}{\bar{F}_1+2}(\lambda_p+1)-\dfrac{1}{kL}\left(1-\dfrac{2\bar{F}_7}{\bar{F}_1+2}\right)\right]\right\}$
$P \leftarrow\!\!-\!\!-\!\!-\!\!\rightarrow P$ (fixed)	$+\dfrac{P}{\phi\bar{k}}\left(\phi_{M2}\dfrac{\bar{F}_2}{\bar{F}_4}+\phi_{M1}\dfrac{\bar{F}_3}{\bar{F}_4}\right)$	$+\dfrac{P}{\phi}\left(\phi_{Q2}\dfrac{\bar{F}_2}{\bar{F}_4}+\phi_{M2}\dfrac{\bar{F}_3}{\bar{F}_4}\right)$
$M \curvearrowleft \cdots \curvearrowright M$ (fixed)	$+\dfrac{M}{\phi}\left(\phi_{M2}\dfrac{\bar{F}_3}{\bar{F}_4}+\phi_{M1}\dfrac{\bar{F}_1+1}{\bar{F}_4}\right)$	$+\dfrac{M\bar{k}}{\phi}\left(\phi_{Q2}\dfrac{\bar{F}_3}{\bar{F}_4}+\phi_{M2}\dfrac{\bar{F}_1+1}{\bar{F}_4}\right)$
$p_v\lambda_p$, $p_v(1+\lambda_p)$ (fixed)	$+\dfrac{p_v}{\phi 2\bar{k}^2}\left\{\phi_{M2}\left[\lambda_p+\dfrac{\bar{F}_1}{\bar{F}_4}(\lambda_p+1)\right]\right.$ $\left.+\phi_{M1}\left[\dfrac{2\bar{F}_8}{\bar{F}_4}(\lambda_p+1)-\dfrac{1}{kL}\right]\right\}$	$+\dfrac{p_v}{\phi 2k}\left\{\phi_{Q2}\left[\lambda_p+\dfrac{\bar{F}_9}{\bar{F}_4}(\lambda_p+1)\right]\right.$ $\left.+\phi_{M2}\left[\dfrac{2\bar{F}_8}{\bar{F}_4}(\lambda_p+1)-\dfrac{1}{kL}\right]\right\}$

For F and k factors see Table 5-10.
For λ_p coefficients see Chapter 3.

Spherical Shell, Any Fixity at the Lower Boundary

General formulas for open or closed spherical shells are presented here in the same manner as for the cylindrical shell. The loading is arbitrary. The boundary conditions along the lower edge can be assumed to be "fixed" or "pinned." The formulas are dependent upon certain factors S_i that are the functions of integration constants C that appeared by derivation of formulas (bending theory). Δr_m is a function of primary loading, as will be explained later. For the fixed lower boundary

$$S_1 = -\frac{Et}{2Rk}\left(\frac{\Delta r_m}{\sin\phi_1}\frac{2F_3}{F_1+2} + \beta_m \frac{R}{k}\frac{F_2}{F_1+2}\right)$$

$$S_2 = -\frac{Et}{Rk}\left(\frac{\Delta r_m}{\sin\phi_1} 2\frac{F_5-1}{F_1+2} + \beta_m \frac{R}{k}\frac{F_4}{F_1+2}\right)$$

$$S_3 = \frac{Et}{2Rk}\left(\frac{\Delta r_m}{\sin\phi_1} 2\frac{F_6+1}{F_1+2} + \beta_m \frac{R}{k}\frac{F_4}{F_1+2}\right)$$

$$S_4 = \frac{Et}{2Rk}\beta_m \frac{R}{k}$$

where k and F are factors as shown in Sec. 5-3 under Definition of F Factors.

For the special case of closed spherical shells, the only constants needed are

$$S_1 = -\frac{Et}{2k^2}\left(\beta_m + \frac{\Delta r_m}{\sin\phi_1}\frac{2k}{R}\right)$$

$$S_2 = -\frac{Et}{2k^2}\beta_m$$

In the preceding formulas Δr_m is the radial movement at the point of interest on the membrane and β_m is the angle of rotation at the point of interest on the membrane, loaded with any primary loading.

For the lower boundary pinned

$$S_1 = \frac{-\Delta r_m Et F_1}{2\sin\phi_1 Rk F_4}$$

$$S_2 = \frac{\Delta r_m Et}{2\sin\phi_1 Rk}$$

$$S_3 = \frac{\Delta r_m Et}{2\sin\phi_1 Rk}$$

$$S_4 = \frac{-\Delta r_m Et}{2\sin\phi_1 Rk}\frac{F_2}{F_4}$$

For the special case of the closed spherical shell, the only constants needed are

$$S_1 = -\frac{\Delta r_m}{2\sin\phi_1}\frac{Et}{Rk}$$

$$S_2 = \frac{\Delta r_m}{2\sin\phi_1}\frac{Et}{Rk}$$

With these indications the general formulas can be given.

GENERAL CASE OF OPEN SPHERICAL SHELL, ANY BOUNDARY AT LOWER EDGE

LONGITUDINAL LOAD:

$$N_\phi = N_{\phi_{\text{membr}}} + \cot\phi[S_1 F_7(\alpha) + S_2 F_{15}(\alpha) + S_3 F_{16}(\alpha) + S_4 F_8(\alpha)]$$

CIRCUMFERENTIAL LOAD:

$$N_\theta = N_{\theta_{\text{membr}}} - k[-S_1 F_9(\alpha) + S_2 F_{14}(\alpha) + S_3 F_{13}(\alpha) + S_4 F_{10}(\alpha)]$$

SHEAR:

$$Q_\phi = S_1 F_7(\alpha) + S_2 F_{15}(\alpha) + S_3 F_{16}(\alpha) + S_4 F_8(\alpha)$$

MOMENT M_ϕ:

$$M_\phi = -\frac{R}{2k}[-S_1 F_{10}(\alpha) + S_2 F_{13}(\alpha) - S_3 F_{14}(\alpha) - S_4 F_9(\alpha)]$$

MOMENT M_θ:

$$M_\theta = \mu\frac{R}{2k}[-S_1 F_{10}(\alpha) + S_2 F_{13}(\alpha) - S_3 F_{14}(\alpha) - S_4 F_9(\alpha)]$$

DEFORMATIONS

HORIZONTAL MOVEMENTS:

$$\Delta r = \Delta r_m - \frac{Rk}{Et}\sin\phi[-S_1 F_9(\alpha) + S_2 F_{14}(\alpha) + S_3 F_{13}(\alpha) + S_4 F_{10}(\alpha)]$$

ROTATION OF MERIDIAN:

$$\beta = \beta_m - \frac{2k^2}{Et}[-S_1 F_8(\alpha) + S_2 F_{16}(\alpha) - S_3 F_{15}(\alpha) + S_4 F_7(\alpha)]$$

General Case of Closed Spherical Shell, Any Boundaries at Lower Edge

Longitudinal Load:

$$N_\phi = N_{\phi\text{membr}} + \cot\phi[S_1 F_{17}(\alpha) + S_2 F_{18}(\alpha)]$$

Circumferential Load:

$$N_\theta = N_{\theta\text{membr}} + k[S_1 F_{19}(\alpha) - S_2 F_{20}(\alpha)]$$

Shear:

$$Q_\phi = S_1 F_{17}(\alpha) + S_2 F_{18}(\alpha)$$

Moment M_ϕ:

$$M_\phi = \frac{-R}{2k}[S_1 F_{20}(\alpha) + S_2 F_{19}(\alpha)]$$

Moment M_θ:

$$M_\theta = -\frac{R}{2k}\left\{-S_1\left[\frac{\cot\phi}{k}F_{18}(\alpha) - \mu F_{20}(\alpha)\right] + S_2\left[\frac{\cot\phi}{k}F_{17}(\alpha) + \mu F_{19}(\alpha)\right]\right\}$$

Deformations

Horizontal Displacement:

$$\Delta r = \Delta r_m + \frac{Rk}{Et}\sin\phi[S_1 F_{19}(\alpha) - S_2 F_{20}(\alpha)]$$

Rotation of the Tangent of the Meridian:

$$\beta = \beta_m - \frac{2k^2}{Et}[S_1 F_{18}(\alpha) - S_2 F_{17}(\alpha)]$$

Definition of F Factors

The general solution of the homogeneous differential equation

$$w^{IV} + 4k^4 w = 0$$

can be represented with the following combination of trigonometric and hyperbolic functions:

$$\cosh kL\xi \cos kL\xi \qquad \sinh kL\xi \cos kL\xi$$
$$\cosh kL\xi \sin kL\xi \qquad \sinh kL\xi \sin kL\xi$$

where kL is a dimensionless parameter and ξ is a dimensionless ordinate. There are numerous combinations which can be expressed by certain factors F, which are given in Table 5-10. In order to avoid calculation of every combination, some are calculated and represented graphically (Figs. 5-2, 5-3, and 5-4). The factors F are plotted as a function of a

TABLE 5-10 $F_i(\xi)$ and F_i **Factors** (Ref. 5-1)

i	$F_i(\xi)$	F_i
1	$\sinh^2 kL\xi - \sin^2 kL\xi$	$\sinh^2 kL - \sin^2 kL$
2	$\sinh^2 kL\xi + \sin^2 kL\xi$	$\sinh^2 kL + \sin^2 kL$
3	$\sinh kL\xi \cosh kL\xi + \sin kL\xi \cos kL\xi$	$\sinh kL \cosh kL + \sin kL \cos kL$
4	$\sinh kL\xi \cosh kL\xi - \sin kL\xi \cos kL\xi$	$\sinh kL \cosh kL - \sin kL \cos kL$
5	$\sin^2 kL\xi$	$\sin^2 kL$
6	$\sinh^2 kL\xi$	$\sinh^2 kL$
7	$\cosh kL\xi \cos kL\xi$	$\cosh kL \cos kL$
8	$\sinh kL\xi \sin kL\xi$	$\sinh kL \sin kL$
9	$\cosh kL\xi \sin kL\xi - \sinh kL\xi \cos kL\xi$	$\cosh kL \sin kL - \sinh kL \cos kL$
10	$\cosh kL\xi \sin kL\xi + \sinh kL\xi \cos kL\xi$	$\cosh kL \sin kL + \sinh kL \cos kL$
11	$\sin kL\xi \cos kL\xi$	$\sin kL \cos kL$
12	$\sinh kL\xi \cosh kL\xi$	$\sinh kL \cosh kL$
13	$\cosh kL\xi \cos kL\xi - \sinh kL\xi \sin kL\xi$	$\cosh kL \cos kL - \sinh kL \sin kL$
14	$\cosh kL\xi \cos kL\xi + \sinh kL\xi \sin kL\xi$	$\cosh kL \cos kL + \sinh kL \sin kL$
15	$\cosh kL\xi \sin kL\xi$	$\cosh kL \sin kL$
16	$\sinh kL\xi \cos kL\xi$	$\sinh kL \cos kL$
17	$\exp(-kL\xi \cos kL\xi)$	$\exp(-kL \cos kL)$
18	$\exp(-kL\xi \sin kL\xi)$	$\exp(-kL \sin kL)$
19	$\exp\left[-kL\xi (\cos kL\xi + \sin kL\xi)\right]$	$\exp\left[-kL (\cos kL + \sin kL)\right]$
20	$\exp\left[-kL\xi (\cos kL\xi - \sin kL\xi)\right]$	$\exp\left[-kL (\cos kL - \sin kL)\right]$

certain parameter η, i.e., $F = F(\eta)$ which is different for cylindrical, conical, and spherical shells. The η is defined as follows:

For a cylindrical shell

$$\eta = kL \text{ or } \eta = kL\xi \qquad \text{where } k = \frac{\sqrt[4]{3(1-\mu^2)}}{\sqrt{Rt}} \text{ and } \xi = \frac{x}{L}$$

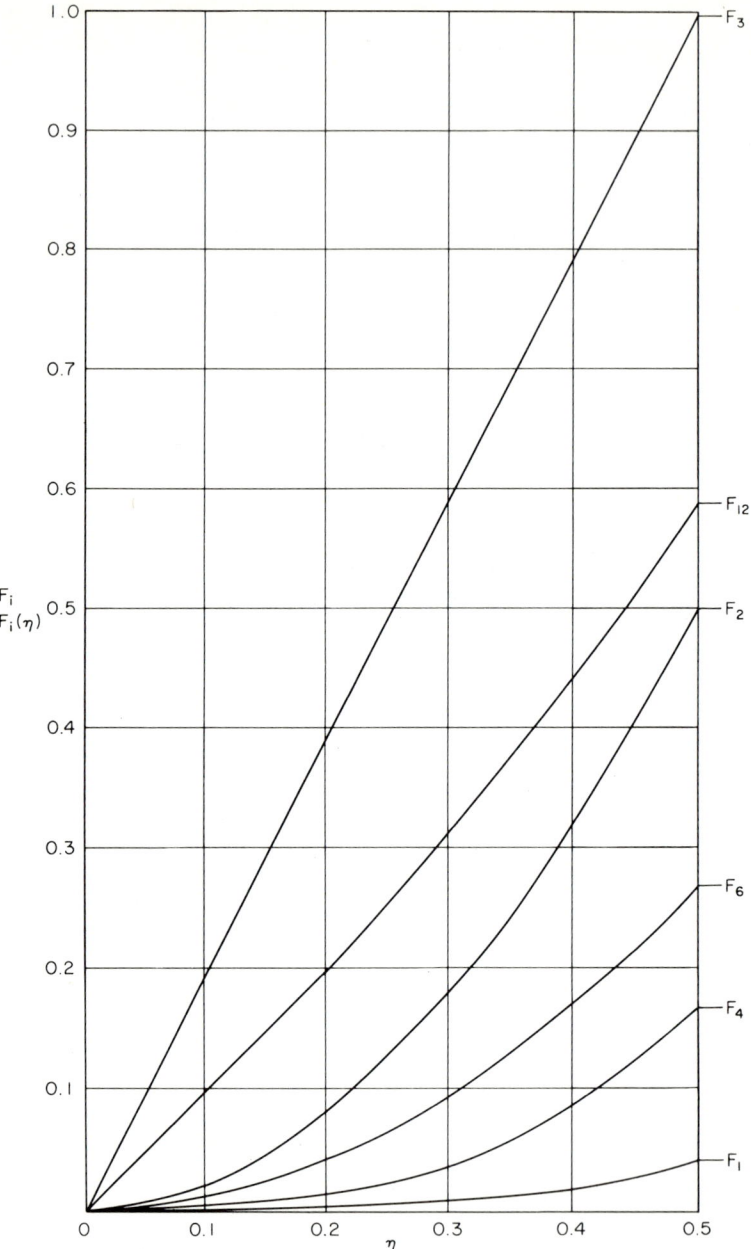

figure 5-2 F_i *factors* ($i = 1, 2, 3, 4, 6, 12$).

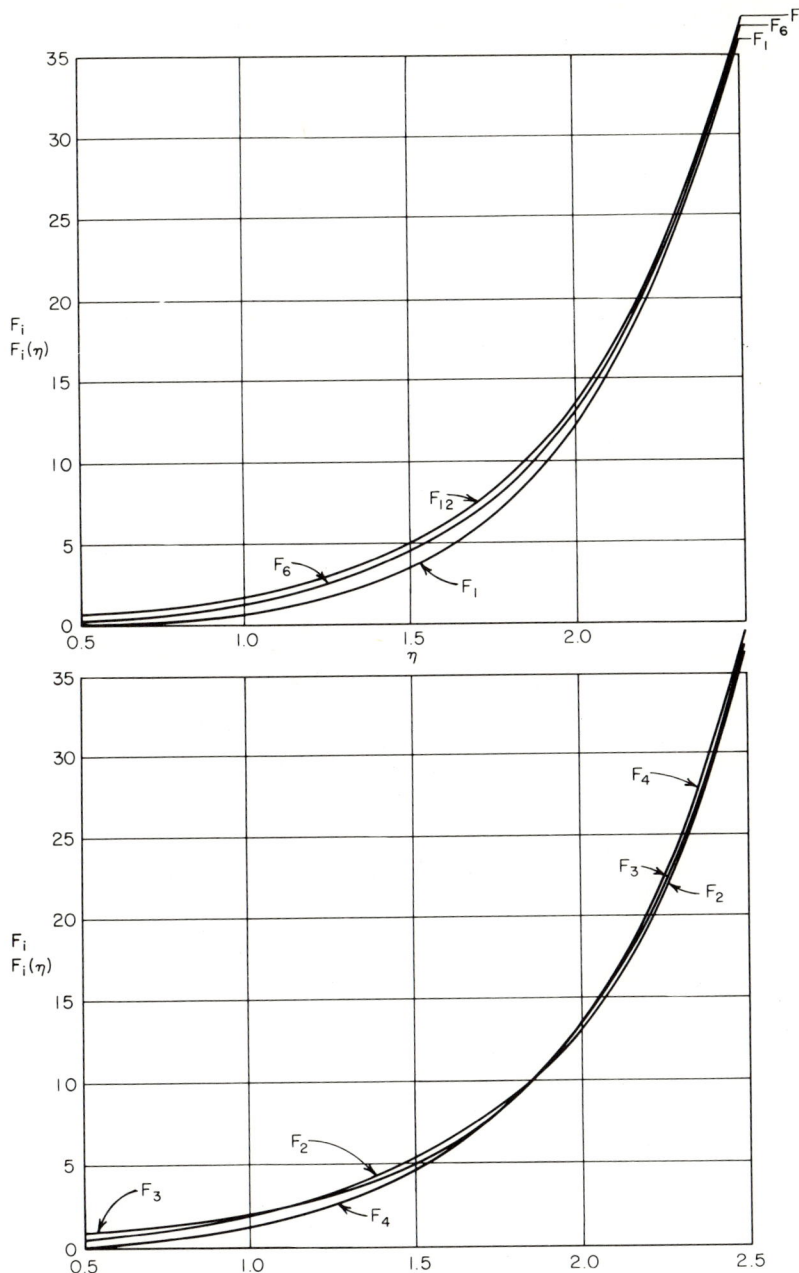

figure 5-2 Continued.

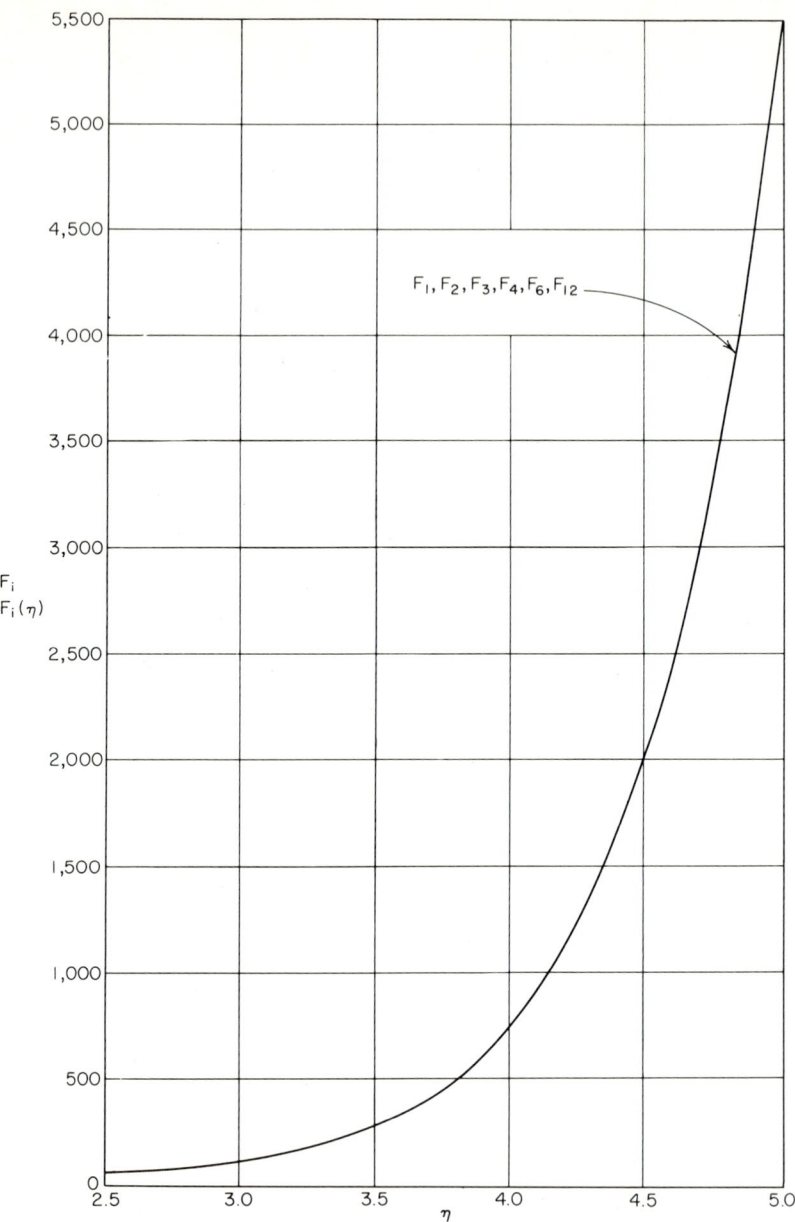

figure 5-2 F_i factors ($i = 1, 2, 3, 4, 6, 12$). *Continued.*

For a conical shell

$$\eta = kL \quad \text{or} \quad \eta = kL\xi \qquad \text{where} \quad k = \frac{\sqrt[4]{3(1-\mu^2)}}{\sqrt{tx_m \cot \alpha_0}} \quad \text{and} \quad \xi = \frac{\bar{x}}{L}$$

Special Solutions   131

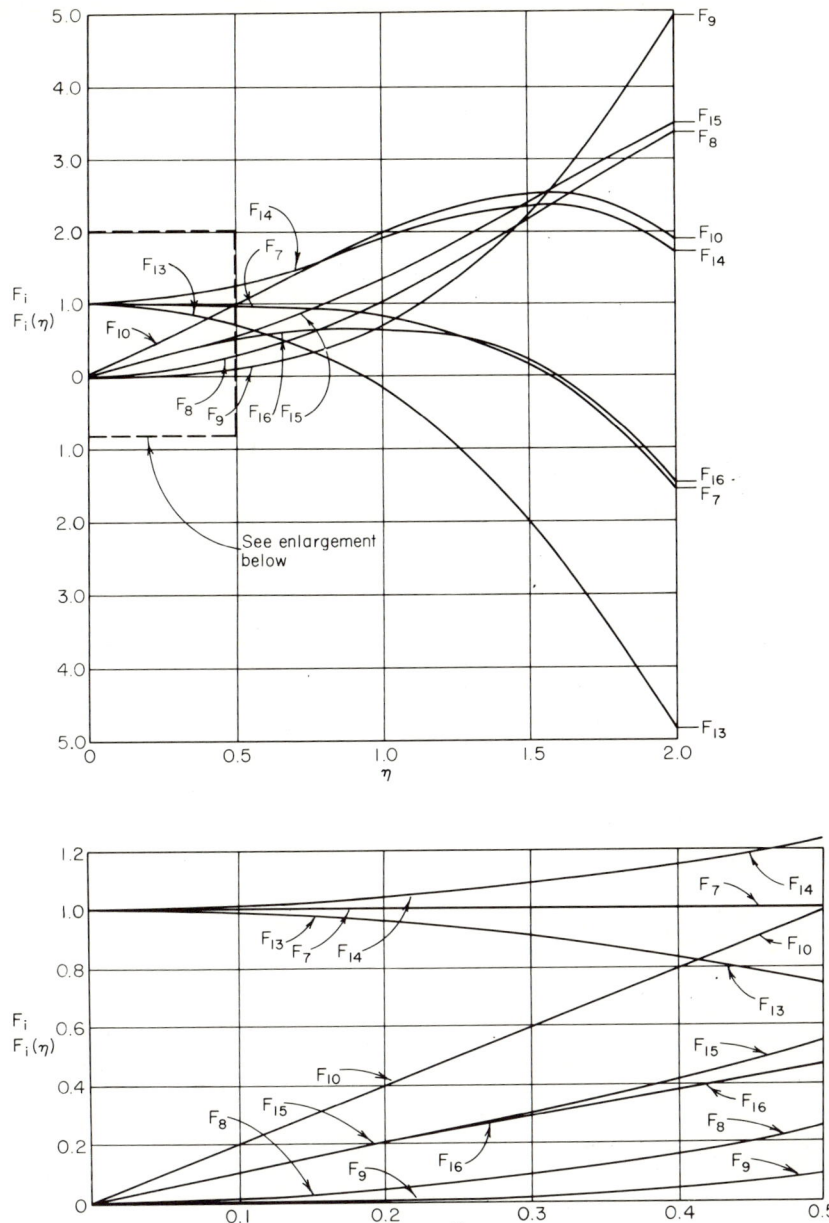

figure 5-3 F_i factors ($i = $ 7, 8, 9, 10, 13, 14, 15, 16).

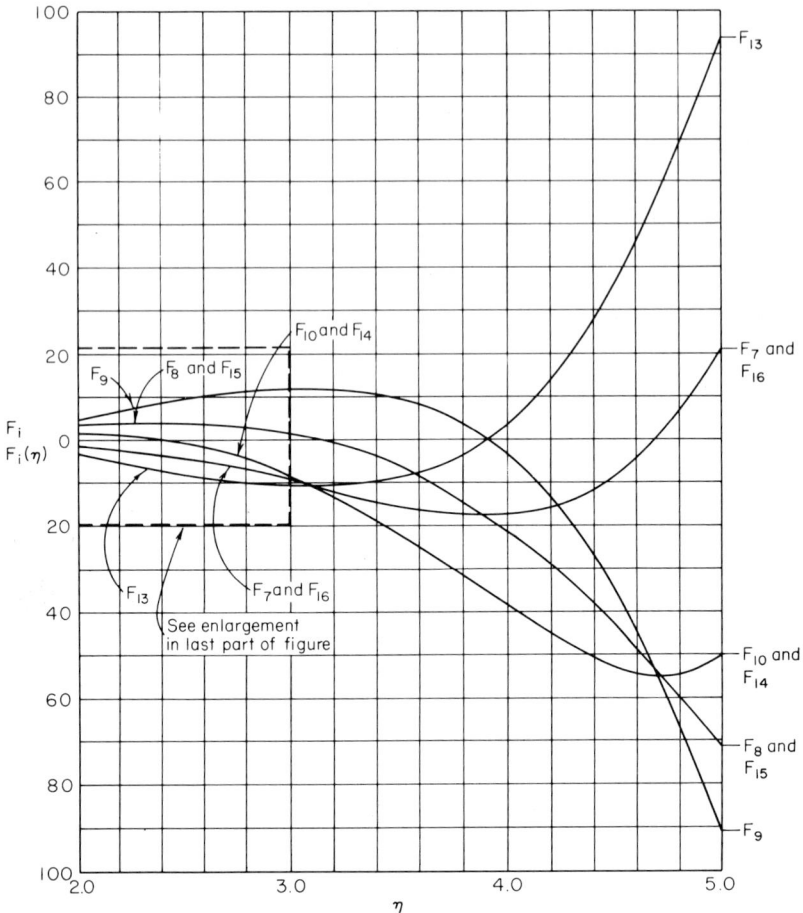

figure 5-3 F_i factors ($i = 7, 8, 9, 10, 13, 14, 15, 16$). *Continued.*

For a spherical shell

$$\eta = k \text{ for } F_i \text{ or } \eta = k\alpha \text{ for } F_i(\alpha) \quad \text{and} \quad k = \sqrt[4]{3(1-\mu^2)\left(\frac{R}{t}\right)^2}$$

because for the spherical shell $\xi = x/L = \alpha/\phi$, (α and ϕ as previously defined), which means that L is proportional to ϕ and x to α.

$$\eta = kL\xi = k\phi\frac{\alpha}{\phi} = k\alpha$$

Definition of the F factors is given in Table 5-10. If high accuracy is required, it is suggested that the F values be calculated as

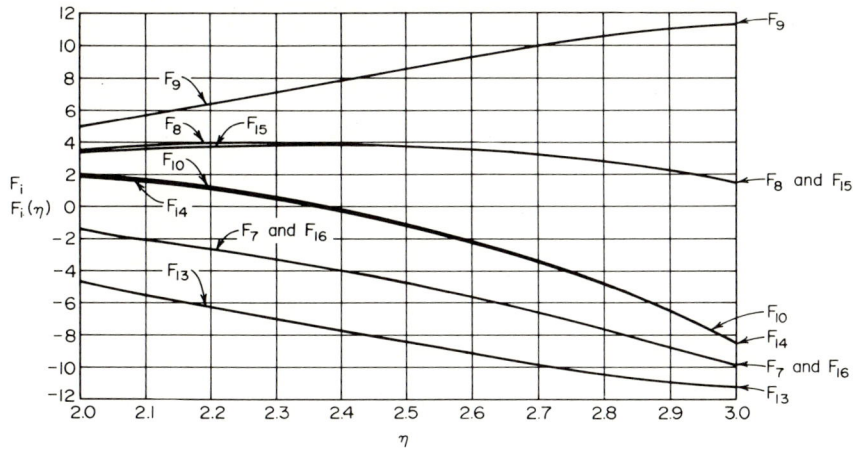

figure 5-3 *Continued.*

presented in Table 5-10. If extreme accuracy is not required, the graphs in Figs. 5-2 through 5-4 should be used. These graphs also present the values F_i, but because of sensitivity of those factors, graphical determination is not as accurate as analytical determination from Table 5-10.

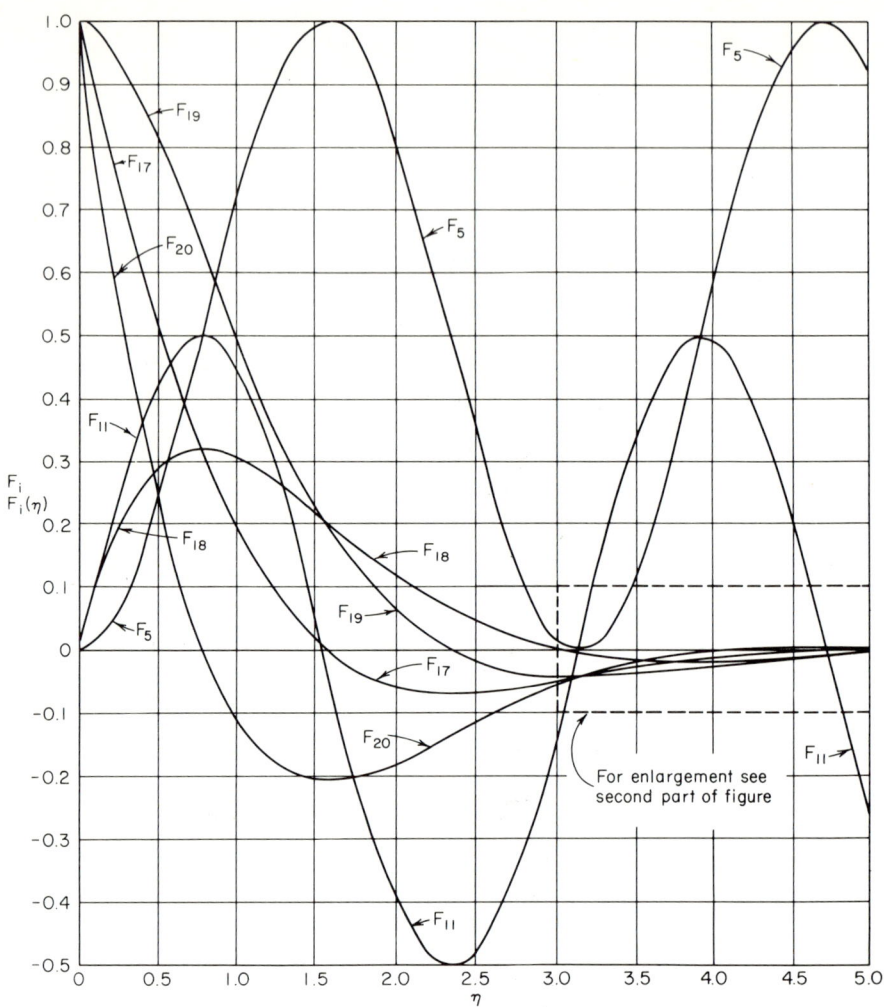

figure 5-4 F_i factors ($i = 5, 11, 17, 18, 19, 20$).

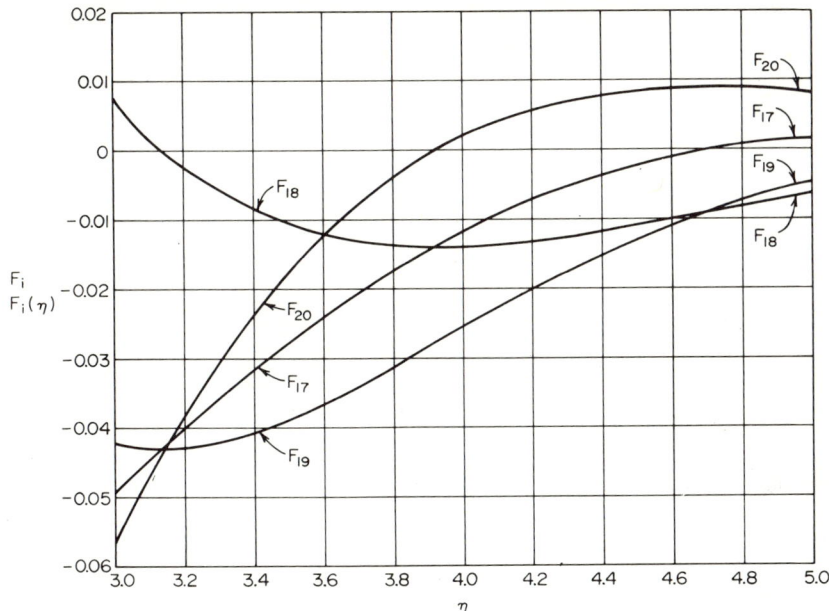

figure 5-4 *Continued.*

Approximate Method for Determination of Location and Maximum Stresses in Cylinders (Ref. 5-1)

The following approximate method is presented for preliminary design. ξ_N and ξ_M are nondimensional values that represent the location of maximum circumferential stress and maximum moment due to any linear loading, characterized by p_v and λ_p (see Chap. 3). $N_{\max}$ is the maximum circumferential force, and $M_{\max}$ is the maximum moment along the meridian. Q_F and M_F are the reaction forces at the boundary (discontinuity forces). The graphs are plotted for different geometries of cylinders represented with parameter kL where $k = \sqrt[4]{3(1-\mu^2)}/\sqrt{Rt}$. There are two similar sets of graphs, one for a fixed and the other for a pinned lower boundary. Consequently,

Graphs in Figs. 5-5 and 5-6 lead to determination of ξ_N and $N_{\max}$.
Graphs in Fig. 5-7 or 5-8 lead to determination of ξ_M.
Graphs in Fig. 5-9 or 5-10 lead to determination of $M_{\max}$.
Graphs in Fig. 5-11 or 5-12 lead to determination of H_F.
Graphs in Fig. 5-13 lead to determination of M_F.

These graphs were developed by Hampe (Ref. 6-1) and are based on his method, which has been described.

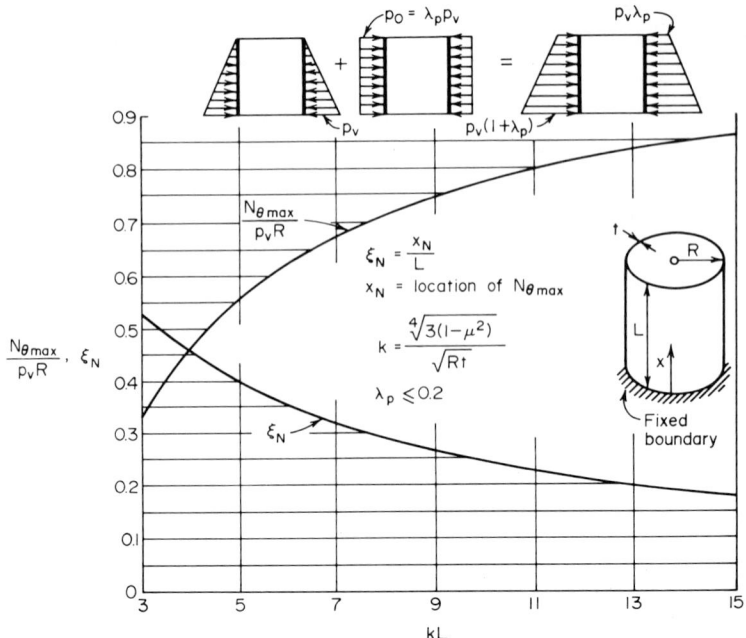

figure 5-5 *Determination of location and value of maximum circumferential load in cylinders loaded linearly, with fixed lower boundary.*

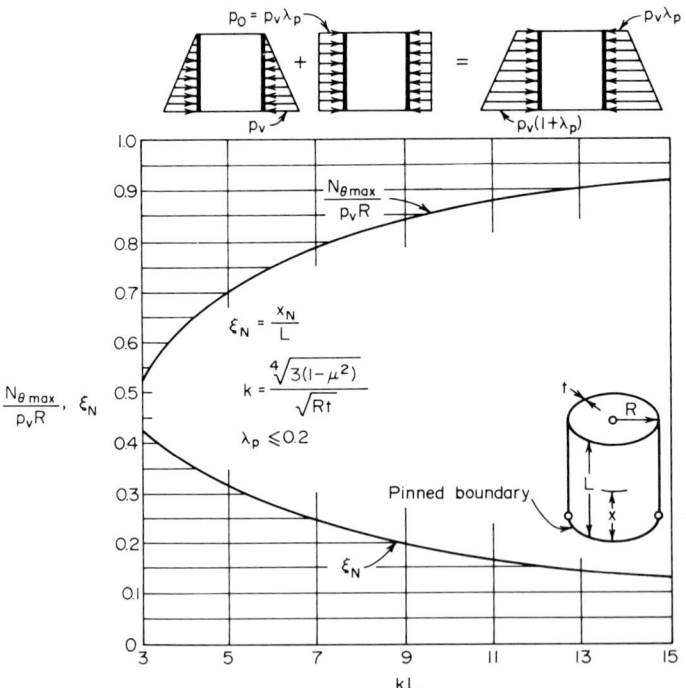

figure 5-6 *Determination of location and value of maximum circumferential load in cylinders loaded linearly, with pinned lower boundary.*

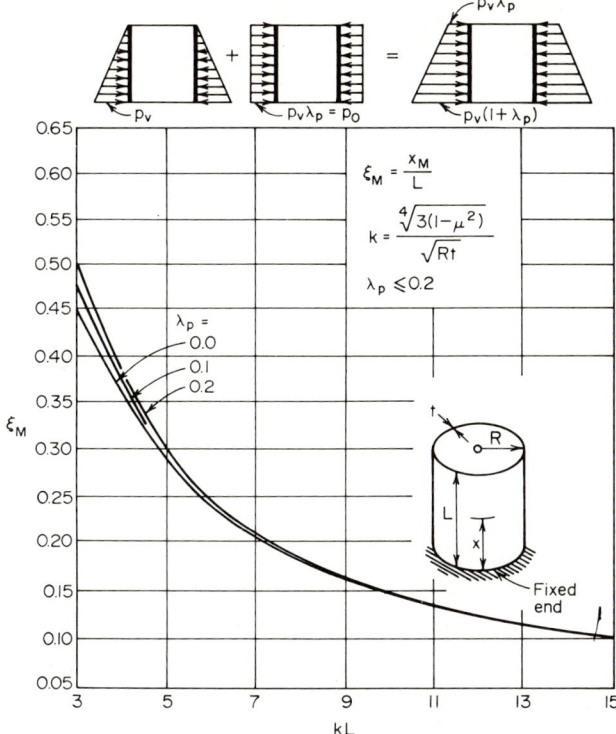

figure 5-7 *Determination of location for maximum moment for cylinder loaded linearly with fixed lower boundary.*

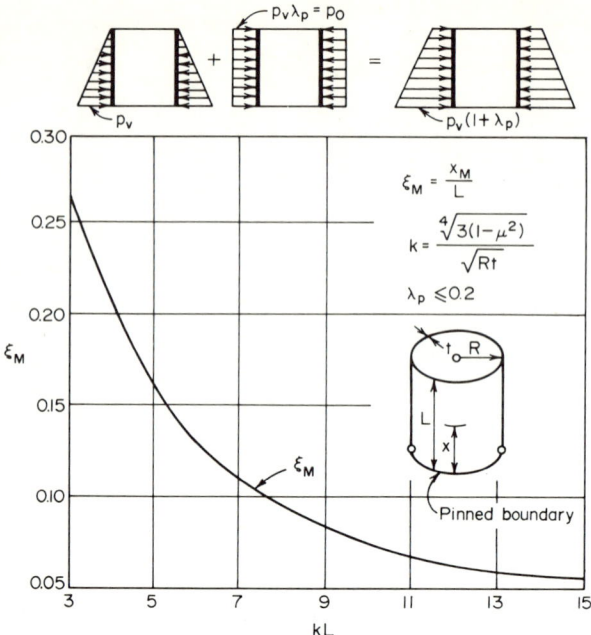

figure 5-8 *Determination of location for maximum moment for cylinders loaded linearly with pinned lower boundary.*

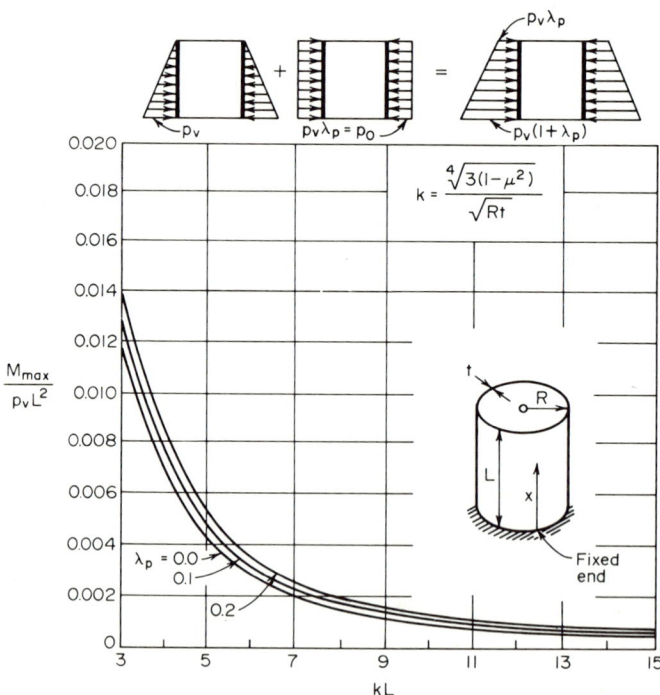

figure 5-9 *Determination of maximum moment for linearly loaded cylinders, fixed at lower boundary.*

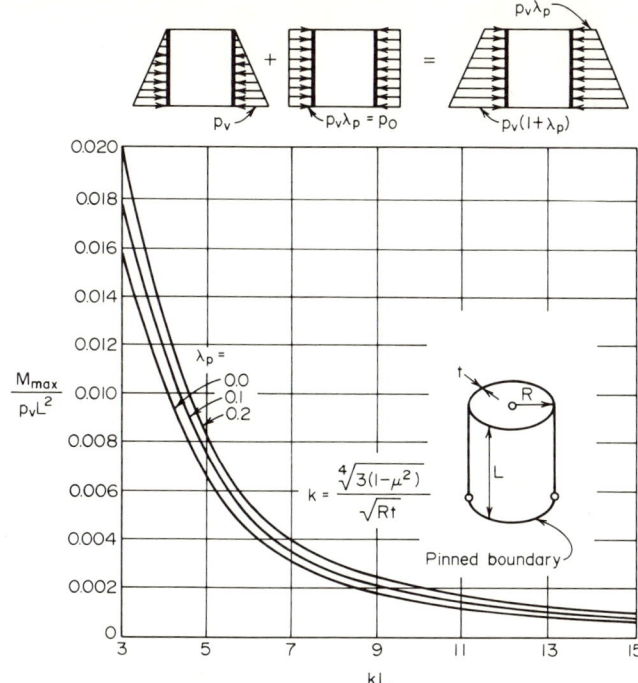

figure 5-10 *Determination of maximum moment for linearly loaded cylinders with pinned lower boundary.*

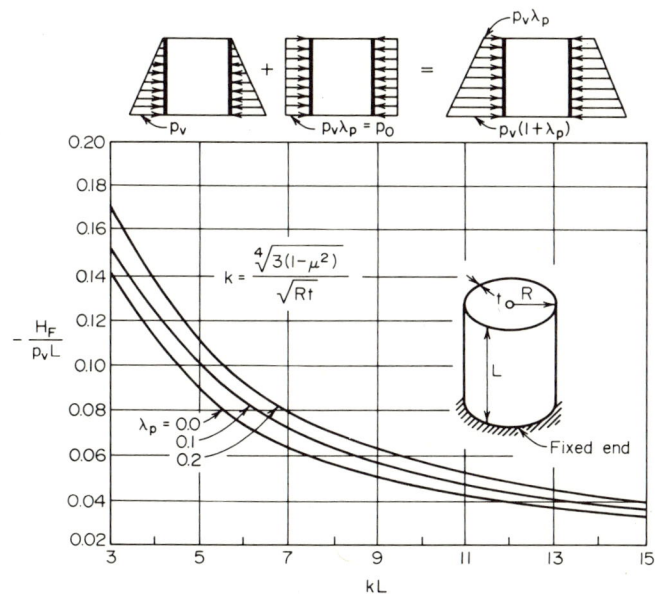

figure 5-11 *Determination of shear at lower fixed boundary of cylinders loaded with linearly distributed pressure.*

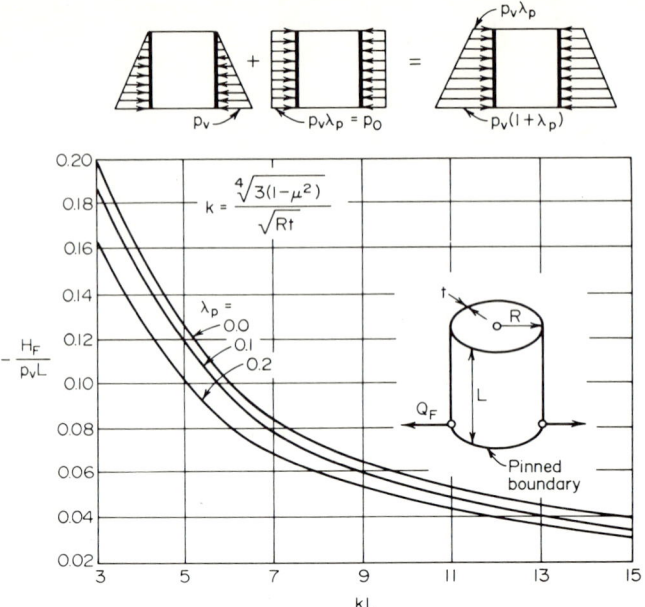

figure 5-12 *Determination of shear at lower pinned boundary of cylinders loaded with linearly distributed pressure.*

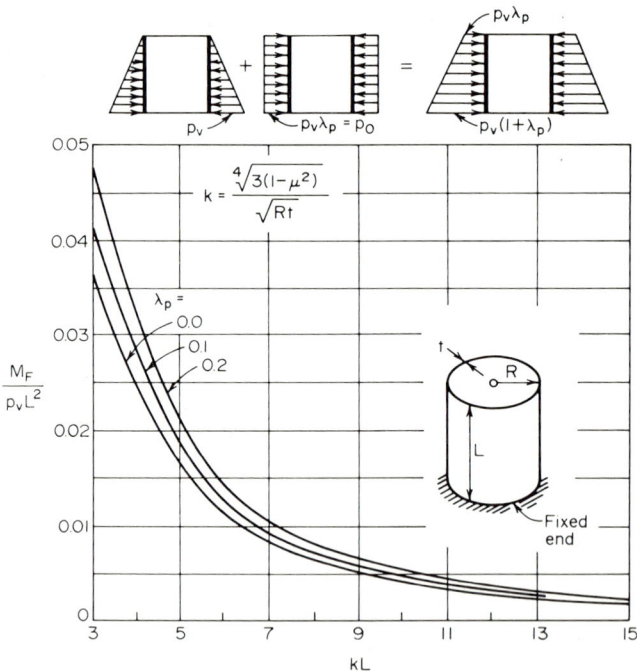

figure 5-13 *Determination of the fixity moment at the bottom of the fixed cylinders for linear loading.*

Special Solutions 141

5-3 Circular Plates

A collection of solutions for circular plates with different axisymmetrical loading conditions is presented in this section. The circular plates with and without a central circular hole are considered. These solutions can be used individually or in the interaction procedure for more complicated structures. The following nomenclature is used:

w = deflection (positive downward)
β = rotation (positive if it causes tension at lower surface)

E = Young's modulus of elasticity
t = thickness of plate
μ = Poisson's ratio
$D = Et^3/12(1 - \mu^2)$
M_r = radial moment (positive if it causes tension at lower surface)
M_t = tangential moment (positive if it causes tension at lower surface)
Q_r = radial shear
A = reaction at support

Other designations are indicated in the tables in this section.

The formulas presented were derived by using the linear bending theory. The primary solution is presented first; then secondary solutions are presented in the same way as for shells. Finally, special cases (fixed boundary conditions) will be given. These solutions were obtained from Ref. 5-2.

Primary Solutions

Primary solutions are assembled in Tables 5-11 and 5-12.

Secondary Solutions

The only unit edge loading of importance is a unit moment loading along the edges (Fig. 5-14). Table 5-13 presents solutions for this loading for different cases of circular plate with a circular opening at the center. Additional formulas for circular plates for determination of stresses due to equally distributed edge loadings are presented in Table 5-14.

Special Cases

Special cases and solutions for circular plates that commonly occur in practice are presented in this section. The geometry, boundary conditions, and loadings for special circular plates (with and without a central hole) are shown in Tables 5-15 and 5-16. The information for these tables was obtained from Ref. 5-2.

TABLE 5-11 Simply Supported Circular Plates (Ref. 5-2)

	Uniform load	Parabolic distribution	Conical distribution	Reverse parabolic distribution
	$p = \text{const}$, $P = pa^2\pi$ $A = \dfrac{P}{2a\pi}$, $Q_r = -\dfrac{Pp}{2a\pi}$	$p = p_0(1-\rho^2)$; $P = p_0 a^2 \dfrac{\pi}{2}$ $A = \dfrac{P}{2a\pi}$, $Q_r = -\dfrac{Pp(2-\rho^2)}{2\pi a}$	$p = p_0(1-\rho)$; $P = p_0 a^2 \dfrac{\pi}{3}$ $A = \dfrac{P}{2a\pi}$, $Q_r = -\dfrac{Pp(3-2\rho)}{2a\pi}$	$p = p_0(1-\rho)^2$, $P = \dfrac{1}{6} p_0 a^2 \pi$ $A = \dfrac{P}{2a\pi}$, $Q_r = -\dfrac{Pp(6-8\rho+3\rho^2)}{2a\pi}$
w	$\dfrac{Pa^2(1-\rho^2)}{64D\pi}\left(\dfrac{5+\mu}{1+\mu} - \rho^2\right)$	$\dfrac{Pa^2}{288D\pi}\left[\dfrac{31+7\mu}{1+\mu} - \dfrac{39+15\mu}{1+\mu}\rho^2 + 9\rho^4 - \rho^6\right]$	$\dfrac{Pa^2}{4800D\pi}\left[\dfrac{3(83+43\mu)}{1+\mu} - \dfrac{10(71+29\mu)}{1+\mu}\rho^2 + 225\rho^4 - 64\rho^5\right]$	$\dfrac{Pa^2}{2400D\pi}\left[\dfrac{323+83\mu}{1+\mu} - \dfrac{5(89+4\mu)}{1+\mu}\rho^2 + 225\rho^4 - 128\rho^5 + 25\rho^6\right]$
β	$\dfrac{Pa}{16D\pi}\left(\dfrac{3+\mu}{1+\mu} - \rho^2\right)$	$\dfrac{Pa\rho}{48D\pi}\left(\dfrac{13+5\mu}{1+\mu} - 6\rho^2 + \rho^4\right)$	$\dfrac{Pa\rho}{240D\pi}\left(\dfrac{71+29\mu}{1+\mu} - 45\rho^2 + 16\rho^3\right)$	$\dfrac{Pa\rho}{240D\pi}\left(\dfrac{89+4\mu}{1+\mu} - 90\rho^2 + 64\rho^3 - 15\rho^4\right)$
M_r	$\dfrac{P}{16\pi}(3+\mu)(1-\rho^2)$	$\dfrac{P}{48\pi}\left[3+5\mu-6(3+\mu)\rho^2+(5+\mu)\rho^4\right]$	$\dfrac{P}{240\pi}\left[71+29\mu-45\rho^2(3+\mu)+16\rho^3(4+\mu)\right]$	$\dfrac{P}{240\pi}\left[89+4\mu-90\rho^2(3+\mu)+64\rho^3(4+\mu)-15\rho^4(5+\mu)\right]$
M_t	$\dfrac{P}{16\pi}\left[3+\mu-(1+3\mu)\rho^2\right]$	$\dfrac{P}{48\pi}\left[3+5\mu-6(1+3\mu)\rho^2+(1+5\mu)\rho^4\right]$	$\dfrac{P}{240\pi}\left[71+29\mu-45\rho^2(1+3\mu)+16\rho^3(1+4\mu)\right]$	$\dfrac{P}{240\pi}\left[89+4\mu-90\rho^2(1+3\mu)+64\rho^3(1+4\mu)-15\rho^4(1+5\mu)\right]$

TABLE 5-11 Continued

	Uniform running load $A = Px$		Uniformly distributed moment $A = 0$	
	For $0 \leq \rho \leq X$	For $X \leq \rho \leq 1$	For $0 \leq \rho \leq X$	For $X \leq \rho \leq 1$
w	$\dfrac{Pa^3}{8D}\dfrac{X}{1+\mu}\left\{(3+\mu)(1-X^2)+2(1+\mu)X^2\ln X - [(1-\mu)(1-X^2)-2(1+\mu)\ln X]\rho^2\right\}$	$\dfrac{Pa^3}{8D}\dfrac{X}{1+\mu}\left\{[(3+\mu)-(1-\mu)X^2](1-\rho^2)+2(1+\mu)(X^2+\rho^2)\ln\rho\right\}$	$\dfrac{Ma^2}{4D}\dfrac{1}{1+\mu}\left\{2X^2[1-(1+\mu)\ln X]-[1+\mu+(1-\mu)X^2]\rho^2\right\}$	$\dfrac{Ma^2}{4D}\dfrac{X^2}{1+\mu}\left\{[1-\mu)(1-\rho^2)-2(1+\mu)\ln\rho\right\}$
β	$\dfrac{Pa^2}{4D}\dfrac{X}{1+\mu}\rho\left[(1-\mu)(1-X^2)-2(1+\mu)\ln X\right]$	$\dfrac{Pa^2}{4D}\dfrac{X}{1+\mu}\rho\left[2-(1-\mu)X^2-(1+\mu)\dfrac{X^2}{\rho^2}-2(1+\mu)\ln\rho\right]$	$\dfrac{Ma\rho}{2D(1+\mu)}\left[1+\mu+(1-\mu)X^2\right]$	$\dfrac{MaX^2}{2D(1+\mu)}\left[(1-\mu)\rho+\dfrac{1+\mu}{\rho}\right]$
M_r	$\dfrac{PaX}{4}\left[(1-\mu)(1-X^2)-2(1+\mu)\ln X\right]$	$\dfrac{PaX}{4}\left[(1-\mu)X^2\left(\dfrac{1}{\rho^2}-1\right)-2(1+\mu)\ln\rho\right]$	$\dfrac{M}{2}\left[1+\mu+(1-\mu)X^2\right]$	$\dfrac{M}{2}(1-\mu)X^2\left(1-\dfrac{1}{\rho^2}\right)$
M_t	$\dfrac{PaX}{4}\left[(1-\mu)(1-X^2)-2(1+\mu)\ln X\right]$	$\dfrac{PaX}{4}\left\{(1-\mu)\left[2-X^2\left(\dfrac{1}{\rho^2}+1\right)\right]-2(1+\mu)\ln\rho\right\}$	$\dfrac{M}{2}\left[1+\mu+(1-\mu)X^2\right]$	$\dfrac{M}{2}(1-\mu)X^2\left(1+\dfrac{1}{\rho^2}\right)$
Q_r	0	$-Px\dfrac{1}{\rho}$	0	0

TABLE 5-11 Simply Supported Circular Plates Continued

	Uniform load over portion of plate	Concentrated load
	$p = $ constant; $P = pb^2\pi = p a^2 \pi X^2$, $A = \dfrac{P}{2a\pi}$ For $0 \leq \rho \leq X$: $Q_r = -\dfrac{P\rho}{2aX^2\pi}$ For $X \leq \rho \leq 1$: $Q_r = -\dfrac{P}{2a\pi\rho}$	$A = \dfrac{P}{2\pi a}$, For $\rho \geq X$, $Q_r = -\dfrac{P}{2a\pi\rho}$ For $\rho = 0$, $Q_r = 0$ (X is small)
w	$\dfrac{Pa^2}{64D\pi}\dfrac{1}{1+\mu}\left\{\left[4(3+\mu) - (7+3\mu)X^2 + 4(1+\mu)X^2 \ln X\right.\right.$ $\left.\left. - 2\left[4-(1-\mu)X^2 - 4(1+\mu)\ln X\right]\rho^2 + (1+\mu)\dfrac{\rho^4}{X^2}\right\}$	$\dfrac{Pa^2}{16D\pi}\left[\dfrac{3+\mu}{1+\mu}(1-\rho^2) + 2\rho^2 \ln\rho\right]$
β	$\dfrac{Pa}{16\pi D}\dfrac{1}{1+\mu}\rho\left[4-(1-\mu)X^2 - 4(1+\mu)\ln X - \dfrac{1+\mu}{X^2}\rho^2\right]$	$\dfrac{Pa}{4D\pi}\rho\left(\dfrac{1}{1+\mu} - \ln\rho\right)$
M_r	$\dfrac{P}{16\pi}\left[(1-\mu)X^2 - 4(1+\mu)\ln X - \dfrac{3+\mu}{X^2}\rho^2\right]$	For $\rho \geq X$, $-\dfrac{P}{4\pi}(1+\mu)\ln\rho$ For $\rho = 0$, $\dfrac{P}{4\pi}[1-(1+\mu)\ln X]$
M_t	$\dfrac{P}{16\pi}\left\{(1-\mu)\left[4-X^2\left(\dfrac{1}{\rho^2}+1\right)\right] - 4(1+\mu)\ln\rho\right\}$	For $\rho \geq X$, $\dfrac{P}{4\pi}[1-\mu-(1+\mu)\ln\rho]$ For $\rho = 0$, $\dfrac{P}{4\pi}[1-(1+\mu)\ln X]$

TABLE 5-12 Simply Supported Circular Plates with Central Hole (Ref. 5-2)

	Uniform load	Uniform running load along free edge
	$Q_r = -\dfrac{pa}{2}\left(\rho - X^2\dfrac{1}{\rho}\right)$, $A = \dfrac{pa}{2}(1-X^2)$; $X<1$; $A = \dfrac{pa}{2}(X^2-1)$; $X>1$; $k_2 = X^2\left[3+\mu+4(1+\mu)\dfrac{X^2}{1-X^2}\ln X\right]$	$Q_r = -\dfrac{PX}{\rho}$, $A = PX$ for $X<1$; $A = -PX$ for $X>1$; $k_4 = (1+\mu)\dfrac{X^2}{1-X^2}\ln X$
w	$\dfrac{pa^4}{64D}\left\{\dfrac{2}{1+\mu}[3+\mu)(1-2X^2)+k_2](1-\rho^2)-(1-\rho^4)-\dfrac{4k_2\ln\rho}{1-\mu}-8X^2\rho^2\ln\rho\right\}$	$\dfrac{Pa^3X}{8D}\left[\dfrac{3+\mu-2k_4}{1+\mu}(1-\rho^2)+4\dfrac{k_4}{1-\mu}\ln\rho+2\rho^2\ln\rho\right]$
β	$\dfrac{pa^3}{16D}\left[\dfrac{1}{1+\mu}(3+\mu-4X^2+k_2)\rho-\rho^3+\dfrac{k_2}{1-\mu}\dfrac{1}{\rho}+4X^2\rho\ln\rho\right]$	$\dfrac{Pa^2X}{2D}\left(\dfrac{1-k_4}{1+\mu}\rho-\dfrac{k_4}{1-\mu}\dfrac{1}{\rho}-\rho\ln\rho\right)$
M_r	$\dfrac{pa^2}{16}\left[(3+\mu)(1-\rho^2)+k_2\left(1-\dfrac{1}{\rho^2}\right)+4(1+\mu)X^2\ln\rho\right]$	$\dfrac{PaX}{2}\left[k_4\left(\dfrac{1}{\rho^2}-1\right)-(1+\mu)\ln\rho\right]$
M_t	$\dfrac{pa^2}{16}\left[2(1-\mu)(1-2X^2)+(1+3\mu)(1-\rho^2)+k_2\left(1+\dfrac{1}{\rho^2}\right)+4(1+\mu)X^2\ln\rho\right]$	$\dfrac{PaX}{2}\left[1-k_4\left(\dfrac{1}{\rho^2}+1\right)-(1+\mu)\ln\rho\right]$

TABLE 5-13 Simply Supported Circular Plates with Central Hole (Ref. 5-2)

	Uniformly distributed moment ($\rho = X$) M is in in.-lb/in.; $Q_r = 0$; $A = 0$; $k_6 = \dfrac{X^2}{(1-X^2)}$	Uniformly distributed moment ($\rho = 1$) M is in in.-lb/in.; $Q_r = 0 = A$; $k_7 = \dfrac{1}{(1-X^2)}$
w	$\dfrac{Ma^2}{2D}\dfrac{k_6}{1+\mu}\left(1-\rho^2 - 2\dfrac{1+\mu}{1-\mu}\ln\rho\right)$	$\dfrac{Ma^2}{2D}\dfrac{k_7}{1+\mu}\left(1-\rho^2 - 2\dfrac{1+\mu}{1-\mu}X^2\ln\rho\right)$
β	$\dfrac{Ma}{D}\dfrac{k_6}{1+\mu}\left(\rho + \dfrac{1+\mu}{1-\mu}\dfrac{1}{\rho}\right)$	$\dfrac{Ma}{D}\dfrac{k_7}{1+\mu}\left(\rho + \dfrac{1+\mu}{1-\mu}X^2\dfrac{1}{\rho}\right)$
M_r	$Mk_6\left(1 - \dfrac{1}{\rho^2}\right)$	$MX^2 k_7\left(\dfrac{1}{X^2} - \dfrac{1}{\rho^2}\right)$
M_t	$Mk_6\left(1 + \dfrac{1}{\rho^2}\right)$	$MX^2 k_7\left(\dfrac{1}{X^2} + \dfrac{1}{\rho^2}\right)$

TABLE 5-14 Circular Plates, Stresses in Plates Due to Edge Elongation (Ref. 5-1)

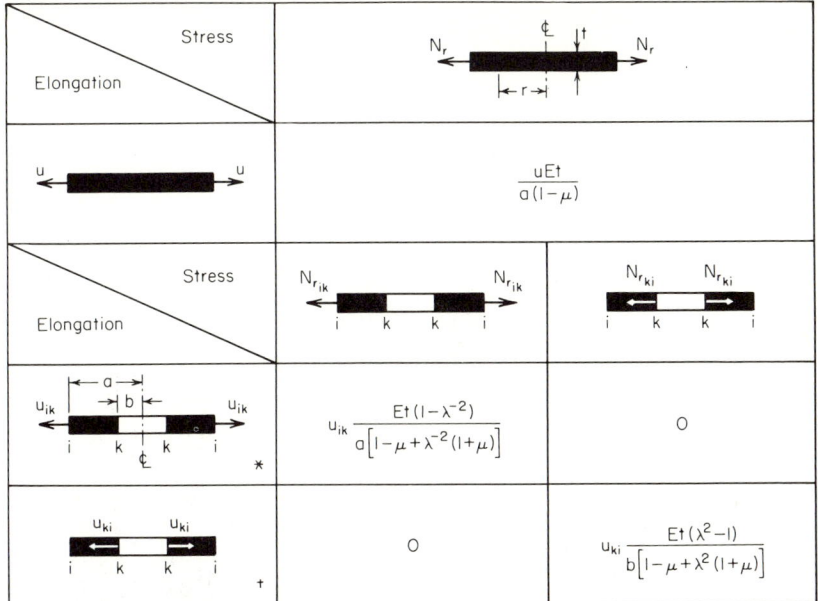

Note: $\lambda = a/b$.   Boundary conditions: *If $r=a$, $u=u_{ik}$; if $r=b$, $N_r=0$.
† If $r=a$, $N_r=0$; if $r=b$, $u=u_{ik}$.

TABLE 5-14a Edge Deformations for Circular Plates (Ref. 5-1)

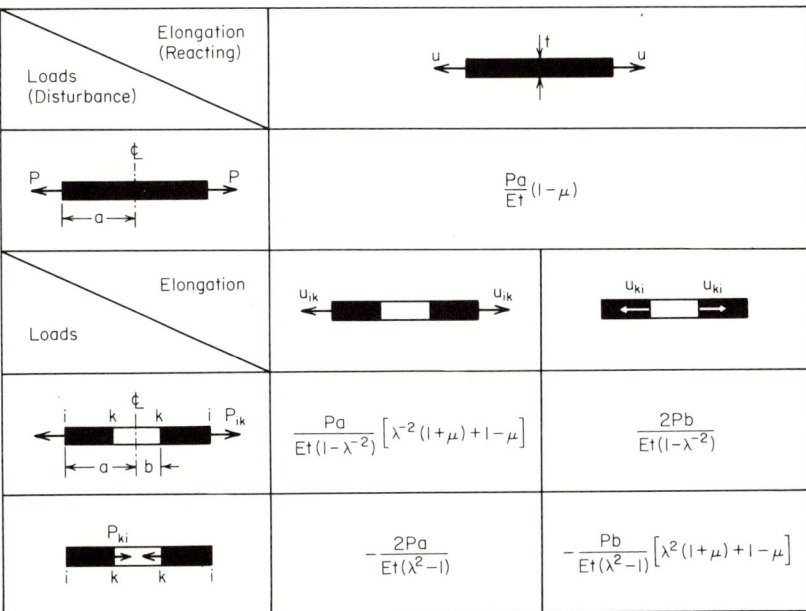

Note: $\lambda = a/b$; vectors as indicated are applied all around the edge.

TABLE 5-15 Circular Plate with Central Hole (Ref. 5-2)

	Uniform load	Uniform running load along free edge	Uniformly distributed moment along free edge
	$k_1 = X^2 \dfrac{(1-\mu)X^2 + (1+\mu)(1+4X^2\ln X)}{1-\mu+(1+\mu)X^2}$	$k_3 = X^2 \dfrac{1+(1+\mu)\ln X}{1-\mu+(1+\mu)X^2}$	$k_5 = \dfrac{X^2}{1-\mu+(1+\mu)X^2}$
	$A = \dfrac{pa}{2}(1-X^2)$ for $X<1$, $A = \dfrac{pa}{2}(X^2-1)$ for $X>1$	$A = PX$ for $X<1$ $A = -PX$ for $X>1$	$A = 0$ (vertical reaction)
		P is in lb/in.	M (in-lb/in.)
	[diagrams: $b = aX$, $r = a\rho$, showing w, M_r, M_t, Q_r]	[diagrams: $b = aX$, $r = a\rho$, showing w, M_r, M_t, Q_r]	[diagrams: $b = aX$, $r = a\rho$, showing w, M_r, M_t; $Q_r = 0$]
w	$\dfrac{pa^4}{64D}\left[-1+2(1-k_1-2X^2)(1-\rho^2)+\rho^4-4k_1\ln\rho - 8X^2\rho^2\ln\rho\right]$	$\dfrac{Pa^3 X}{8D}\left[(1+2k_3)(1-\rho^2)+4k_3\ln\rho + 2\rho^2\ln\rho\right]$	$\dfrac{Ma^2}{2D}k_5\left(-1+\rho^2-2\ln\rho\right)$
β	$\dfrac{pa^3}{16D}\left[(1-k_1)\rho - \rho^3 + k_1\dfrac{1}{\rho} + 4X^2\rho\ln\rho\right]$	$\dfrac{Pa^2 X}{2D}\left[k_3\left(\rho-\dfrac{1}{\rho}\right)-\rho\ln\rho\right]$	$\dfrac{Ma}{D}k_5\left(\dfrac{1}{\rho}-\rho\right)$
M_r	$\dfrac{pa^2}{16}\left[(1+\mu)(1-k_1)+4X^2(1-k_1)-(1-\mu)k_1\dfrac{1}{\rho^2}-(3+\mu)\rho^2+4X^2(1+\mu)\ln\rho\right]$	$\dfrac{Pa X}{2}\left[-1+(1+\mu)k_3+(1-\mu)k_3\dfrac{1}{\rho^2}-(1+\mu)\ln\rho\right]$	$-Mk_5\left[1+\mu+(1-\mu)\dfrac{1}{\rho^2}\right]$
M_t	$\dfrac{pa^2}{16}\left[(1+\mu)(1-k_1)+4\mu X^2-(1+3\mu)\rho^2+(1-\mu)\dfrac{k_1}{\rho^2}+4(1+\mu)X^2\ln\rho\right]$	$\dfrac{Pa X}{2}\left[-\mu+(1+\mu)k_3-(1-\mu)\dfrac{k_3}{\rho^2}-(1+\mu)\ln\rho\right]$	$-Mk_5\left[1+\mu-(1-\mu)\dfrac{1}{\rho^2}\right]$
Q_r	$-\dfrac{pa}{2}\left(\rho - X^2\dfrac{1}{\rho}\right)$	$-PX\dfrac{1}{\rho}$	0

TABLE 5-16 Circular Plates with Clamped Edges (Ref. 5-2)

	Uniform load	Parabolic distribution	Conical distribution	Reverse parabolic distribution
	$p = \text{const}$ $\quad P = pa^2\pi$ $\quad A = \dfrac{P}{2a\pi}$	$p = p_0(1-\rho^2)$ $\quad P = \dfrac{1}{2}p_0 a^2\pi$ $\quad A = \dfrac{P}{2a\pi}$	$p = p_0(1-\rho)$ $\quad P = \dfrac{1}{3}p_0 a^2\pi$ $\quad A = \dfrac{P}{2a\pi}$	$p = p_0(1-\rho)^2$ $\quad P = \dfrac{1}{6}p_0 a^2\pi$ $\quad A = \dfrac{P}{2a\pi}$
w	$\dfrac{Pa}{64\pi D}(1-\rho^2)^2$	$\dfrac{Pa^2}{288D\pi}(7-15\rho^2+9\rho^4-\rho^6)$	$\dfrac{Pa^2}{4800D\pi}(129-290\rho^2+225\rho^4-64\rho^5)$	$\dfrac{Pa^2}{2400D\pi}(83-205\rho^2+225\rho^4-128\rho^5+25\rho^6)$
β	$\dfrac{Pa}{16D\pi}\rho(1-\rho^2)$	$\dfrac{Pa\rho}{48D\pi}(5-6\rho^2+\rho^4)$	$\dfrac{Pa\rho}{240D\pi}(29-45\rho^2+16\rho^3)$	$\dfrac{Pa\rho}{240D\pi}(41-90\rho^2+64\rho^3-15\rho^4)$
M_r	$\dfrac{P}{16\pi}\left[1+\mu-(3+\mu)\rho^2\right]$	$\dfrac{P}{48\pi}\left[5(1+\mu)-6(3+\mu)\rho^2+\rho^4(5+\mu)\right]$	$\dfrac{P}{240\pi}\left[29(1+\mu)-45(3+\mu)\rho^2+16(4+\mu)\rho^3\right]$	$\dfrac{P}{240\pi}\left[41(1+\mu)-90(3+\mu)\rho^2+64(4+\mu)\rho^3-15(5+\mu)\rho^4\right]$
M_t	$\dfrac{P}{16\pi}\left[1+\mu-(1+3\mu)\rho^2\right]$	$\dfrac{P}{48\pi}\left[5(1+\mu)-6(1+3\mu)\rho^2+(1+5\mu)\rho^4\right]$	$\dfrac{P}{240\pi}\left[29(1+\mu)-45(1+3\mu)\rho^2+16(1+4\mu)\rho^3\right]$	$\dfrac{P}{240\pi}\left[41(1+\mu)-90(1+3\mu)\rho^2+64(1+4\mu)\rho^3-15(1+5\mu)\rho^4\right]$
Q_r	$-\dfrac{P}{2a\pi}\rho$	$-\dfrac{P}{2a\pi}\rho(2-\rho^2)$	$-\dfrac{P}{2a\pi}\rho(3-2\rho)$	$-\dfrac{P}{2a\pi}\rho(6-8\rho+3\rho^2)$

149

TABLE 5-16 Continued

	Uniform load over portion of plate $P=pb^2\pi=pa^2\pi X^2$	Concentrated load P (lb)	Uniform running load P (lb/in)	Uniformly distributed moment M (in.-lb/in)
	$0 \leq \rho \leq X$; $X \leq \rho \leq 1$; $A = \dfrac{P}{2a\pi}$	$\rho \geq X$; X is small ; $\rho = 0$; $A = \dfrac{P}{2a\pi}$	$0 \leq \rho \leq X$; $X \leq \rho \leq 1$; $A = PX$	$X \leq \rho \leq 1$; $0 \leq \rho \leq X$; $A = 0$
w	$\dfrac{Pa^2}{64D\pi}[4-3X^2+4X^2\ln X -2(X^2-4\ln X)\rho^2+\dfrac{1}{X^2}\rho^4]$; $\dfrac{Pa^2}{32D\pi}[(2-X^2)(1-\rho^2)+2(X^2+2\rho^2)\ln\rho]$	$\dfrac{Pa^2}{16D\pi}(1-\rho^2+2\rho^2\ln\rho)$	$\dfrac{Pa^3X}{8D}[1-X^2+2X^2\ln X +(1-X^2+2\ln X)\rho^2]$; $\dfrac{Pa^3X}{8D}[(1+X^2)(1-\rho^2)+2(X^2+\rho^2)\ln\rho]$	$-\dfrac{Ma^2X^2}{4D}(1-\rho^2)+2\ln\rho)$; $-\dfrac{Ma^2}{4D}[2X^2\ln X +\rho^2(1-X^2)]$
β	$\dfrac{Pa}{16D\pi}(X^2-4\ln X)\rho -\dfrac{1}{X^2}\rho^3$	$-\dfrac{Pa}{4D\pi}\rho\ln\rho$	$-\dfrac{Pa^2X\rho}{4D}(1-X^2+2\ln X)$	$-\dfrac{Pa^2X\rho}{4D}[X^2(1-\dfrac{1}{\rho^2})-2\ln\rho]$	$\dfrac{Max^2}{2D}(\dfrac{1}{\rho}-\rho)$; $\dfrac{Ma\rho}{2D}$
M_r	$\dfrac{P}{16\pi}[(1+\mu)(X^2-4\ln X)-\dfrac{3+\mu}{X^2}\rho^2]$; $\dfrac{P}{16\pi}[-4+(1+\mu)\dfrac{X^2}{\rho^2}+(1+\mu)(X^2-4\ln\rho)]$	$-\dfrac{P}{4\pi}[1+(1+\mu)\ln\rho]$	$-\dfrac{PaX}{4}(1+\mu)(1-X^2+2\ln X)$	$-\dfrac{PaX}{4}[2-(1-\mu)\dfrac{X^2}{\rho^2}-(1+\mu)(X^2-4\ln\rho)]$	$-\dfrac{MX^2}{2}[1+\mu+(1-\mu)\dfrac{1}{\rho^2}]$; $\dfrac{M}{2}(1+\mu)(1-X^2)$
M_t	$\dfrac{P}{16\pi}[(1+\mu)(X^2-4\ln X)-\dfrac{1+3\mu}{X^2}\rho^2]$; $\dfrac{P}{16\pi}[-4\mu-(1-\mu)\dfrac{X^2}{\rho^2}+(1+\mu)(X^2-4\ln\rho)]$	$-\dfrac{P}{4\pi}[\mu+(1+\mu)\ln\rho]$	$-\dfrac{PaX}{4}(1+\mu)(1-X^2+2\ln X)$	$-\dfrac{PaX}{4}[2\mu+(1-\mu)\dfrac{X^2}{\rho^2}-(1+\mu)(X^2-2\ln\rho)]$	$-\dfrac{MX^2}{2}[1+\mu-(1-\mu)\dfrac{1}{\rho^2}]$; $\dfrac{M}{2}(1+\mu)(1-X^2)$
Q_r	$-\dfrac{P}{2a\pi}\dfrac{\rho}{X^2}$; $-\dfrac{P}{2a\pi}\dfrac{1}{\rho}$	$-\dfrac{P}{2a\pi}\dfrac{1}{\rho}$	0 ; $-PX\dfrac{1}{\rho}$	0 ; 0

150

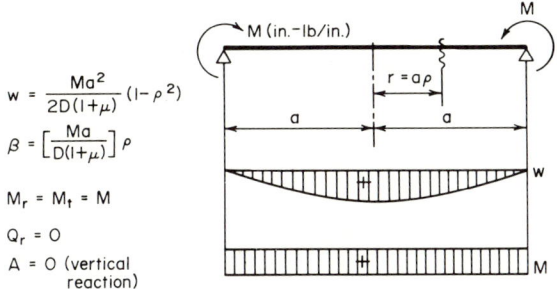

figure 5-14 *Formulas of influences for a simply supported circular plate loaded with a uniformly distributed end moment.*

5-4 Circular Rings

Circular rings are important structural elements which often interact with shells. In this section, such information is summarized and presented for loading symmetrical with respect to the center of the ring.

Nomenclature employed is as follows:

$A =$ area of cross section

$I_1, I_2 =$ moment of inertia for centroidal axis in the plane or normal to the plane of the ring

$J =$ torsional rigidity factor of section

Table 5-17 presents the solutions for different loads on rings. This information was obtained from Ref. 5-3.

REFERENCES

5-1. Hampe, E.: *Statik Rotationssymmetrischer Flächentragwerke*, VEB Verlag für Bauwesen, Berlin, 1964.

5-2. Worch, G.: *Elastische Platten* (in German), *Beton-Kalender*, Wilhelm Ernst & Sohn KG, Berlin, 1943.

5-3. Flügge, W.: *Stresses in Shells*, Springer-Verlag OHG, Berlin, 1960.

TABLE 5-17 Symmetrically Loaded Circular Rings (Ref. 5-3)

Item	Loading condition	n > 1 Stresses	n > 1 Displacements	n = 1 Stresses	n = 1 Displacements	n = 0 Stresses	n = 0 Displacements
1	Radial loading $p = p_n \cos n\theta$, where n is an integer: n = 1, 2, 3, ...	$N = -\dfrac{p_n a}{n^2-1}\cos n\theta$ $M = -\dfrac{p_n a^2}{n^2-1}\cos n\theta$	$v = -\dfrac{p_n a^2}{n(n^2-1)}\left(\dfrac{a^2}{(n^2-1)EI_2} + \dfrac{1}{EA}\right)\cos n\theta$ $w = -\dfrac{p_n a^4}{(n^2-1)^2 EI_2}\cos n\theta$	\multicolumn{2}{c	}{Problem does not exist because the loading is not in equilibrium}	$N = pa$ $M = 0$	$v = 0$ $w = \dfrac{pa^2}{EA}$
2	Tangential loading $p = p_n \sin n\theta$	$N = \dfrac{np_n a}{n^2-1}\cos n\theta$ $M = \dfrac{p_n a^2}{n(n^2-1)}\cos n\theta$	$v = \dfrac{p_n a^2}{n^2-1}\left[\dfrac{a^2}{n^2(n^2-1)EI_2} + \dfrac{1}{EA}\right]\sin n\theta$ $w = \dfrac{p_n a^4}{n(n^2-1)^2 EI_2}\cos n\theta$	\multicolumn{2}{c	}{Problem does not exist because the loading is not in equilibrium}	\multicolumn{2}{c	}{Problem does not exist because the loading is not in equilibrium}
3	Loading normal to the plane of the ring $p = p_n \cos n\theta$	$M_l = \dfrac{p_n a^2}{n^2-1}\cos n\theta$ $M_T = -\dfrac{p_n a^2}{n(n^2-1)}\sin n\theta$	$u = \dfrac{p_n a^4}{(n^2-1)^2}\left[\dfrac{1}{EI_1} + \dfrac{1}{n^2 GJ}\right]\cos n\theta$ $\beta = \dfrac{p_n a^3}{(n^2-1)^2}\left[\dfrac{1}{EI_1} + \dfrac{1}{GJ_T}\right]\cos n\theta$	\multicolumn{2}{c	}{Problem does not exist because the loading is not in equilibrium}	\multicolumn{2}{c	}{Problem does not exist because the loading is not in equilibrium}
4	External moments, about the ring axis $m = m_n \cos n\theta$	$M_l = \dfrac{m_n a}{n^2-1}\cos n\theta$ $M_T = \dfrac{n m_n a}{n^2-1}\cos n\theta$	$u = -\dfrac{m_n a^3}{(n^2-1)^2}\left[\dfrac{1}{EI_1} + \dfrac{1}{GJ}\right]\cos n\theta$ $\beta = \dfrac{m_n a^2}{(n^2-1)^2}\left[\dfrac{1}{EI_1} + \dfrac{n^2}{GJ}\right]\cos n\theta$	\multicolumn{2}{c	}{Problem does not exist because the loading is not in equilibrium}	$M_l = ma$ $M_T = 0$	u may assume any desired value $\beta = \dfrac{ma^2}{EI_1}$

Chapter 6

MULTISHELL STRUCTURES

6-1 Introduction

A multishell structure consists of a combination of simple shell elements, as is shown in Fig. 6-1. The structure may be composed of spherical, cylindrical, and conical shells, circular plates, and axisymmetrically loaded circular rings.

A combined shell is a shell of revolution which can be obtained by the rotation of a curved line about an axis of revolution. Such a line may be replaced by several segments of a regular shape such as circular, straight, or parabolical segments. The segments are "connected in series." Such a shell is a special case of a multishell, which is produced by the rotation of a branched curved or straight line about an axis of revolution. Such shells include elements which are connected both "in parallel" and "in series."

Axisymmetrical loadings only are considered. From the point of view of the analyst there is no difference in the analysis procedure for combined shells or multishells. Consequently, in the following section, no distinction is made between them, and both shells are called multishells.

154 Structural Analysis of Shells

In Chap. 2 the simplest interaction of one junction of a two-shell element was shown. If the shell elements are of such a geometry that all disturbances due to the corrective loading die out before reaching the opposite edge of the shell element, each junction may be interacted separately regardless of the rest of the structure. Then there are for each junction a simple set of two (or three, if the vertical corrective load is considered) linear algebraic equations to determine the corrective loading at this junction.

If, however, the length of the shell element is such that the influences due to the corrective loading do not die out before reaching the opposite edge of the shell element, it is not possible to solve one junction separately, but the rest of the structure must be considered. Then, as will be shown later, the discontinuity equation at the junction under consideration involves also the discontinuity loadings at the opposite edges of shell elements.

Previously, when considering only one imaginary cut, only two equations with two unknowns were obtained. If n imaginary cuts are introduced simultaneously, $2n$ linear equations with $2n$ unknowns can be obtained.

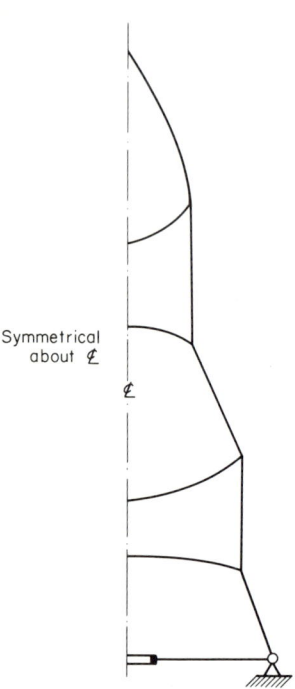

figure 6-1 *Multishell.*

6-2 Equation of Deformations of Shell Elements

A typical shell element is presented in Fig. 6-2. The edge loadings are indicated at the corresponding edges. The primary loading is denoted by p.

The following relation can be written for the shell element ①-②:

$$\begin{bmatrix} \beta_1 \\ \delta_1 \\ \beta_2 \\ \delta_2 \end{bmatrix} = \alpha_{ik} \begin{bmatrix} M_1 \\ H_1 \\ M_2 \\ H_2 \end{bmatrix} + \begin{bmatrix} \beta_1^0 \\ \delta_1^0 \\ \beta_2^0 \\ \delta_2^0 \end{bmatrix}$$

where β_1, β_2 = final rotations at ①, ②
δ_1, δ_2 = final horizontal displacement at ①, ②
H_1, H_2 = horizontal corrective shear at ①, ②
M_1, M_2 = corrective moments at ①, ②
p = primary loading
β_1^0, β_2^0 = rotations at ①, ② due to primary loading (usually zero, if the primary loading is pressure loading)

$\delta_1{}^0, \delta_2{}^0$ = horizontal displacements at ①, ② due to primary loading

$$\alpha_{ik} = \begin{bmatrix} \beta_1^{M1} & \beta_1^{H1} & \beta_1^{M2} & \beta_1^{H2} \\ \delta_1^{M1} & \delta_1^{H1} & \delta_1^{M2} & \delta_1^{H2} \\ \beta_2^{M1} & \beta_2^{H1} & \beta_2^{M2} & \beta_2^{H2} \\ \delta_2^{M1} & \delta_2^{H1} & \delta_2^{M2} & \delta_2^{H2} \end{bmatrix} = \text{flexibility matrix}$$

The flexibility coefficients in the matrix are the deformations at the edges ① or ② (as indicated by the lower subscript) due to the unit edge loadings $H = 1$ and $M = 1$ acting at edges ① or ② (as indicated by the upper index). The rotations are indicated by β, and the horizontal displacements are indicated by δ.

The coefficients β and δ are dependent on the elastic characteristics of shell elements only, and are obtained from Tables 6-3 to 6-14.

Table 6-1 represents special cases of shell elements with the corresponding equations for deformations.

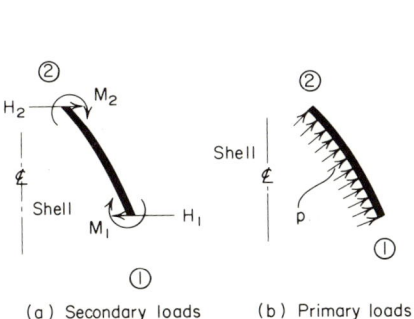

figure 6-2 *Segment of interacting shell (typical shell element).*

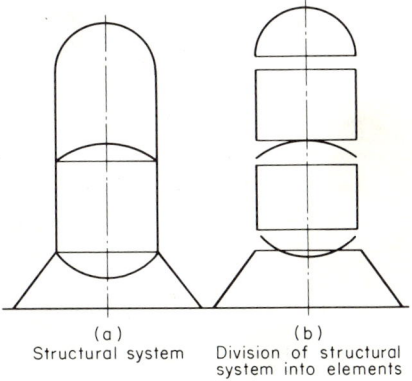

figure 6-3 *Multishell separated in single shell elements.*

156 Structural Analysis of Shells

TABLE 6-1 Shell-element Equations

No.	Geometry	Equations	Matrix
1	(shell/plate diagrams)	$\begin{bmatrix} \beta_1 \\ \delta_1 \\ \beta_2 \\ \delta_2 \end{bmatrix} = \alpha \begin{bmatrix} M_1 \\ H_1 \\ M_2 \\ H_2 \end{bmatrix} + \begin{bmatrix} \beta_1^0 \\ \delta_1^0 \\ \beta_2^0 \\ \delta_2^0 \end{bmatrix}$	$\begin{bmatrix} \beta_1^{M_1} & \beta_1^{H_1} & \beta_1^{M_2} & \beta_1^{H_2} \\ \delta_1^{M_1} & \delta_1^{H_1} & \delta_1^{M_2} & \delta_1^{H_2} \\ \beta_2^{M_1} & \beta_2^{H_1} & \beta_2^{M_2} & \beta_2^{H_2} \\ \delta_2^{M_1} & \delta_2^{H_1} & \delta_2^{M_2} & \delta_2^{H_2} \end{bmatrix}$
2	(shell/plate diagrams)	$\begin{bmatrix} \beta_1 \\ \delta_1 \end{bmatrix} = \alpha \begin{bmatrix} M_1 \\ H_1 \end{bmatrix} + \begin{bmatrix} \beta_1^0 \\ \delta_1^0 \end{bmatrix}$	$\begin{bmatrix} \beta_1^M & \beta_1^H \\ \delta_1^M & \delta_1^H \end{bmatrix}$

6-3 Equilibrium of Junctions

An additional set of equations is needed for every junction to make the structure work as a unit. For example, if junction ② is combined from two shell elements, the following equation can be written:

$$\begin{bmatrix} M_2 \\ H_2 \\ \beta_2 \\ \delta_2 \end{bmatrix} - \begin{bmatrix} M_2' \\ H_2' \\ 0 \\ 0 \end{bmatrix} = \begin{bmatrix} 0 \\ 0 \\ \beta_2' \\ \delta_2' \end{bmatrix}$$

Table 6-2 presents similar sets of equations for two- and three-branch junctions and also shows how the boundary conditions of the supports will be considered.

Multishell Structures 157

Positive sign convention for shell element—right-hand side
a. Upper edge of a shell element:
   Moment in the clockwise direction
   Horizontal load directed outward
   Vertical load directed up
b. Lower edge of a shell element:
   Moment in the counterclockwise direction
   Horizontal load directed inward
   Vertical load directed downward

Positive sign convention for circular plates—right-hand side
a. Inside edge of circular plate with central hole: As upper edge of shell element under *a*
b. Outside edge of a circular plate with or without central hole: As lower edge of shell element under *b*

In this figure all corrective loadings are indicated by X_i.

TABLE 6-2 Junction Equations

No.	Junction	Equations
I	Junction of two shell elements	$\begin{bmatrix} M_2 \\ H_2 \\ \beta_2 \\ \delta_2 \end{bmatrix} - \begin{bmatrix} M'_2 \\ H'_2 \\ 0 \\ 0 \end{bmatrix} = \begin{bmatrix} 0 \\ 0 \\ \beta'_2 \\ \delta'_2 \end{bmatrix}$
II	Junction of three shell elements	$\begin{bmatrix} M_2 \\ H_2 \\ \beta_2 \\ \beta_2 \\ \delta_2 \\ \delta_2 \end{bmatrix} - \begin{bmatrix} M'_2 \\ H'_2 \\ 0 \\ 0 \\ 0 \\ 0 \end{bmatrix} - \begin{bmatrix} M''_2 \\ H''_2 \\ 0 \\ 0 \\ 0 \\ 0 \end{bmatrix} = \begin{bmatrix} 0 \\ 0 \\ \beta'_2 \\ \beta''_2 \\ \delta'_2 \\ \delta''_2 \end{bmatrix}$
III	Fixed edge	Use formulas 1 or 2 (Table 6-1) and I or II (this table) considering: $\beta_1 = 0 \quad \delta_1 = 0$
IV	Pinned edge	$M_1 = 0 \quad \delta_1 = 0$
V	Simple support edge	$M_1 = 0 \quad H_1 = 0$

158 Structural Analysis of Shells

6-4 Interaction Procedure

1. Separate the multishell structure into single shell elements as is shown in Fig. 6-3. Each shell element of this type can be designated as a "statically determinate element," as was explained in the membrane-theory discussion.

2. Enter at every junction the corrective loadings, considering the values on the right side of Fig. 6-4 as positive for the purpose of interaction.

3. Depending on the geometry of each shell element, prescribe to each the corresponding set of equilibrium equations as shown in Table 6-1 (See Fig. 6-5. Enter the influence coefficients β and δ in

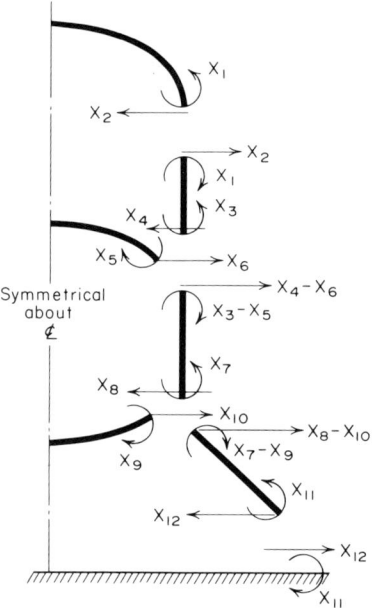

figure 6-4 *The corrective loadings at each junction of multishell.*

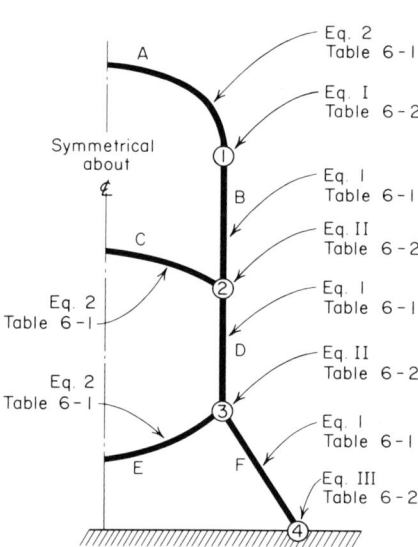

figure 6-5 *The set of relations to be used in order to find the statically indeterminate loadings.*

accordance with the adapted sign convention for deformations and with Chaps. 3 and 4.) To simplify the application of the method, Tables 6-3 through 6-14 contain the formulas for the edge deformations due to primary and secondary loadings. The formulas in this section are shorter and simpler than the general formulas given in Chaps. 3 and 4 because they cover the special cases of loads and displacements at the edge only.

4. In addition for every junction enter a corresponding set of equations as shown in Table 6-2. Consider all boundary conditions where needed (Fig. 6-5).

5. All equations of sets of types 1, 2, and I, II, ... from Tables 6-1 and 6-2 mentioned in steps 3 and 4 reduce to one final set of algebraic equations with as many unknowns as there are equations. This set is indicated in Fig. 6-6.

6. Solve the set of equations and find all corrective loadings (discontinuity loadings), which are indicated in Figs. 6-4 and 6-6 by X_i instead of M_i and H_i.

7. The formulas in Chaps. 3 and 4 must be used to find the stresses at any point of the elementary shell. In this way the discontinuity loadings and the stresses, where needed, are found.

i	X_1	X_2	X_3	X_4	X_5	X_6	X_7	X_8	X_9	X_{10}	X_{11}	X_{12}	C_{i0}
1	=	—	—	—	—								C_{10}
2	—	=	—	—	—								C_{20}
3	—	—	=	—	—	—	—	—					C_{30}
4	—	—	—	=	—	—	—	—					C_{40}
5		—	—	—	=	—	—	—					C_{50}
6		—	—	—	—	=	—	—					C_{60}
7					—	—	=	—	—	—	—	—	C_{70}
8					—	—	—	=	—	—	—	—	C_{80}
9							—	—	=	—	—	—	C_{90}
10							—	—	—	=	—	—	$C_{10\,0}$
11									—	—	=	—	$C_{11\,0}$
12									—	—	—	=	$C_{12\,0}$

figure 6-6 *The matrix of unknowns if the boundary disturbances at opposite boundaries are influencing each other. (Force method, with the statically determinate elements.)*

Example

Determine the corrective loading as shown in Fig. 6-7a. The statically indeterminate corrective loading is entered at every junction as indicated. Figure 6-7b shows sets of equations from Tables 6-1 and 6-2 prescribed to each shell element A, B, C, and D and to each junction ①, ②, and ③. First sets (1) and (2) from Table 6-1:

$$\begin{bmatrix} \beta_1 \\ \delta_1 \end{bmatrix} = \alpha_A \begin{bmatrix} M_1 \\ H_1 \end{bmatrix} + \begin{bmatrix} \beta_{10} \\ \delta_{10} \end{bmatrix} \tag{2}$$

$$\begin{bmatrix} \beta_1' \\ \delta_1' \\ \beta_2 \\ \delta_2 \end{bmatrix} = \alpha_B \begin{bmatrix} M_1' \\ H_1' \\ M_2 \\ H_2 \end{bmatrix} + \begin{bmatrix} \beta_{10}' \\ \delta_{10}' \\ \beta_{20} \\ \delta_{20} \end{bmatrix} \tag{1}$$

160 *Structural Analysis of Shells*

$$\begin{bmatrix} \beta_2' \\ \delta_2' \end{bmatrix} = \alpha_C \begin{bmatrix} M_2' \\ H_2' \end{bmatrix} + \begin{bmatrix} \beta_{20}' \\ \delta_{20}' \end{bmatrix} \tag{2}$$

$$\begin{bmatrix} \beta_2'' \\ \delta_2'' \\ \beta_3 \\ \delta_3 \end{bmatrix} = \alpha_D \begin{bmatrix} M_2'' \\ H_2'' \\ M_3 \\ H_3 \end{bmatrix} + \begin{bmatrix} \beta_{20}'' \\ \delta_{20}'' \\ \beta_{30} \\ \delta_{30} \end{bmatrix} \tag{1}$$

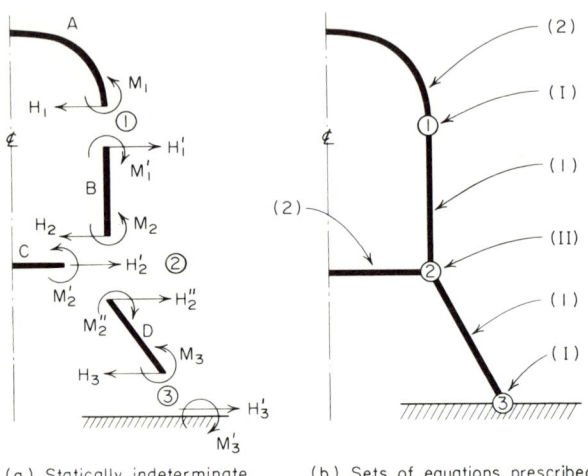

(a) Statically indeterminate corrective loading

(b) Sets of equations prescribed to each shell element and to each junction.

figure 6-7 *Multishell conventions.*

So far there are 24 unknown loads and deformations. Additional equations I and II from Table 6-2 will reduce these unknowns to the required number of eight.

$$\begin{bmatrix} M_1 \\ H_1 \\ \beta_1 \\ \delta_1 \end{bmatrix} - \begin{bmatrix} M_1' \\ H_1' \\ 0 \\ 0 \end{bmatrix} = \begin{bmatrix} 0 \\ 0 \\ \beta_1' \\ \delta_1' \end{bmatrix} \tag{I}$$

$$\begin{bmatrix} M_2 \\ H_2 \\ \beta_2 \\ \delta_2 \\ \beta_2 \\ \delta_2 \end{bmatrix} - \begin{bmatrix} M_2' \\ H_2' \\ 0 \\ 0 \\ 0 \\ 0 \end{bmatrix} - \begin{bmatrix} M_2'' \\ H_2'' \\ 0 \\ 0 \\ 0 \\ 0 \end{bmatrix} = \begin{bmatrix} 0 \\ 0 \\ \beta_2' \\ \delta_2' \\ \beta_2'' \\ \delta_2'' \end{bmatrix} \quad \text{(II)}$$

$$\begin{bmatrix} M_3 \\ H_3 \\ \beta_3 \\ \delta_3 \end{bmatrix} - \begin{bmatrix} M_3' \\ H_3' \\ 0 \\ 0 \end{bmatrix} = \begin{bmatrix} 0 \\ 0 \\ \beta_3' \\ \delta_3' \end{bmatrix} \quad \text{(I)}$$

The boundary conditions of junction ③ are to be considered in the above sets.

$$\beta_3 = \beta_3' = 0 \qquad \delta_3 = \delta_3' = 0$$

All the above sets of equations can be algebraically reduced to one set with eight unknowns:

	M_1	H_1	M_2	H_2	M_2'	H_2'	M_3	H_3	C_{i0}
1									C_{10}
2									C_{20}
3									C_{30}
4									C_{40}
5									C_{50}
6									C_{60}
7									C_{70}
8									C_{80}

C_{i0} are known numbers which are functions of the membrane influences β_{i0} and δ_{i0}. The values M_1', H_1', M_2'', H_2'', M_3', and H_3' do not appear because these values are functions of other unknowns as listed in the above table:

$$M_1' = M_1 \quad M_2'' = M_2 - M_2' \quad M_3' = M_3$$
$$H_1' = H_1 \quad H_2'' = H_2 - H_2' \quad H_3' = H_3$$

6-5 Final Sets of Equations for Corrective Loadings

Structural System Combined from Statically Determinate Elements

Two cases which are considered when the structure is built up from statically determinate elements are shown in Fig. 6-4.

Case 1: The length of each element is such that the disturbances due to the unit loading overlap. In other words, the stresses due to the unit loading depend in every joint on both edge unit loads. For such a case, the matrix to determine X_i has the appearance shown in Fig. 6-6.

Case 2: The length of the element is such that the disturbances caused by the unit loadings will die at short distances from the edge and will not overlap the disturbances due to the unit loading on the opposite edge of a statically determinate shell element. In this case the matrix will have the shape shown in Fig. 6-8.

It can be seen that case 2 can be split into four independent systems of linear equations. The solution will be much faster and simpler than under case 1.

i	X_1	X_2	X_3	X_4	X_5	X_6	X_7	X_8	X_9	X_{10}	X_{11}	X_{12}	C_{i0}
1	=	—											C_{10}
2	—	=											C_{20}
3			=	—	—	—							C_{30}
4			—	=	—	—							C_{40}
5			—	—	=	—							C_{50}
6			—	—	—	=							C_{60}
7							=	—	—	—			C_{70}
8							—	=	—	—			C_{80}
9							—	—	=	—			C_{90}
10							—	—	—	=			$C_{10\,0}$
11											=	—	$C_{11\,0}$
12											—	=	$C_{12\,0}$

figure 6-8 *The matrix of unknowns if the boundary disturbances at opposite boundaries are not influencing each other. (Force method with statically determinate elements.)*

Structural System Combined from Statically Indeterminate Elements

The separation of the multishell into simpler components does not necessarily need to be made as shown in Fig. 6-3. The separation can be performed as shown in Fig. 6-9, but then the elements are statically indeterminate. It does not change the philosophy, but the analysis will be much faster if formulas are available for the statically indeterminate elements subjected to the primary and secondary loadings because fewer elements are present. These loads can also be obtained by calculation if the collection of formulas does not cover the particular case. As before, two cases are considered.

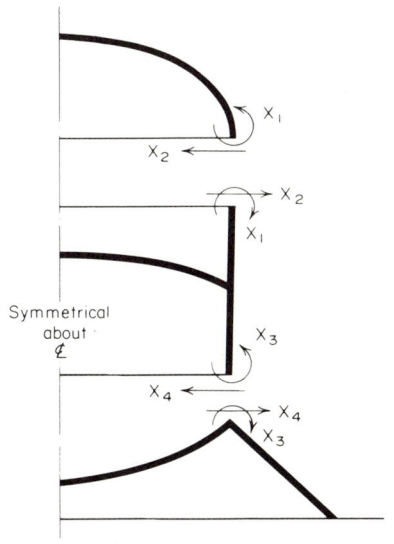

figure 6-9 *Free body diagram. The system is divided into the statically indeterminate elements.*

Case 1: The length of the element is such that the disturbances due to the unit loads on opposite edges overlap. In other words, stresses due to the unit loadings will depend, at every point of the shell, on both edge unit loadings. For this case the corresponding matrix is shown in Fig. 6-10.

Case 2: The lengths of the elements are such that the disturbances due to the unit loading do not depend on the unit loading of the opposite edge. For such a case, the corresponding matrix is shown in Fig. 6-11.

In case 2, the system of four linear equations is split into two independent systems of only two linear equations, which makes the solution simpler. Also, use of the externally statically indeterminate elements greatly reduces the number of unknowns, making the solution much simpler.

164 Structural Analysis of Shells

i	X_1	X_2	X_3	X_4	C_{io}
1	=	—	—	—	C_{10}
2	—	=	—	—	C_{12}
3	—	—	=	—	C_{13}
4	—	—	—	=	C_{14}

figure 6-10 *The matrix of unknowns if the boundary disturbances at opposite boundaries are influencing each other. (Force method with the statically indeterminate elements.)*

i	X_1	X_2	X_3	X_4	C_{io}
1	=	—			C_{10}
2	—	=			C_{20}
3			=	—	C_{30}
4			—	=	C_{40}

figure 6-11 *The matrix of unknowns if the boundary disturbances are not influencing each other. (Force method with statically indeterminate elements.)*

6-6 Application of Tables

Tables 6-3 to 6-14 are used to obtain the influence coefficients at the junctions due to the various primary loadings and due to the unit secondary loadings.

TABLE 6-3 Cylindrical Shell with Free Edges. Edge Distortions Due to Primary Loadings (Ref. 6-1)

Shell geometry	Boundary distortions	
[diagram: cylinder of length L, coordinate ξ, $p = p(\xi)$]	[diagram showing Δr_{ki} at both edges]	[diagram showing β_{ki} angles at both edges]
Loading condition	ΔH_{ki}	β_{ki}
[diagram] $p(\xi) = p_0 = \text{const}$	$+\dfrac{p_0 R^2}{Et}$	0
[diagram] $p(\xi) = p_v(1-\xi)$	0	$-\dfrac{p_v R^2}{EtL}$
[diagram] $p(\xi) = p_0 \sin \alpha \xi$	$\dfrac{p_0 R^2}{Et} \dfrac{4(kL)^4}{\alpha^4 - 4(kL)^4} \left[\sin \alpha - \dfrac{\alpha^2}{2(kL)^2} \right.$ $\left. \cdot \left(\dfrac{\alpha}{kL} \dfrac{F_9}{F_1} - \dfrac{F_2}{F_1} \sin \alpha + \dfrac{\alpha}{kL} \dfrac{F_4}{F_1} \cos \alpha \right) \right]$	$\dfrac{p_0 R^2}{Et} \dfrac{4(kL)^4}{\alpha^4 + 4(kL)^4} k \left[\dfrac{\alpha}{kL} \cos \alpha + \dfrac{\alpha^2}{2(kL)^2} \right.$ $\left. \cdot \left(\dfrac{2F_3}{F_1} \sin \alpha - \dfrac{\alpha}{kL} \dfrac{2F_8}{F_1} - \dfrac{\alpha}{kL} \dfrac{F_2}{F_1} \cos \alpha \right) \right]$
[diagram] $p(\xi) = p_0 \cos \alpha \xi$	$\dfrac{p_0 R^2}{Et} \dfrac{4(kL)^4}{\alpha^4 + 4(kL)^4} \left[\cos \alpha - \dfrac{\alpha^2}{2(kL)^2} \right.$ $\left. \cdot \left(\dfrac{2F_8}{F_1} - \dfrac{F_3}{F_1} \cos \alpha - \dfrac{\alpha}{kL} \dfrac{F_4}{F_1} \sin \alpha \right) \right]$	$-\dfrac{p_0 R^2}{Et} \dfrac{4(kL)^4}{\alpha^4 + 4(kL)^4} k \left[\dfrac{\alpha}{kL} \sin \alpha + \dfrac{\alpha^2}{2(kL)^2} \right.$ $\left. \cdot \left(\dfrac{2F_{10}}{F_1} - \dfrac{2F_3}{F_1} \cos \alpha - \dfrac{\alpha}{kL} \dfrac{F_2}{F_1} \sin \alpha \right) \right]$
[diagram] $p(\xi) = p_0 \exp(-\alpha \xi)$	$\dfrac{p_0 R^2}{Et} \dfrac{4(kL)^4}{\alpha^4 + 4(kL)^4} \left[\exp(-\alpha) \right.$ $+ \dfrac{\alpha^2}{2(kL)^2} \left(\dfrac{2F_8}{F_1} - \dfrac{\alpha}{kL} \dfrac{F_9}{F_1} \right)$ $\left. - \dfrac{\alpha^2}{2(kL)^2} \left(\dfrac{F_2}{F_1} + \dfrac{\alpha}{kL} \dfrac{F_4}{F_1} \right) \exp(-\alpha) \right]$	$-\dfrac{p_0 R^2}{Et} \dfrac{4(kL)^4}{\alpha^4 + 4(kL)^4} k \left[\dfrac{\alpha}{kL} \exp(-\alpha) \right.$ $- \dfrac{\alpha^2}{2(kL)^2} \left(\dfrac{2F_{10}}{F_1} - \dfrac{\alpha}{kL} \dfrac{2F_8}{F_1} \right)$ $\left. + \dfrac{\alpha^2}{2(kL)^2} \left(\dfrac{2F_3}{F_1} + \dfrac{\alpha}{kL} \dfrac{F_2}{F_1} \right) \exp(-\alpha) \right]$

For F and k factors see under "Definition of F Factors" in Sec. 5-2.

166 *Structural Analysis of Shells*

TABLE 6-3 Continued

Shell geometry	Boundary distortions	
$p = p(\xi)$ (with length L, coordinate ξ, indices i, k)	Δr_{ik}	β_{ik}
Loading condition	ΔH_{ik}	β_{ik}
$p(\xi) = p_0 = \text{const}$	$+\dfrac{p_0 R^2}{Et}$	0
$p(\xi) = p_v(1-\xi)$	$+\dfrac{p_v R^2}{Et}$	$-\dfrac{p_v R^2}{EtL}$
$p(\xi) = p_0 \sin \alpha \xi$	$\dfrac{p_0 R^2}{Et}\dfrac{\alpha^2}{2(kL)^2} - \dfrac{4(kL)^4}{\alpha^4 + 4(kL)^4}\left(\dfrac{\alpha}{kL}\dfrac{F_4}{F_1} - \dfrac{2F_8}{F_1}\sin\alpha + \dfrac{\alpha}{kL}\dfrac{F_9}{F_1}\cos\alpha\right)$	$\dfrac{p_0 R^2}{Et}\dfrac{4(kL)^4}{\alpha^4 + 4(kL)^4}k\left[\dfrac{\alpha}{kL} - \dfrac{\alpha^2}{2(kL)^2}\cdot\left(\dfrac{\alpha}{kL}\dfrac{F_2}{F_1} - \dfrac{2F_{10}}{F_1}\sin\alpha + \dfrac{\alpha}{kL}\dfrac{2F_8}{F_1}\cos\alpha\right)\right]$
$p(\xi) = p_0 \cos \alpha \xi$	$\dfrac{p_0 R^2}{Et}\dfrac{4(kL)^4}{\alpha^4 + 4(kL)^4}\left[1 + \dfrac{\alpha^2}{2(kL)^2}\cdot\left(\dfrac{F_2}{F_1} - \dfrac{2F_8}{F_1}\cos\alpha - \dfrac{\alpha}{kL}\dfrac{F_9}{F_1}\sin\alpha\right)\right]$	$-\dfrac{p_0 R^2}{Et}\dfrac{4(kL)^4}{\alpha^4 + 4(kL)^4}k\dfrac{\alpha^2}{2(kL)^2}\cdot\left(\dfrac{2F_3}{F_1} - \dfrac{2F_{10}}{F_1}\cos\alpha - \dfrac{\alpha}{kL}\dfrac{2F_8}{F_1}\sin\alpha\right)$
$p(\xi) = p_0 \exp(-\alpha\xi)$	$\dfrac{p_0 R^2}{Et}\dfrac{4(kL)^4}{\alpha^4 + 4(kL)^4}\left[1 - \dfrac{\alpha^2}{2(kL)^2}\cdot\left(\dfrac{F_2}{F_1} - \dfrac{\alpha}{kL}\dfrac{F_4}{F_1}\right) + \dfrac{\alpha^2}{2(kL)^2}\cdot\left(\dfrac{2F_8}{F_1} + \dfrac{\alpha}{kL}\dfrac{F_9}{F_1}\right)\exp(-\alpha)\right]$	$-\dfrac{p_0 R^2}{Et}\dfrac{4(kL)^4}{\alpha^4 + 4(kL)^4}k\left[\dfrac{\alpha}{kL} - \dfrac{\alpha^2}{2(kL)^2}\cdot\left(\dfrac{2F_3}{F_1} - \dfrac{\alpha}{kL}\dfrac{F_2}{F_1}\right) + \dfrac{\alpha^2}{2(kL)^2}\cdot\left(\dfrac{2F_{10}}{F_1} + \dfrac{\alpha}{kL}\dfrac{2F_8}{F_1}\right)\exp(-\alpha)\right]$

For F and k factors see under "Definition of F Factors" in Sec. 5-2.

TABLE 6-4 Cylindrical Shells. Edge Distortions Due to Edge Loadings (Ref. 6-1)

Loading condition \ Edge distortions	Δr_{ik} (i) Δr_{ik}	β_{ik} (i) β_{ik}	Δr_{ki} (i) Δr_{ki}	β_{ki} (i) β_{ki}
H_{ik} (i) H_{ik}	$-H_{ik}\dfrac{2R^2 k}{Et}\dfrac{F_4}{F_I}$	$H_{ik}\dfrac{2R^2 k^2}{Et}\dfrac{F_2}{F_I}$	$H_{ik}\dfrac{2R^2 k}{Et}\dfrac{F_9}{F_I}$	$H_{ik}\dfrac{2R^2 k^2}{Et}\dfrac{2F_8}{F_I}$
M_{ik} (i) M_{ik}	$M_{ik}\dfrac{2R^2 k^2}{Et}\dfrac{F_2}{F_I}$	$-M_{ik}\dfrac{2R^2 k^3}{Et}\dfrac{2F_3}{F_I}$	$-M_{ik}\dfrac{2R^2 k^2}{Et}\dfrac{2F_8}{F_I}$	$-M_{ik}\dfrac{2R^2 k^3}{Et}\dfrac{2F_{10}}{F_I}$
H_{ki} (i) H_{ki}	$-H_{ik}\dfrac{2R^2 k}{Et}\dfrac{F_9}{F_I}$	$H_{ki}\dfrac{2R^2 k^2}{Et}\dfrac{2F_8}{F_I}$	$H_{ki}\dfrac{2R^2 k}{Et}\dfrac{F_4}{F_I}$	$H_{ki}\dfrac{2R^2 k^2}{Et}\dfrac{F_2}{F_I}$
M_{ki} (i) M_{ki}	$-M_{ki}\dfrac{2R^2 k^2}{Et}\dfrac{2F_8}{F_I}$	$M_{ki}\dfrac{2R^2 k^3}{Et}\dfrac{2F_{10}}{F_I}$	$M_{ki}\dfrac{2R^2 k^2}{Et}\dfrac{F_2}{F_I}$	$M_{ki}\dfrac{2R^2 k^3}{Et}\dfrac{2F_3}{F_I}$

For F and k factors see under "Definition of F Factors" in Sec. 5-2.

168 *Structural Analysis of Shells*

TABLE 6-5 Conical Shell with Free Edges. Edge Distortions Due to Primary Loadings (Ref. 6-1)

Loading condition / Edge distortions	$\Delta r \leftarrow \triangle \rightarrow \Delta r$	$\beta \leftarrow \triangle \rightarrow \beta$
(q, triangular load)	$-\dfrac{qx_2^2}{Et}\cot\alpha_0\left(\cos^2\alpha_0 - \dfrac{\mu}{2}\right)$	$\dfrac{qx_2}{Et}\dfrac{\cos\alpha_0}{\sin^2\alpha_0}\left[\cos^2\alpha_0(2+\mu) - \dfrac{1}{2} - \mu\right]$
(p, uniform)	$-\dfrac{px_2^2}{Et}\cos\alpha_0\cot\alpha_0\left(\cos^2\alpha_0 - \dfrac{\mu}{2}\right)$	$\dfrac{px_2}{Et}\cot^2\alpha_0\left[\cos^2\alpha_0(2+\mu) - \dfrac{1}{2} - \mu\right]$
Hydrostatic	$\dfrac{\rho x_2^3}{Et}\cos^2\alpha_0\left(\dfrac{\mu}{3} - 1\right)$	$\rho\,\dfrac{8}{3}\dfrac{x_2^2}{Et}\dfrac{\cos^2\alpha_0}{\sin\alpha_0}$
(p, normal)	$-\dfrac{px_2^2}{Et}\cos\alpha_0\cot\alpha_0\left(1 - \dfrac{\mu}{2}\right)$	$\dfrac{px_2}{Et}\dfrac{3}{2}\cot^2\alpha_0$

For loadings see Table 3-5.

Loading condition / Edge distortions	$\Delta r \leftarrow \triangledown \rightarrow \Delta r$	$\beta \; \triangledown \; \beta$
(q)	$\dfrac{qx_2^2}{Et}\cot\alpha_0\left(\cos^2\alpha_0 - \dfrac{\mu}{2}\right)$	$\dfrac{qx_2}{Et}\dfrac{\cos\alpha_0}{\sin^2\alpha_0}\left[\cos^2\alpha_0(2+\mu) - \dfrac{1}{2} - \mu\right]$
(p)	$\dfrac{px_2^2}{Et}\cos\alpha_0\cot\alpha_0\left(\cos^2\alpha_0 - \dfrac{\mu}{2}\right)$	$\dfrac{px_2}{Et}\cot^2\alpha_0\left[\cos^2\alpha_0(2+\mu) - \dfrac{1}{2} - \mu\right]$
Hydrostatic	$-\dfrac{1}{6}\rho\,\dfrac{\mu x_2^3}{Et}\cos^2\alpha_0$	$-\dfrac{7}{6}\rho\,\dfrac{x_2^2}{Et}\dfrac{\cos^2\alpha_0}{\sin\alpha_0}$
(p)	$\dfrac{px_2^2}{Et}\cos\alpha_0\cot\alpha_0\left(1 - \dfrac{\mu}{2}\right)$	$\dfrac{px_2}{Et}\dfrac{3}{2}\cot^2\alpha_0$

TABLE 6-6 Conical Segment with Free Edges. Edge Distortions Due to Primary Loadings (Ref. 6-1)

Loading condition \ Edge distortion	Δr_{ik}	β_{ik}	Δr_{ki}	β_{ki}
(linear q)	$-\dfrac{qx_2^2}{Et}\cot\alpha_0\left[2\cos^2\alpha_0 - \mu\left(1 - \dfrac{x_1^2}{x_2^2}\right)\right]$	$\dfrac{qx_2}{Et}\dfrac{\cos\alpha_0}{2\sin^2\alpha_0}\left[2\cos^2\alpha_0(2+\mu) - 1 - 2\mu + \left(\dfrac{x_1}{x_2}\right)^2\right]$	$-\dfrac{qx_1^2}{Et}\dfrac{\cos^3\alpha_0}{2\sin\alpha_0}$	$\dfrac{qx_1}{Et}\dfrac{\cos\alpha_0}{\sin^2\alpha_0}\left[\cos^2\alpha_0(2+\mu) - \mu\right]$
(uniform p)	$-\dfrac{px_2^2}{Et}\dfrac{\cos^2\alpha_0}{2\sin\alpha_0}\left[2\cos^2\alpha_0 - \mu\left(1 - \dfrac{x_1^2}{x_2^2}\right)\right]$	$\dfrac{px_2}{Et}\dfrac{\cot^2\alpha_0}{2}\left[2\cos^2\alpha_0(2+\mu) - 1 - 2\mu + \left(\dfrac{x_1}{x_2}\right)^2\right]$	$-\dfrac{px_1^2}{Et}\dfrac{\cos^4\alpha_0}{\sin\alpha_0}$	$\dfrac{px_1}{Et}\cot^2\alpha_0\left[\cos^2\alpha_0(2+\mu) - \mu\right]$
Hydrostatic	$-\dfrac{px_2^2}{Et}\cos^2\alpha_0\cot\alpha_0\left[x_2 - x_1 - \dfrac{\mu x_2}{3} + \dfrac{\mu x_1}{2}\left(1 - \dfrac{x_1^2}{3x_2^2}\right)\right]$	$\dfrac{px_2}{Et}\dfrac{\cos^2\alpha_0}{\sin\alpha_0}\left[\dfrac{8}{3}x_2 - \dfrac{x_1}{2}\left(3 + \dfrac{1}{3}\dfrac{x_1^2}{x_2^2}\right)\right]$	0	$\dfrac{px_1^2}{Et}\dfrac{\cos^2\alpha_0}{\sin\alpha_0}$
(distributed q)	$-\dfrac{px_2^2}{Et}\cos\alpha_0\cot\alpha_0\left[1 - \dfrac{\mu}{2}\left(1 - \dfrac{x_1^2}{x_2^2}\right)\right]$	$\dfrac{px_2}{Et}\dfrac{\cos^2\alpha_0}{2}\left[3 + \left(\dfrac{x_1}{x_2}\right)^2\right]$	$-\dfrac{px_1^2}{Et}\cos\alpha_0\cot\alpha_0$	$2\dfrac{px_1}{Et}\cot^2\alpha_0$
(P loads)	$\dfrac{P}{Et}x_1\mu\cot\alpha_0$	$\dfrac{P}{Et}\dfrac{\cot\alpha_0\, x_1}{\sin\alpha_0\, x_2}$	$\dfrac{P}{Et}x_1\mu\cot\alpha_0$	$-\dfrac{P}{Et}\dfrac{\cot\alpha_0}{\sin\alpha_0}$

TABLE 6-6 Continued

Loading condition	Edge distortion: $\Delta r_{ki} \leftrightarrow \Delta r_{ki}$ (k)(i)	β_{ki} (k) x x (i) β_{ki}	Δr_{ik} (k)(i) Δr_{ik}	β_{ki} (k)(i) β_{ki}
(trapezoidal load, α_0, q, x)	$\dfrac{qx_2^2}{Et}\dfrac{\cot\alpha_0}{2}\left[2\cos^2\alpha_0 - \mu\left(1-\dfrac{x_1^2}{x_2^2}\right)\right]$	$\dfrac{qx_2}{Et}\dfrac{\cos\alpha_0}{2\sin^2\alpha_0}\left[2\cos^2\alpha_0(2+\mu)-1-2\mu+\left(\dfrac{x_1}{x_2}\right)^2\right]$	$\dfrac{qx_2^2}{Et}\dfrac{\cos^3\alpha_0}{\sin\alpha_0}$	$\dfrac{qx_1}{Et}\dfrac{\cos\alpha_0}{\sin^2\alpha_0}\left[\cos^2\alpha(2+\mu)-\mu\right]$
(uniform load p)	$\dfrac{px_2^2}{Et}\dfrac{\cos^2\alpha_0}{2\sin\alpha_0}\left[2\cos^2\alpha_0 - \mu\left(1-\dfrac{x_1^2}{x_2^2}\right)\right]$	$\dfrac{px_2}{Et}\dfrac{\cot^2\alpha_0}{2}\left[2\cos^2\alpha_0(2+\mu)-1-2\mu+\left(\dfrac{x_1}{x_2}\right)^2\right]$	$\dfrac{px_1^2}{Et}\dfrac{\cos^4\alpha_0}{\sin\alpha_0}$	$\dfrac{px_1}{Et}\cot^2\alpha_0\left[\cos^2\alpha_0(2+\mu)-\mu\right]$
Hydrostatic	$-p\dfrac{\cos^2\alpha_0}{Et}\dfrac{x_2^2}{2}\left[\dfrac{x_2}{2}\left(1-\dfrac{x_1^2}{x_2^2}\right)+\dfrac{x_1^2}{3}\dfrac{x_2}{x_2^2}\dfrac{x_2}{3}\right]$	$\dfrac{px_2}{Et}\dfrac{\cos^2\alpha_0}{\sin\alpha_0}\left[-\dfrac{7}{6}+\left(\dfrac{x_1}{x_2}\right)^2\left(\dfrac{x_2}{2}-\dfrac{x_1}{3}\right)\right]$	$\dfrac{px_1^2}{Et}\cos^2\alpha_0(x_2-x_1)$	$\dfrac{px_1}{Et}\cos^2\alpha_0(2x_2-3x_1)$
(trapezoidal, p)	$\dfrac{px_2}{Et}\cos\alpha_0\cot\alpha_0\left[1-\dfrac{1}{2}\left(1-\dfrac{x_1^2}{x_2^2}\right)\right]$	$\dfrac{px_2}{Et}\dfrac{\cot^2\alpha_0}{2}\left[3+\left(\dfrac{x_1}{x_2}\right)^2\right]$	$\dfrac{px_1^2}{Et}\cos\alpha_0\cot\alpha_0$	$2\dfrac{px_1}{Et}\cot^2\alpha_0$
(P, P)	$-\dfrac{P}{Et}x_1\mu\cot\alpha_0$	$-\dfrac{P}{Et}\dfrac{\cot\alpha_0}{\sin\alpha_0}\dfrac{x_1}{x_2}$	$-\dfrac{P}{Et}x_1\mu\cot\alpha_0$	$-\dfrac{P}{Et}\dfrac{\cot\alpha_0}{\sin\alpha_0}$

TABLE 6-7 Conical Segment with Free Edges. Edge Distortions Due to Secondary Loadings (Ref. 6-1)

Loading condition \ Edge distortion	Δr_{ik} ... Δr_{ik}	β_{ik} ... β_{ik}	Δr_{ki} ... Δr_{ki}	β_{ki} ... β_{ki}
H_{ik}, α_0, H_{ik}	$-H_{ik}\dfrac{\sin^2\alpha_0}{2Dk^3}\dfrac{F_4}{F_1}$	$+H_{ik}\dfrac{\sin\alpha_0}{2Dk^2}\dfrac{F_2}{F_1}$	$+H_{ik}\dfrac{\sin^2\alpha_0}{2Dk^3}\dfrac{F_9}{F_1}$	$+H_{ik}\dfrac{\sin\alpha_0}{2Dk^2}\dfrac{2F_8}{F_1}$
M_{ik} ... M_{ik}	$+M_{ik}\dfrac{\sin\alpha_0}{2Dk^2}\dfrac{F_2}{F_1}$	$-M_{ik}\dfrac{1}{2Dk}\dfrac{2F_3}{F_1}$	$-M_{ik}\dfrac{\sin\alpha_0}{2Dk^2}\dfrac{2F_8}{F_1}$	$-M_{ik}\dfrac{1}{2Dk}\dfrac{2F_{10}}{F_1}$
H_{ki} ... H_{ki}	$-H_{ki}\dfrac{\sin^2\alpha_0}{2Dk^3}\dfrac{F_9}{F_1}$	$+H_{ki}\dfrac{\sin\alpha_0}{2Dk^2}\dfrac{2F_8}{F_1}$	$+H_{ki}\dfrac{\sin^2\alpha_0}{2Dk^3}\dfrac{F_4}{F_1}$	$+H_{ki}\dfrac{\sin\alpha_0}{2Dk^2}\dfrac{F_2}{F_1}$
M_{ki} ... M_{ki}	$-M_{ki}\dfrac{\sin\alpha_0}{2Dk^2}\dfrac{2F_8}{F_1}$	$+M_{ki}\dfrac{1}{2Dk}\dfrac{2F_{10}}{F_1}$	$+M_{ki}\dfrac{\sin\alpha_0}{2Dk^2}\dfrac{F_2}{F_1}$	$+M_{ki}\dfrac{1}{2Dk}\dfrac{2F_3}{F_1}$

For F and k factors see under "Definition of F Factors" in Sec. 5-2.

TABLE 6-8 Spherical Shell with Free Edges. Edge Distortions Due to Primary Loadings (Ref. 6-1)

Loading condition	Edge distortions Δr	β
(dome, load q, angle ϕ_1)	$\dfrac{R^2 q}{Et} \sin\phi_1 \left(-\cos\phi_1 + \dfrac{1+\mu}{1+\cos\phi_1} \right)$	$-\dfrac{Rq}{Et}(2+\mu)\sin\phi_1$
(dome, load p)	$\dfrac{R^2 p}{Et} \sin\phi_1 \left(-\cos^2\phi_1 + \dfrac{1+\mu}{2} \right)$	$-\dfrac{Rp}{Et}(3+\mu)\sin\phi_1 \cos\phi_1$
Hydrostatic (dome, p_v)	$\dfrac{-\rho R^3}{6Et} \sin\phi_1 \left[6(1-\cos\phi_1) - (1+\mu)\left(1 - 2\dfrac{\cos^2\phi_1}{1+\cos\phi_1}\right) \right]$	$\dfrac{\rho R^2}{Et}\sin\phi_1$
(dome, load p)	$-\dfrac{R^2 p}{2Et}(1-\mu)\sin\phi_1$	0

Loading condition	Edge distortions Δr	β
(bowl, load q, angle ϕ_1)	$\dfrac{R^2 q}{Et} \sin\phi_1 \left(\cos\phi_1 - \dfrac{1+\mu}{1+\cos\phi_1} \right)$	$-\dfrac{Rq}{Et}(2+\mu)\sin\phi_1$
(bowl, load p)	$-\dfrac{R^2 p}{Et} \sin\phi_1 \left(-\cos^2\phi_1 + \dfrac{1+\mu}{2} \right)$	$-\dfrac{Rp}{Et}(3+\mu)\sin\phi_1 \cos\phi_1$
Hydrostatic (bowl, p_v)	$\dfrac{-\rho R^3}{6Et} \sin\phi_1 \left(2 - 3\cos\phi_1 + 2\dfrac{\cos^2\phi_1}{1+\cos\phi_1}(1+\mu) \right)$	$-\dfrac{\rho R^2}{Et}\sin\phi_1$
(bowl, load p)	$\dfrac{R^2 p}{2Et}(1-\mu)\sin\phi_1$	0

TABLE 6-9 Spherical Segment with Free Edges. Edge Distortions Due to Primary Loadings (Ref. 6-1)

Loading condition	Edge distortion Δr_{ik}	β_{ik}	Δr_{ki}	β_{ki}
(q, ϕ_1, ϕ_2)	$\dfrac{Rq}{Et}\sin\phi_1\left[-\cos\phi_1+\dfrac{1+\mu}{\sin^2\phi_1}(-\cos\phi_2+\cos\phi_2)\right]$	$-\dfrac{Rq}{Et}(2+\mu)\sin\phi_1$	$-\dfrac{R^2 q}{Et}\sin\phi_2\cos\phi_2$	$-\dfrac{Rq}{Et}(2+\mu)\sin\phi_2$
(p, uniform)	$\dfrac{R^2 p}{Et}\sin\phi_1\left[-\cos^2\phi_1+\dfrac{1+\mu}{2}\left(1-\dfrac{\sin^2\phi_2}{\sin^2\phi_1}\right)\right]$	$-\dfrac{Rp}{Et}(3+\mu)\sin\phi_1\cos\phi_1$	$-\dfrac{R^2 p}{Et}\sin\phi_2\cos^2\phi_2$	$-\dfrac{Rp}{Et}(3+\mu)\sin\phi_2\cos\phi_2$
Hydrostatic	$-\dfrac{\rho R^3}{Et}\sin\phi_1\left\{\cos\phi_2-\cos\phi_1-\dfrac{1+\mu}{6}\left[\dfrac{3\cos\phi_2}{}\right.\right.$ $\left.\left.\cdot\left(1-\dfrac{\sin^2\phi_2}{\sin^2\phi_1}\right)-2\dfrac{\cos^3\phi_2-\cos^3\phi_1}{\sin^2\phi_1}\right]\right\}$	$+\dfrac{\rho R^2}{Et}\sin\phi_1$	0	$+\dfrac{\rho R^2}{Et}\sin\phi_2$
(p)	$-\dfrac{R^2 p}{Et}\sin\phi_1\left[1-\dfrac{1+\mu}{2}\left(1-\dfrac{\sin^2\phi_2}{\sin^2\phi_1}\right)\right]$	0	$\dfrac{R^2 p}{Et}\sin\phi_2$	0
(P)	$\dfrac{PR}{Et}(1+\mu)\dfrac{\sin\phi_2}{\sin\phi_1}$	0	$\dfrac{PR}{Et}(1+\mu)$	0

TABLE 6-9 Continued

Loading condition	Edge distortion Δr_{ki}	β_{ki}	Δr_{ik}	β_{ik}
(conical with ϕ_1, ϕ_2, q)	$-\frac{R^2 q}{Et}\sin\phi_1\left[-\cos\phi_1+\frac{1+\mu}{\sin^2\phi_1}(-\cos\phi_1+\cos\phi_2)\right]$	$-\frac{Rq}{Et}(2+\mu)\sin\phi_1$	$+\frac{R^2 q}{Et}\sin\phi_2\cos\phi_2$	$-\frac{Rq}{Et}(2+\mu)\sin\phi_2$
(uniform p)	$\frac{R^2 p}{Et}\sin\phi_1\left[-\cos^2\phi_1+\frac{1+\mu}{2}\left(1-\frac{\sin^2\phi_2}{\sin^2\phi_1}\right)\right]$	$-\frac{Rp}{Et}(3+\mu)\sin\phi_1\cos\phi_1$	$\frac{R^2 p}{Et}\sin\phi_2\cos^2\phi_2$	$-\frac{Rp}{Et}(3+\mu)\sin\phi_2\cos\phi_2$
Hydrostatic	$\frac{\rho R^3}{6Et}\sin\phi_2(1+\mu)\left[3\cos\phi_1\left(1-\frac{\sin^2\phi_2}{\sin^2\phi_1}\right)-2\frac{\cos^3\phi_2-\cos^3\phi_1}{\sin^2\phi_1}\right]$	$-\frac{\rho R^2}{Et}\sin\phi_1$	$+\frac{\rho R^3}{Et}\sin\phi_2(\cos\phi_1-\cos\phi_2)$	$-\frac{\rho R^2}{Et}\sin\phi_2$
(p)	$\frac{R^2 p}{Et}\sin\phi_1\left[1-\frac{1+\mu}{2}\left(1-\frac{\sin^2\phi_2}{\sin^2\phi_1}\right)\right]$	0	$\frac{R^2 p}{Et}\sin\phi_2$	0
(P)	$-\frac{PR}{Et}(1+\mu)\frac{\sin\phi_2}{\sin\phi_1}$	0	$-\frac{PR}{Et}(1+\mu)$	0

TABLE 6-10 Spherical Segment, Free Edges. Edge Distortions Due to Secondary Loadings (Ref. 6-1)

Loading condition	Edge distortion Δr_{ik}	β_{ik}	Δr_{ki}	β_{ki}
H_{ik}	$+H_{ik}\dfrac{2Rk}{Et}\sin^2\phi_1\dfrac{F_4}{F_1}$	$-H_{ik}\dfrac{2k^2}{Et}\sin\phi_1\dfrac{F_2}{F_1}$	$-H_{ik}\dfrac{2Rk}{Et}\sin\phi_1\sin\phi_2\dfrac{F_9}{F_1}$	$-H_{ik}\dfrac{2k^2}{Et}\sin\phi_1\dfrac{2F_8}{F_1}$
M_{ik}	$-M_{ik}\dfrac{2k^2}{Et}\sin\phi_1\dfrac{F_2}{F_1}$	$+M_{ik}\dfrac{2k^3}{EtR}\dfrac{2F_3}{F_1}$	$+M_{ik}\dfrac{2k^2}{Et}\sin\phi_2\dfrac{2F_8}{F_1}$	$+M_{ik}\dfrac{2k^3}{EtR}\dfrac{2F_{10}}{F_1}$
H_{ki}	$+H_{ki}\dfrac{2Rk}{Et}\sin\phi_1\sin\phi_2\dfrac{F_9}{F_1}$	$-H_{ki}\dfrac{2k^2}{Et}\sin\phi_2\dfrac{2F_8}{F_1}$	$-H_{ik}\dfrac{2Rk}{Et}\sin^2\phi_2\dfrac{F_4}{F_1}$	$-H_{ki}\dfrac{2k^2}{Et}\sin\phi_2\dfrac{F_2}{F_1}$
M_{ki}	$-M_{ki}\dfrac{2k^2}{Et}\sin\phi_1\dfrac{2F_8}{F_1}$	$-M_{ki}\dfrac{2k^3}{EtR}\dfrac{2F_{10}}{F_1}$	$+M_{ki}\dfrac{2k^2}{Et}\sin\phi_2\dfrac{F_2}{F_1}$	$+M_{ki}\dfrac{2k^3}{EtR}\dfrac{2F_3}{F_1}$

For F and k factors see under "Definition of F Factors" in Sec. 5-2.

176 *Structural Analysis of Shells*

TABLE 6-11 Cylindrical Shell with Edge Free, Other Edge fixed. Edge Distortions Due to Primary Loadings (Ref. 6-1)

Shell geometry		
Loading condition	Δr = deflection	β = rotation of edge
$p(\xi) = p_0$ = const	$p_0 \dfrac{R^2}{Et}\left(1 - \dfrac{2F_7}{F_1+2}\right)$	$p_0 \dfrac{R^2 k}{Et}\dfrac{2F_9}{F_1+2}$
$p(\xi) = p_v(1-\xi)$	$p_v \dfrac{R^2}{Et}\left(\dfrac{1}{kL}\dfrac{F_{10}}{F_1+2} - \dfrac{2F_7}{F_1+2}\right)$	$-\dfrac{R^2 p_v}{EtL}\left(1 - kL\dfrac{2F_9}{F_1+2} - \dfrac{2F_7}{F_1+2}\right)$
$p(\xi) = p_0 \sin\alpha\xi$	$\dfrac{p_0 R^2}{Et}\dfrac{4(kL)^4}{\alpha^4+4(kL)^4}\left[\sin\alpha - \dfrac{\alpha}{kL}\dfrac{F_{10}}{F_1+2} \right.$ $\left. + \dfrac{\alpha^2}{2(kL)^2}\left(\dfrac{F_2}{F_1+2}\sin\alpha - \dfrac{\alpha}{kL}\dfrac{F_4}{F_1+2}\cos\alpha\right)\right]$	$\dfrac{p_0 R^2 k}{Et}\dfrac{4(kL)^4}{\alpha^4+4(kL)^4}\left[\dfrac{\alpha}{kL}\cos\alpha - \dfrac{\alpha}{kL}\dfrac{2F_7}{F_1+2}\right.$ $\left. + \dfrac{\alpha^2}{2(kL)^2}\left(\dfrac{2F_3}{F_1+2}\sin\alpha - \dfrac{\alpha}{kL}\dfrac{F_2}{F_1+2}\cos\alpha\right)\right]$
$p(\xi) = p_0 \cos\alpha\xi$	$\dfrac{p_0 R^2}{Et}\dfrac{4(kL)^4}{\alpha^4+4(kL)^4}\left[\cos\alpha - \dfrac{2F_7}{F_1+2} \right.$ $\left. + \dfrac{\alpha^2}{2(kL)^2}\left(\dfrac{F_2}{F_1+2}\cos\alpha + \dfrac{\alpha}{kL}\dfrac{F_4}{F_1+2}\sin\alpha\right)\right]$	$-\dfrac{p_0 R^2 k}{Et}\dfrac{4(kL)^4}{\alpha^4+4(kL)^4}\left[\dfrac{\alpha}{kL}\sin\alpha - \dfrac{2F_9}{F_1+2}\right.$ $\left. - \dfrac{\alpha^2}{2(kL)^2}\left(\dfrac{2F_3}{F_1+2}\cos\alpha + \dfrac{\alpha}{kL}\dfrac{F_2}{F_1+2}\sin\alpha\right)\right]$
$p(\xi) = p_0 \exp(-\alpha\xi)$	$p_0 \dfrac{R^2}{Et}\dfrac{4(kL)}{\alpha^4+4(kL)^4}\left[\exp(-\alpha) - \dfrac{2F_7}{F_1+2} + \dfrac{\alpha}{kL}\right.$ $\left. \cdot \dfrac{F_{10}}{F_1+2} - \dfrac{\alpha^2}{2(kL)^2}\left(\dfrac{F_2}{F_1+2} + \dfrac{\alpha}{kL}\dfrac{F_4}{F_1+2}\right)\exp(-\alpha)\right]$	$-\dfrac{p_0 R^2 k}{Et}\dfrac{4(kL)^4}{\alpha^4+4(kL)^4}\left[\dfrac{\alpha}{kL}\exp(-\alpha) - \dfrac{2F_9}{F_1+2} - \dfrac{\alpha}{kL}\right.$ $\left. \cdot \dfrac{2F_7}{F_1+2} + \dfrac{\alpha^2}{2(kL)^2}\left(-\dfrac{2F_3}{F_1+2} + \dfrac{\alpha}{kL}\dfrac{F_2}{F_1+2}\right)\exp(-\alpha)\right]$

TABLE 6-12 Cylindrical Shell with One Edge Free, Other Edge Fixed. Edge Influence Due to Primary Loadings (Ref. 6-1)

Shell geometry		
Loading condition	Δr = deflection	β = rotation
$p(\xi) = p_0 = \text{const}$	$+\dfrac{p_0 R^2}{Et}\left(1 + \dfrac{F_9}{F_4}\right)$	$+\dfrac{p_0 R^2}{Et} k \dfrac{2F_8}{F_4}$
$p(\xi) = p_v(1-\xi)$	$+\dfrac{p_v R^2}{Et}\dfrac{F_9}{F_4}$	$-\dfrac{p_v R^2}{Et} k\left(\dfrac{1}{kL} - \dfrac{2F_8}{F_4}\right)$
$p(\xi) = p_0 \sin\alpha\xi$	$+\dfrac{p_0 R^2}{Et}\dfrac{4(kL)^4}{\alpha^4 + 4(kL)^4}\left[\sin\alpha + \dfrac{\alpha^2}{2(kL)^2}\right.$ $\left.\cdot\left(\sin\alpha\dfrac{F_3}{F_4} - \dfrac{\alpha}{kL}\cos\alpha\dfrac{F_2}{F_4}\right)\right]$	$+\dfrac{p_0 R^2}{Et} k \dfrac{4(kL)^4}{\alpha^4 + 4(kL)^4}\left[\dfrac{\alpha}{kL}\cos\alpha + \dfrac{\alpha^2}{2(kL)^2}\right.$ $\left.\cdot\left(\sin\alpha\dfrac{2(F_1+1)}{F_4} - \dfrac{\alpha}{kL}\cos\alpha\dfrac{F_3}{F_4}\right)\right]$
$p(\xi) = p_0 \cos\alpha\xi$	$+\dfrac{pR^2}{Et}\dfrac{4(kL)^4}{\alpha^4 + 4(kL)^4}\left[\cos\alpha + \dfrac{F_9}{F_4} - \dfrac{\alpha^2}{2(kL)^2}\right.$ $\left.\cdot\dfrac{F_{10}}{F_4} + \dfrac{\alpha^2}{2(kL)^2}\left(\cos\alpha\dfrac{F_3}{F_4} + \dfrac{\alpha}{kL}\sin\alpha\dfrac{F_2}{F_4}\right)\right]$	$-\dfrac{pR^2}{Et} k \dfrac{4(kL)^4}{\alpha^4 + 4(kL)^4}\left[\dfrac{\alpha}{kL}\sin\alpha - \dfrac{2F_8}{F_4} + \dfrac{\alpha^2}{2(kL)^2}\right.$ $\left.\cdot\dfrac{2F_7}{F_4} - \dfrac{\alpha^2}{2(kL)^2}\left(\cos\alpha\dfrac{2(F_1+1)}{F_4} + \dfrac{\alpha}{kL}\sin\alpha\dfrac{F_3}{F_4}\right)\right]$
$p(\xi) = p_0 \exp(-\alpha\xi)$	$+\dfrac{p_0 R^2}{Et}\dfrac{4(kL)^4}{\alpha^4 + 4(kL)^4}\left[\exp(-\alpha) + \dfrac{F_9}{F_4} + \dfrac{\alpha^2}{2(kL)^2}\right.$ $\left.\cdot\dfrac{F_{10}}{F_4} - \dfrac{\alpha^2}{2(kL)^2}\left(\dfrac{F_3}{F_4} + \dfrac{\alpha}{kL}\dfrac{F_2}{F_4}\right)\exp(-\alpha)\right]$	$-\dfrac{p_0 R^2}{Et} k \dfrac{4(kL)^4}{\alpha^4 + 4(kL)^4}\left[\dfrac{\alpha}{kL}\exp(-\alpha) - \dfrac{2F_8}{F_4} - \dfrac{\alpha^2}{2(kL)^2}\right.$ $\left.\cdot\dfrac{2F_7}{F_4} + \dfrac{\alpha^2}{2(kL)^2}\left(\dfrac{2(F_1+1)}{F_4} + \dfrac{\alpha}{kL}\dfrac{F_3}{F_4}\right)\exp(-\alpha)\right]$

178 *Structural Analysis of Shells*

TABLE 6-13 Distortions Due to Secondary Loadings (Ref. 6-1)

Shell geometry and loading condition	Edge distortion Δr	β
Hemisphere with H loads at base, angle ϕ_1	$+H \dfrac{2Rk}{Et} \sin^2 \phi_1$	$-H \dfrac{2k^2}{Et} \sin \phi_1$
Hemisphere with M moments at base	$-M \dfrac{2k^2}{Et} \sin \phi_1$	$+M \dfrac{4k^3}{EtR}$
Shell loading condition	ΔH = deflection	β = rotation
Cylinder fixed base, M at top	$M \dfrac{R^2 k^2}{Et} \dfrac{2F_2}{F_1+2}$	$M \dfrac{R^2 k^3}{Et} \dfrac{4F_3}{F_1+2}$
Cylinder fixed base, H at top	$H \dfrac{R^2 k}{Et} \dfrac{2F_4}{F_1+2}$	$H \dfrac{R^2 k^2}{Et} \dfrac{2F_2}{F_1+2}$
Cylinder pinned base, M at top	$M \dfrac{R^2 k^2}{Et} \dfrac{2F_3}{F_4}$	$M \dfrac{R^2 k^3}{Et} \dfrac{4(F_1+1)}{F_4}$
Cylinder pinned base, H at top	$H \dfrac{R^2 k}{Et} \dfrac{2F_2}{F_4}$	$H \dfrac{R^2 k^2}{Et} \dfrac{2F_3}{F_4}$

For F and k factors see under "Definition of F Factors" in Sec. 5-2.

TABLE 6-14 Cylindrical Shell with Various Edge Conditions. Edge Reactions Due to Primary and Secondary Loadings (Ref. 6-1)

Loading condition \ Edge reactions	M_{ik} (both ends)	H_{ik} (both ends)	M_{ki} (top)	H_{ki} (top)
$p(\xi) = \text{const} = p_0$	$-\dfrac{p_0}{2k^2}\left(\dfrac{F_2}{F_1} - \dfrac{2F_8}{F_1}\right)$	$-\dfrac{p_0}{k}\left(\dfrac{F_3}{F_1} - \dfrac{F_{10}}{F_1}\right)$	$+\dfrac{p_0}{2k^2}\left(\dfrac{F_2}{F_1} - \dfrac{2F_8}{F_1}\right)$	$-\dfrac{p_0}{k}\left(\dfrac{F_3}{F_1} - \dfrac{F_{10}}{F_1}\right)$
$p(\xi) = p_v(1-\xi)$	$-\dfrac{p_v}{2k^2}\left[\dfrac{F_2}{F_1} - \dfrac{1}{kL}\left(\dfrac{F_4}{F_1} + \dfrac{F_9}{F_1}\right)\right]$	$-\dfrac{p_v}{2k}\left[\dfrac{2F_3}{F_1} - \dfrac{1}{kL}\left(\dfrac{F_2}{F_1} + \dfrac{2F_8}{F_1}\right)\right]$	$-\dfrac{p_v}{2k^2}\left[\dfrac{2F_8}{F_1} - \dfrac{1}{kL}\left(\dfrac{F_4}{F_1} + \dfrac{F_9}{F_1}\right)\right]$	$+\dfrac{p_v}{2k}\left[\dfrac{2F_{10}}{F_1} - \dfrac{1}{kL}\left(\dfrac{F_2}{F_1} + \dfrac{2F_8}{F_1}\right)\right]$

Loading condition \ Edge reactions	H_{ik} (both ends, fixed top)	M_{ki} (top)	H_{ki} (top)
$p(\xi) = \text{const} = p_0$	$+\dfrac{p_0}{2k}\left(\dfrac{2F_7}{F_4} - \dfrac{F_1+2}{F_4}\right)$	$+\dfrac{p_0}{2k^2}\left(\dfrac{F_3}{F_4} - \dfrac{F_{10}}{F_4}\right)$	$-\dfrac{p_0}{2k}\left(\dfrac{2(F_1+1)}{F_4} - \dfrac{2F_7}{F_4}\right)$
$p(\xi) = p_v(1-\xi)$	$-\dfrac{p_v}{2k}\left(\dfrac{F_1+2}{F_4} - \dfrac{1}{kL}\dfrac{F_{10}}{F_4}\right)$	$-\dfrac{p_v}{2k^2}\left(\dfrac{F_{10}}{F_4} - \dfrac{1}{kL}\dfrac{F_2}{F_4}\right)$	$+\dfrac{p_v}{2k}\left(\dfrac{2F_7}{F_4} - \dfrac{1}{kL}\dfrac{F_3}{F_4}\right)$

Loading condition \ Edge reactions	H_{ik} (both ends, fixed top)	M_{ik} (both ends, fixed top)
$p(\xi) = \text{const} = p_0$	$-\dfrac{p_0}{2k}\dfrac{2F_3}{F_1+2}$	$-\dfrac{p_0}{2k^2}\dfrac{F_2}{F_1+2}$
$p(\xi) = p_v(1-\xi)$	$-\dfrac{p_v}{2k}\left(\dfrac{2F_3}{F_1+2} - \dfrac{1}{kL}\dfrac{F_2}{F_1+2}\right)$	$-\dfrac{p_v}{2k^2}\left(\dfrac{F_2}{F_1+2} - \dfrac{1}{kL}\dfrac{F_4}{F_1+2}\right)$

For F and k factors see under "Definition of F Factors" in Sec. 5-2.

REFERENCES

6-1. Hampe, E.: *Statik Rotationssymmetrischer Flächentragwerke*, vol. 4, VEB Verlag für Bauwesen, Berlin, 1964.

Chapter 7

SHELLS WITH COMPOSITE OR STIFFENED WALLS

7-1 Introduction

Previous chapters have considered primarily homogeneous isotropic shells. However, certain arrangements of material in the shell-wall cross section increase the rigidity, and consequently, less material is needed. In order to obtain a lighter structure, the material in the cross section can be arranged to make the cross section most resistant to the stresses that are predominant. This chapter considers the stress and deflection analysis of shells with composite or stiffened wall construction. Only shells of revolution subjected to axisymmetric loads will be discussed.

If the load on the shell causes bending or if stability considerations are included, the load-carrying capacity of the shell can be increased by increasing the shell-wall bending stiffness. The stiffness of the shell-wall can be increased by attaching meridional and circumferential stiffeners to a relatively thin sheet which forms the shell. The shell wall would then be more efficient in resisting bending than would homogeneous wall construction. In addition, if the bending is primarily in the meridional direction, the stiffeners added in the meridional direction would be made much larger than the circumferential stiffeners. This would give the shell different stiffness properties in the meridional and circumferential directions. A shell with stiffness properties which are not the same in the meridional and circumferential directions is called an orthotropic shell. Many types of orthotropic constructions are used to make shell structures, for examples, reinforced concrete, wood, stiffened, sandwich, waffle, and multilayered filament matrix composites. The stress and deflection analysis for composite and stiffened orthotropic shells will be discussed in this chapter. The stability analyses are presented in Chaps. 11, 12, and 13.

7-2 Extensional and Bending Stiffness

The relationship between the force resultants N_ϕ, N_θ and the centroidal surface strains ϵ_ϕ, ϵ_θ for an orthotropic composite shell can be written as

$$N_\phi = B_\phi(\epsilon_\phi + \mu_\theta' \epsilon_\theta)$$
$$N_\theta = B_\theta(\epsilon_\theta + \mu_\phi' \epsilon_\phi)$$
(7-1)

where B_ϕ, B_θ = the extensional stiffness of the shell wall in the meridional and circumferential directions, respectively (lb/in.) and μ_ϕ', μ_θ' = Poisson's ratio associated with extension in the meridional and circumferential directions, respectively.

Subscripts ϕ and θ refer to the meridional and circumferential direction, respectively.

Equation 7-1 is similar to Eq. 1-21, which is presented for homogeneous isotropic shell walls.

The relationship between the moments M_ϕ, M_θ and the curvatures χ_ϕ, χ_θ for an orthotropic composite shell can be written as

$$M_\phi = -D_\phi[\chi_\phi + \mu_\theta \chi_\theta]$$
$$M_\theta = D_\theta[\chi_\theta + \mu_\phi \chi_\phi]$$
(7-2)

where D_ϕ, D_θ = the bending stiffness of the shell wall in the ϕ and θ directions, respectively (in.-lb)

μ_ϕ, μ_θ = Poisson's ratio associated with bending

Equation 7-2 is similar to Eq. 1-21, which is presented for homogeneous isotropic shell walls. Equations 7-1 and 7-2 are valid if the location of the centroid plane in the ϕ direction coincides approximately with the centroid plane in the θ direction. The Poisson ratios may be slightly different for the cases of bending stresses and axial stresses, but for many cases they can be assumed equal. From the reciprocity theorem it has been shown that the following relationships exist:

$$B_\phi \mu_\theta' = B_\theta \mu_\phi' \quad \text{and} \quad D_\phi \mu_\theta = D_\theta \mu_\phi$$

The physical meaning of B and D can be seen by comparing Eqs. 7-1 and 7-2 with Eq. 1-21. If Young's modulus E is constant through the thickness of the shell wall, B is proportional to the area A and D is proportional to the moment of inertia I of a 1-inch-wide strip of the shell wall.

$$B_i = \frac{E_i A_i}{1 - \mu_\phi' \mu_\theta'} \qquad D_i = \frac{E_i I_i}{1 - \mu_\phi \mu_\theta}$$

where i refers to ϕ or θ. It can be seen that B is a measure of the shell wall's resistance to extension and D is a measure of the shell wall's resistance to bending.

Formulas for the extensional and bending stiffnesses for typical types of construction are presented in Sec. 11-2.

For a shell of revolution subjected to axisymmetrical loads, Eqs. 7-1 and 7-2 are the only changes that are necessary in the theory presented in Chap. 1 to obtain the governing equations for orthotropic composite shells.

7-3 Primary Solutions

It was previously stated that in most cases, the primary solutions are membrane solutions. The primary solutions for axisymmetrically loaded shells of revolution consist of the following quantities:

N_θ = membrane load in circumferential direction
N_ϕ = membrane load in meridional direction
u = displacement in the direction of tangent to the meridian
w = displacement in the direction of the normal-to-the-middle surface

Having u and w, any displacement components can be obtained from geometric relations if only axisymmetrical cases are considered. Consequently, for this purpose, it is adequate to investigate u and w.

All formulas that were presented for the homogeneous isotropic case in Chap. 3 can be used to determine N_θ and N_ϕ for the orthotropic composite case because the membrane is a statically determinate structure and the loads do not depend on the stiffness properties.

When N_θ and N_ϕ are obtained, u and w can be obtained in the following manner (see Ref. 7-1).

The strain components ϵ_ϕ and ϵ_θ can be obtained from Eq. 7-1.

$$\epsilon_\phi = \frac{1}{B_\phi(1 - \mu_\phi' \mu_\theta')}(N_\phi - \mu_\phi' N_\theta)$$

$$\epsilon_\theta = \frac{1}{B_\theta(1 - \mu_\phi' \mu_\theta')}(N_\theta - \mu_\theta' N_\phi)$$

(7-3)

Equation 7-3 is similar to Eq. 3-2, which is presented for homogeneous isotropic shell walls.

Displacements can now be obtained from the following differential equation:

$$\frac{du}{d\phi} - u \cot \phi = R_\phi \epsilon_\phi - R_\theta \epsilon_\theta = f(\phi)$$

The solution of the above equation is

$$u = \sin \phi \left[\int \frac{f(\phi)}{\sin \phi} d\phi + K \right]$$

where K is the constant of integration to be determined from the condition at the support (see Chap. 3). Then, the displacement w is obtained from the following equation:

$$\epsilon_\theta = \frac{u}{R_\theta} \cot \phi - \frac{w}{R_\theta}$$

Consequently, the stresses and deformations for every symmetrically loaded shell of revolution can be determined for the orthotropic composite case. When u and w are determined, Δr and β, needed for interaction purposes, are obtained from simple geometrical relation (see Sec. 3-3).

If the extensional stiffness of a composite shell wall is the same in the meridional and circumferential directions, the formulas for the shell-wall displacements and rotations presented in Chap. 3 can be used directly by setting

$$Et = B(1 - \mu^2) \qquad (7\text{-}4)$$

where $B = B_\phi = B_\theta$
$\mu = \mu_\phi' = \mu_\theta'$

The displacements and rotations for the composite shell will be obtained directly by substituting the Et computed by Eq. 7-4 into the formulas presented in Chap. 3.

7-4 Secondary Solutions

Very few closed-form solutions exist in the literature at the present time for shells with orthotropic composite wall construction. However, the solutions presented for homogeneous isotropic shells can be used in many instances to obtain solutions for orthotropic composite shells. If the stiffnesses for a composite shell wall can be expressed in the following manner:

$$B = B_\phi = B_\theta \qquad D = D_\phi = D_\theta \qquad \mu = \mu_\phi' = \mu_\theta' = \mu_\phi = \mu_\theta \qquad (7\text{-}5)$$

all the solutions presented in the preceding chapters for shells of revolution subjected to axisymmetric loads may be used to obtain the forces, moments, shears, displacements, and rotations for the composite shell by using the following t and E:

$$t = \sqrt{\frac{12D}{B}}$$

$$E = \frac{B(1-\mu^2)}{t} \quad (7\text{-}6)$$

A homogeneous isotropic shell with a thickness t and a Young's modulus of E as computed from Eq. 7-6 will have the same extensional and bending stiffness as the composite shell; therefore, the forces, moments, displacements, and rotations must be the same for the same shell geometry and loading. The stresses, however, for the equivalent homogeneous isotropic shell are not necessarily the same as for the composite shell.

Several types of composite constructions which satisfy Eq. 7-5 are presented in Sec. 11-2 under special cases. The properties of the shell wall must remain constant in the meridional and circumferential direction if Eq. 7-6 is used.

Cylinders

Reference 7-2 derives the governing differential equation for orthotropic composite cylinders subjected to axisymmetric loads. The only stiffness properties which occur in these equations for cylinder walls with large transverse shear stiffness are D_x, the bending stiffness in the axial direction, and B_θ, the extensional stiffness in the circumferential direction. Therefore, the homogeneous isotropic cylinder which has the same stiffness properties as the composite orthotropic cylinder can be obtained from the following equations:

$$t = \sqrt{\frac{12D_x}{B_\theta}}$$

$$E = \frac{B_\theta(1 - \mu_x'\mu_\theta')}{t} \quad (7\text{-}7)$$

$$\mu^2 = \mu_x'\mu_\theta'$$

$$\mu = \mu_x$$

The forces, moments, displacements, and rotations for any axisymmetric orthotropic composite cylinder subjected to axisymmetric loads can be obtained from the previous chapters by substituting Eq. 7-7 for

186 *Structural Analysis of Shells*

the values of t, E, μ^2, and μ which appear in the equations for homogeneous isotropic cylinders. Formulas for computing the constants B_θ, D_x, μ_x', and μ_y' are presented in Sec. 11-2 for typical types of construction.

Most types of cylinder-wall constructions have a relatively large transverse shear stiffness; therefore, this method is valid for most cylinders. One possible exception is the sandwich-type construction which will be discussed in the following paragraph.

Cylinders, Transverse Shear Distortion Included

In the case of a cylinder of sandwich construction with a relatively low transverse shear rigidity, the shear distortion may not be negligible; therefore, the analysis from Ref. 7-2 is presented, which included shear distortions for a symmetrically loaded orthotropic sandwich cylinder.

NOMENCLATURE

D_{Q_x} is the shear stiffness in xz plane per inch of width, lb/in. (see Sec. 11.2)

x is measured in the axial direction starting at edge of cylinder

z is normal to cylinder wall

x, θ subscripts refer to axial and circumferential directions, respectively

The secondary solutions as obtained from Ref. 7-2 are presented in Tables 7-1 and 7-2.

TABLE 7-1 Influence Coefficients Due to Edge Loading M_0 on Orthotropic Cylinder

Constants:
$$a^2 = \frac{B_\theta(1-\mu'_x\mu'_y)}{4D_{Q_x}R^2} \qquad \overline{\beta}^4 = \frac{B_\theta(1-\mu'_x\mu'_y)}{4D_xR^2} \qquad D = D_x$$

	$a^4 - \overline{\beta}^4 > 0$	No.*	$a^4 - \overline{\beta}^4 = 0$	No.*	$a^4 - \overline{\beta}^4 < 0$	No.*
Δr	$\dfrac{M_0}{D(A_2-A_1)}\left(\dfrac{A_2}{4a^2-A_1^2}e^{-A_1x} + \dfrac{A_1}{A_2^2-4a^2}e^{-A_2x}\right)$	1	$\left(\dfrac{e^{mx}}{2a^2}+\dfrac{xe^{mx}}{m}\right)\left(-\dfrac{M_0}{D}\right)$	4	$\dfrac{e^{-Vx}}{2D\overline{\beta}^4}M_0\left(-\overline{\beta}^2\cos Px + \dfrac{V}{P}\overline{\beta}^2\sin Px\right)$	7
M_x	$M_0\left(\dfrac{A_2e^{-A_1x}-A_1e^{-A_2x}}{A_2-A_1}\right)$	2	$e^{mx}M_0(1-mx)$	5	$\dfrac{e^{-Vx}}{P}M_0(P\cos Px + V\sin Px)$	8
Q_x	$M_0\left(\dfrac{-A_1A_2e^{-A_1x}+A_1A_2e^{-A_2x}}{A_2-A_1}\right)$	3	$-2a^2e^{mx}M_0$	6	$-\dfrac{e^{-Vx}}{P}M_0\sin Px(V^2+P^2)$	9
Constants	$A_1 = \left[2a^2+2(a^4-\overline{\beta}^4)^{1/2}\right]^{1/2}$ $A_2 = \left[2a^2-2(a^4-\overline{\beta}^4)^{1/2}\right]^{1/2}$		$m = -(2)^{1/2}a$		$V = (\overline{\beta}^2+a^2)^{1/2}$ $P = (\overline{\beta}^2-a^2)^{1/2}$	
	Influence coefficients for the displacements and rotation at the edge of the shell (x = 0)					
	Δr is positive inward, β is positive if edge rotates in clockwise direction (observing the right side of shell)					
Δr			$-\dfrac{M_0}{2\overline{\beta}^2 D}$			10
β			$M_0\left[\dfrac{(a^2+\overline{\beta}^2)^{1/2}}{D\overline{\beta}^2}\right]$			11

*Formula modifications are presented in Table 7-3.

TABLE 7-2 Influence Coefficients Due to Edge Loading Q_0 on Orthotropic Cylinder

Constants:
$$a^2 = \frac{B_\theta(1-\mu'_x\mu'_y)}{4D_{Q_x}R^2} \qquad \overline{\beta}^4 = \frac{B_\theta(1-\mu'_x\mu'_y)}{4D_xR^2} \qquad D = D_x$$

	$a^4 - \overline{\beta}^4 > 0$	No.*	$a^4 - \overline{\beta}^4 = 0$	No.*	$a^4 - \overline{\beta}^4 < 0$	No.*
Δr	$\dfrac{Q_0}{D(A_2-A_1)}\left(\dfrac{e^{-A_1x}}{4a^2-A_1^2}+\dfrac{e^{-A_2x}}{A_2^2-4a^2}\right)$	12	$\dfrac{Q_0}{Da^2}\left(\dfrac{e^{mx}}{m}+\dfrac{xe^{mx}}{2}\right)$	15	$\dfrac{e^{-Vx}}{2D\overline{\beta}^4}Q_0\left(\dfrac{a^2}{P}\sin Px - V\cos Px\right)$	18
M_x	$Q_0\left(\dfrac{e^{-A_1x}-e^{-A_2x}}{A_2-A_1}\right)$	13	$e^{mx}xQ_0$	16	$\dfrac{e^{-Vx}}{P}Q_0\sin Px$	19
Q_x	$Q_0\left(\dfrac{-A_1e^{-A_1x}+A_2e^{-A_2x}}{A_2-A_1}\right)$	14	$e^{mx}Q_0(1+mx)$	17	$\dfrac{e^{-Vx}}{P}Q_0(P\cos Px - V\sin Px)$	20
Constants	$A_1 = \left[2a^2+2(a^4-\overline{\beta}^4)^{1/2}\right]^{1/2}$ $A_2 = \left[2a^2-2(a^4-\overline{\beta}^4)^{1/2}\right]^{1/2}$		$m = -(2)^{1/2}a$		$V = (\overline{\beta}^2+a^2)^{1/2}$ $P = (\overline{\beta}^2-a^2)^{1/2}$	
	Influence coefficients for the displacements and rotation at the edge of the shell (x = 0)					
	Δr is positive inward, β is positive if edge rotates in clockwise direction (observing the right side of shell)					
Δr			$-Q_0\dfrac{(\overline{\beta}^2+a^2)^{1/2}}{2\overline{\beta}^4 D}$			21
β			$Q_0\dfrac{2a^2+\overline{\beta}^2}{2D\overline{\beta}^4}$			22

*Formula modifications are presented in Table 7-3.

Cylinders, Influence of Axial Force Included

It is usually assumed that the contribution of a uniformly distributed axial force N_0 to the bending deflection is negligible; however, for a cylinder with a relatively large ratio of radius to thickness, the axial force may significantly contribute to the bending deflection. The preceding analysis was therefore extended (Ref. 7-2) to include the effect of the axial force on the deflections. This leads to the modification of the formulas (Tables 7-1 and 7-2) in the manner shown in Table 7-3. The constants are modified as follows:

$$\alpha^2 = \frac{\dfrac{B_\theta(1 - \mu_x' \mu_y')}{D_{Q_x} R^2} + \dfrac{N_0}{D_x}}{4\left(1 + \dfrac{N_0}{D_{Q_x}}\right)}$$

$$\bar{\beta}^4 = \frac{B_\theta(1 - \mu_x' \mu_y')}{4 D_x R^2 \left(1 + \dfrac{N_0}{D_{Q_x}}\right)}$$

$$\gamma^2 = \frac{B_\theta(1 - \mu_x' \mu_y')}{4 D_{Q_x} R^2} \qquad V = 4\gamma^4 - 4\gamma^2 \alpha^2 + \bar{\beta}^4$$

$$S = (\bar{\beta}^2 + \alpha^2)^{1/2}$$

$$D = D_x$$

If the axial force which is uniformly distributed around the circumference of the shell is a compression force, the sign of N_0 should be negative. The formulas presented for orthotropic composite cylinders can be reduced to the case of homogeneous isotropic cylinders by setting $D_{Q_x} = \infty$. The axial force is usually important for very thin-walled, large-radius, isotropic cylinders.

TABLE 7-3 Modification of Tables 7-1 and 7-2 to Include the Effects of Axial Forces on Bending (Ref. 7-2)

Formula	Quantities (formulas in Tables 7-1 and 7-2)	Substitute
1	$(4a^2 - A_1^2)$ $(A_2^2 - 4a^2)$	$(4\gamma^2 - A_1^2)$ $(A_2^2 - 4\gamma^2)$
4	Entire formula	$\Delta r = \dfrac{e^{mx} M_0}{D(m^2 - 4\gamma^2)^2}\left[(4\gamma^2 - 3m^2) + (m^2 - 4\gamma^2)mx\right]$
6	$2a^2 M_0 x$	$-m^2 M_0$
7	Entire formula	$\Delta r = \dfrac{M_0 e^{-sx}}{2RD}\left[(S^2 + \gamma^2)\cos Px + \dfrac{S}{P}(\gamma^2 + p^2)\sin Px\right]$
8	V	S
9	V	S
10	Entire formula	$\Delta r = (-M_0/2VD)(2a^2 + \overline{\beta}^2 - 2\gamma^2)$
11	Entire formula	$\beta = (M_0 \overline{\beta}^2 / 2VD)(a^2 + \overline{\beta}^2)^{1/2}$
12	$(4a^2 - A_1^2)$ $(A_2^2 - 4a^2)$	$(4\gamma^2 - A_1^2)$ $(A_2^2 - 4\gamma^2)$
15	Entire formula	$\Delta r = \dfrac{e^{mx} Q_0}{D(m^2 - 4\gamma^2)^2}\left[2m - (m^2 - 4\gamma^2)x\right]$
18	Entire formula	$\Delta r = \dfrac{Q_0 e^{-sx}}{2VD}\left(\dfrac{\gamma^2}{P}\sin Px - S\cos Px\right)$
19	V	S
20	V	S
21	Entire formula	$\Delta r = (-Q_0/2VD)(a^2 + \overline{\beta}^2)^{1/2}$
22	Entire formula	$\beta = \dfrac{Q_0}{2VD}(\overline{\beta}^2 + 2\gamma^2)$

Spheres and Cones

The secondary solutions for orthotropic composite spheres or cones are not available in closed form. An approximate solution may be obtained by treating the shell as a beam on an elastic foundation. This type of analysis shows that the equivalent homogeneous isotropic shell would have the following properties:

$$t = \sqrt{\frac{12D_\phi}{B_\theta}}$$

$$E = \frac{B_\theta(1 - \mu_x'\mu_\theta')}{t} \tag{7-8}$$

$$\mu^2 = \mu_\phi'\mu_\theta'$$

$$\mu = \mu_\phi$$

The secondary solutions for orthotropic composite spheres and cones can be obtained from the tables presented in Chap. 4 by substituting Eq. 7-8 for the values of t, E, μ^2, and μ which appear in the equations for homogeneous isotropic spheres and cones. Formulas for computing the elastic constant for the orthotropic composite shell are presented in Sec. 11-2 for typical types of construction. This approximation is particularly good for spheres that are close to hemispheres and cones that are close to cylinders, provided that the differences between the stiffness properties in the two directions are not exceptionally large. The range of applicability of this approximation represented by Eq. 7-8 is not known at the present time.

7-5 Stiffened Shells

Stiffened shells are commonly used in the aerospace and civil engineering fields. The stiffened shell is stiffer than an unstiffened one if a meridional, a circumferential, or a combination of both systems of stiffeners is used. The stiffeners usually have all the characteristics of beams and are designed to take the compression and bending loads more effectively than the unstiffened monocoque shell.

Basically, two approaches are possible. If the stiffeners cover a considerable cross-sectional area and are arranged at a wide spacing, the whole construction behaves as a three-dimensional frame. The plates between the stiffeners distribute and transfer loading to the frame, which is analyzed as a space frame. This problem is beyond the scope of this book.

If the stiffeners are located close together, it is convenient to replace the stiffened section with an equivalent monocoque section with the corresponding elastic constants for both directions D_ϕ, D_θ, B_ϕ, B_θ, μ_ϕ', μ_θ', μ_ϕ, μ_θ. Then the shell can be analyzed as an orthotropic shell.

One typical type of stiffened shell has stringers in the longitudinal direction and frames in the circumferential direction. Young's modulus E of the stiffeners and sheet is assumed to be the same for simplicity.

Shells with Composite or Stiffened Walls

Figure 7-1 shows schematically the geometry of the wall of a cylinder. The given geometry is replaced by the idealized one. The idealized thickness of the cross section for longitudinal direction is:

$$t_s = t + \frac{A_s}{b}$$

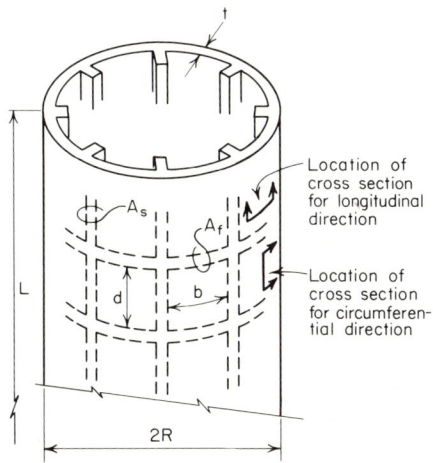

figure 7-1 Stiffened shell.

The idealized thickness of the cross section for the circumferential direction is:

$$t_f = t + \frac{A_f}{d}$$

where t = thickness of original sheet
A_f, A_s = areas of stiffeners in circumferential and longitudinal directions
d, b = distances between the centerline of the stiffeners in longitudinal and circumferential directions

From the above values for t_s and t_f we obtain the extensional stiffness

$$B_\phi = E t_s \qquad B_\theta = E t_f$$

Poisson's ratio is neglected because this effect is small for most stiffened construction.

The bending stiffnesses are

$$D_\phi = E \left(\frac{I_s}{b} + \frac{e_s^2 A_s t}{tb + A_s} + \frac{t^3}{12} \right)$$

$$D_\theta = E \left(\frac{I_f}{d} + \frac{e_f^2 A_f t}{td + A_f} + \frac{t^3}{12} \right)$$

where I_s, I_f = moment of inertia of stringer and frame, respectively
e_s, e_f = distance from center of sheet to centroid of stringer and frame, respectively

These are the only stiffnesses needed for the stress and deflection analysis of axisymmetrically loaded, stiffened shells of revolution. The stiffness properties of shells with skewed stiffeners (waffle) are presented in Sec. 11-2.

7-6 Sandwich Shells

Introduction

Structural sandwich is a layered composite formed by bonding two thin facings to a thick core. It is a type of stressed-skin construction in which the facings resist nearly all the applied edgewise (inplane) loads and flatwise bending moments. The thin facings provide nearly all the bending rigidity to the construction. The core spaces the facings and transmits shear between them so that they are effective about a common neutral axis. The core also provides most of the shear rigidity of the sandwich construction. By proper choice of materials for facings and core, constructions with high ratios of stiffness to weight can be achieved.

An exploded view of honeycomb sandwich construction is shown in Fig. 7-2.

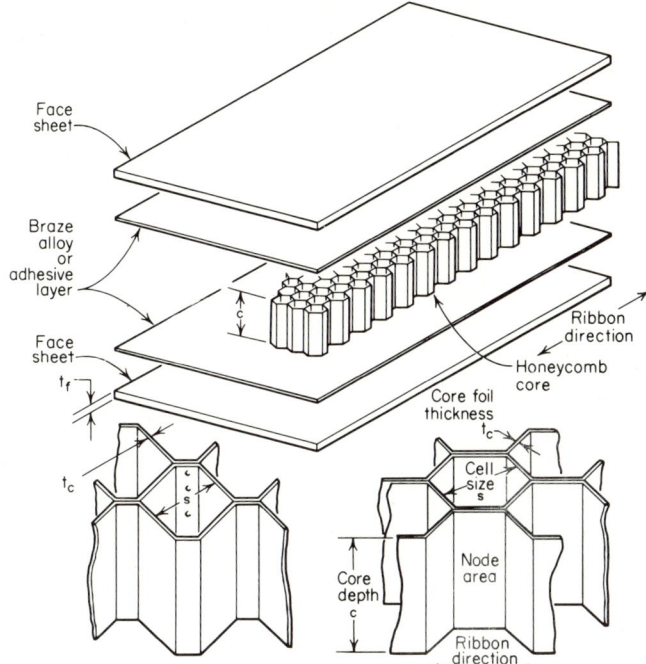

figure 7-2 *Honeycomb sandwich construction.*

Modes of Failure of Sandwich Elements

Generally, two types of allowable data exist. The first type is determined by simple material tests and is associated with material more than with geometry; the second type is dependent on the geometry of the element. If, in a sandwich construction, the materials of construction are considered to be the core, the facings, and the bonding media, the basic material properties would be associated with the properties of these three independent elements.

The second class of allowable data is those which are dependent on configuration as well as the basic properties of the facings, core, and bond media. This class of failure modes may be further subdivided into modes of failure that include the entire configuration, and those which are localized to a portion of the structure, but still limit the overall load-carrying capacity.

The most important local modes of failure are dimpling, wrinkling, and crimping. These modes of failure are dependent on the local geometry and on the basic properties of the materials of the sandwich. The general modes of failure generally are associated with the buckling strength of sandwich structural elements.

Another local mode of failure which may occur is associated with failure at local details such as edge members and close-outs or at points of introduction of concentrated loads. Unfortunately, exact analysis of all these many types of elements that must be introduced into the sandwich by the design requirements is seldom possible. As a consequence, the analyst must either determine the allowables experimentally or must make simplifying, conservative analytical approximations.

Types of Sandwich Cores

As far as the analyst is concerned, the core is a basic material of the sandwich and is treated as such. It has a shear strength, a shear modulus, and a flatwise tensile and compressive strength and modulus, and it may have some bending strength.

Cores may be classified either by their properties or by their geometrical configuration. Analytically, cores are generally categorized by their shear properties in the xz and yz planes. Cores in which the properties on these two planes are equal are termed isotropic cores, whereas those with different properties in the two planes are termed orthotropic cores. Cores may be orthotropic with respect to bending and/or shear rigidity and strength parameters. The foam-type cores generally are isotropic, at least in the xz and yz planes. The properties in the third plane, the xy plane, may be different. The honeycomb-type cellular cores are generally nearly isotropic with respect to shear (about 3:1 ratio maximum), although there are some configurations in which efforts are made to increase properties in one of the two planes. The corrugated cores are highly orthotropic, having orthotropic properties with respect to strength, bending rigidity, and shear rigidity.

Practically speaking, three types of cores are commercially available. There are solid cores, cellular cores, and corrugated cores. Solid cores are typified by the balsa wood or plywood cores in which the core can be used by itself as a structural member. Such cores generally have bending and shear rigidities and strengths in three planes. Consequently, the properties of the resulting sandwich may be greater with the same facing sheets than one in which one of the cellular cores is used.

The most important thing to remember about cores in the design of sandwich structures is that, analytically, they have basic mechanical and physical properties. Some of these properties are included in the basic analytical parameters of the entire sandwich and govern the behavior of the resulting sandwich in stability, stress, and deflection modes.

Analysis of Sandwich Shells

The stress and deflection analysis of sandwich shells can be performed using Secs. 7-3 and 7-4. The sandwich-stiffness properties can be obtained from Sec. 11-2. In most cases the facing sheets of the sandwich are isotropic so that Eq. 7-6 can be used to obtain the equivalent homogeneous isotropic shell, and the solutions presented in previous chapters can be used directly, provided that the transverse shear stiffness of the core is relatively large. A good first approximation is to assume a large transverse shear stiffness. Transverse shear deformations contribute to the shell deflection only in areas of large transverse shear stress. However, because of the large shearing stress, a strengthening of the core in this area is necessary so that the transverse shear stiffness becomes very large.

The local and general instability analysis of sandwich shells can be performed using the method presented in Chap. 13.

REFERENCES

7-1. Timoshenko, S.: *Theory of Plates and Shells*, McGraw-Hill Book Company, New York, 1940.
7-2. Baker, E. H.: Analysis of Symmetrically Loaded Sandwich Cylinder, *J. Am. Inst. Aeronautics Astronautics*, January, 1964.

Chapter 8

UNSYMMETRICALLY LOADED SHELLS

8-1 Introduction

Previous chapters have treated only axisymmetrical geometry, material, and loading cases for shells. In this chapter solutions for shells subjected to unsymmetrical loads are presented.

The types of shells considered are shells of revolution with or without symmetrical boundaries as well as shells which are nonaxisymmetric.

The shells are assumed to be thin enough to use membrane theory. Therefore, the following tables of solutions provide the membrane forces and deflections for a given loading and geometry.

8-2 Shells of Revolution

For an axisymmetric shell loaded unsymmetrically, the membrane forces can be obtained by solving Eq. 1-6.

Table 8-1 presents solutions for certain loadings for spherical, conical, and cylindrical shells loaded unsymmetrically.

Table 8-2 presents the solutions for spherical shells with nonsymmetrical boundaries.

TABLE 8-1 Sphere, Cone, and Cylinder Loaded by Wind Loading (Ref. 8-1)

	Sphere	Cone	Umbrella shell / Conical shell supported at the vertex of the cone with free edge	Cylinder
	$Z = p_w \sin\phi \cos\theta$ Supported at bottom	$Z = p_w \sin\phi \cos\theta$ Supported at bottom Direction of wind	$Z = p_w \sin\phi \cos\theta$	$Z = p_w \cos\theta$ Supported at bottom / Supported at left boundary Differential element
N_ϕ (or N_x)	$-p_w \dfrac{R}{3} \dfrac{\cos\theta \cos\phi}{\sin^3\phi}\left[3(\cos\phi_0) - (\cos^3\phi_0 - \cos^3\phi)\right]$ For $\phi_0 = 0$ (no opening) $-p_w \dfrac{R}{3} \dfrac{\cos\theta \cos\phi}{\sin^3\phi}(2 - 3\cos\phi + \cos^3\phi)$	$-p_w \dfrac{x}{2}\left[\cos\phi - \dfrac{1}{3\cos\phi} - \dfrac{x_0^2}{x^2}\cdot\left(\cos\phi - \dfrac{1}{\cos\phi}\right) - \dfrac{x_0^3}{x^3}\dfrac{2}{3\cos\phi}\right]\cos\theta$ For $x_0 = 0$ (complete cone) $-p_w \dfrac{x}{2}\left(\cos\phi - \dfrac{1}{3\cos\phi}\right)\cos\theta$	$p_w \left[\dfrac{l^3 - x^3}{3x^2} - \dfrac{l^2 - x^2}{2x}\sin^2\phi\right]\dfrac{\cos\theta}{\cos\phi}$	$p_w \dfrac{x^2}{2R}\cos\theta$
$N_{\phi\theta}$	$-p_w \dfrac{R}{3} \dfrac{\sin\theta}{\sin^3\phi}\left[3(\cos\phi_0) - \cos^3\phi_0 + \cos^3\phi\right]$ For $\phi_0 = 0$ (no opening) $-p_w \dfrac{R}{3} \dfrac{\sin\theta}{\sin^3\phi}(2 - 3\cos\phi + \cos^3\phi)$	$-p_w \dfrac{x^3 - x_0^3}{3x^2}\sin\theta$ For $x_0 = 0$ (complete cone) $-p_w \dfrac{x}{3}\sin\theta$	$p_w \dfrac{l^3 - x^3}{3x^2}\sin\theta$	$-p_w x \sin\theta$
N_θ	$p_w \dfrac{R}{3} \dfrac{\cos\theta}{\sin^3\phi}\left[\cos\phi(3\cos\phi_0) - \cos^3\phi_0 - 3\sin^2\phi - 2\cos^4\phi\right]$ For $\phi_0 = 0$ (no opening) $p_w \dfrac{R}{3} \dfrac{\cos\theta}{\sin^3\phi}(2\cos\phi - 3\sin^2\phi - 2\cos^4\phi)$	$-p_w x \cos\phi \cos\theta$	$-p_w x \cos\phi \cos\theta$	$-p_w R \cos\theta$

TABLE 8-2 Spherical Shells with Unsymmetrical Boundaries (Ref. 8-1)

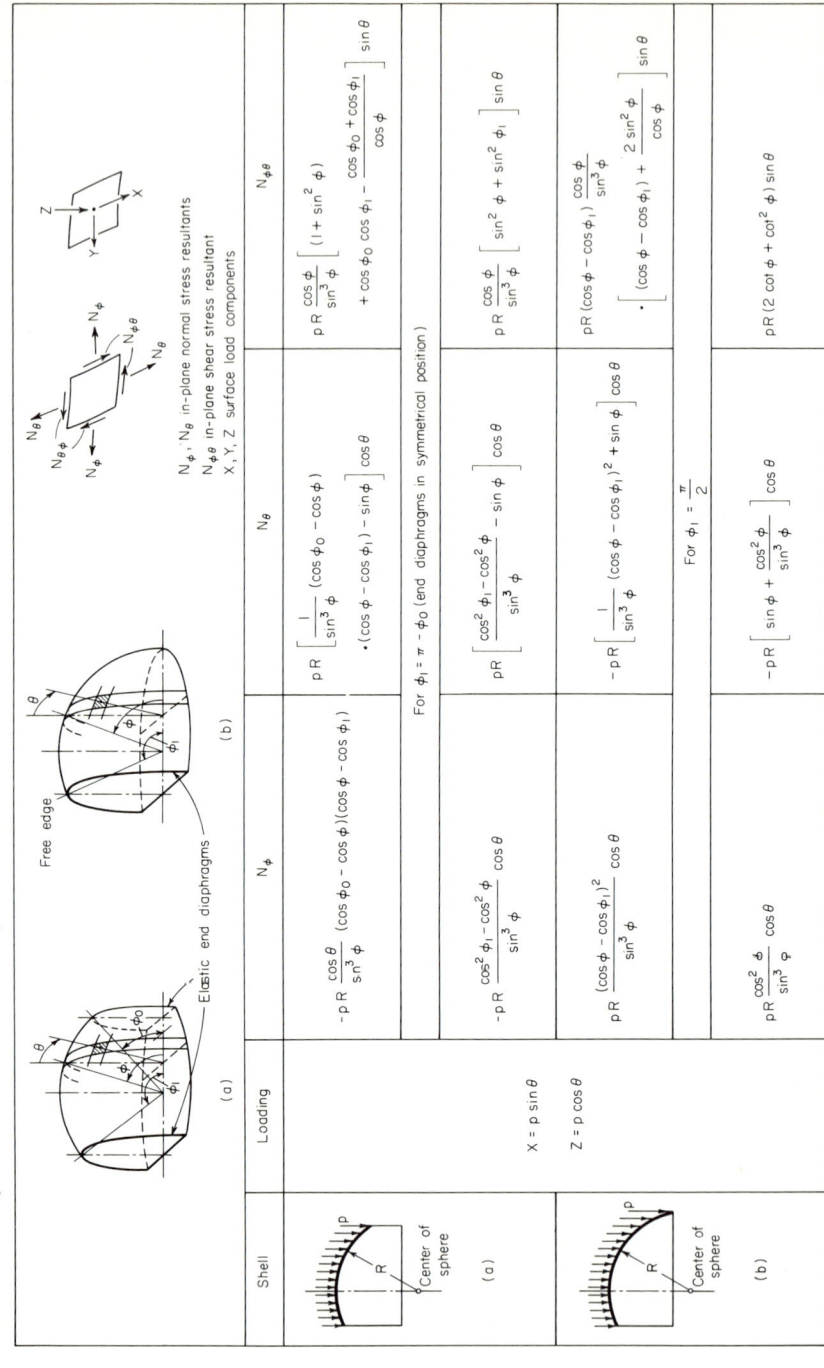

Shell	Loading	N_ϕ	N_θ	$N_{\phi\theta}$
(a)	$X = p\sin\theta$ $Z = p\cos\theta$	$-pR\dfrac{\cos\theta}{\sin^3\phi}(\cos\phi_0 - \cos\phi)(\cos\phi - \cos\phi_1)$	$pR\left[\dfrac{1}{\sin^3\phi}(\cos\phi_0 - \cos\phi) \cdot (\cos\phi - \cos\phi_1) - \sin\phi\right]\cos\theta$	$pR\dfrac{\cos\phi}{\sin^3\phi}\left[(1+\sin^2\phi) + \cos\phi_0\cos\phi_1 - \dfrac{\cos\phi_0 + \cos\phi_1}{\cos\phi}\right]\sin\theta$
		$-pR\dfrac{\cos^2\phi_1 - \cos^2\phi}{\sin^3\phi}\cos\theta$	$pR\left[\dfrac{\cos^2\phi_1 - \cos^2\phi}{\sin^3\phi} - \sin\phi\right]\cos\theta$	$pR\dfrac{\cos\phi}{\sin^3\phi}\left[\sin^2\phi + \sin^2\phi_1\right]\sin\theta$
		For $\phi_1 = \pi - \phi_0$ (end diaphragms in symmetrical position)		
		$pR\dfrac{(\cos\phi - \cos\phi_1)^2}{\sin^3\phi}\cos\theta$	$-pR\left[\dfrac{1}{\sin^3\phi}(\cos\phi - \cos\phi_1)^2 + \sin\phi\right]\cos\theta$	$pR(\cos\phi - \cos\phi_1)\dfrac{\cos\phi}{\sin^3\phi}$ $\cdot \left[(\cos\phi - \cos\phi_1) + \dfrac{2\sin^2\phi}{\cos\phi}\right]\sin\theta$
(b)		$pR\dfrac{\cos^2\phi}{\sin^3\phi}\cos\theta$	For $\phi_1 = \dfrac{\pi}{2}$ $-pR\left[\sin\phi + \dfrac{\cos^2\phi}{\sin^3\phi}\right]\cos\theta$	$pR(2\cot\phi + \cot^2\phi)\sin\theta$

N_ϕ, N_θ in-plane normal stress resultants
$N_{\phi\theta}$ in-plane shear stress resultant
X, Y, Z surface load components

8-3 Shells of Beam Systems

Presented here are some solutions for thin shells with various boundary conditions: cantilever, simply supported, and continuous. The loadings are deadweight, equally distributed loading over the base, end moments, concentrated loads, etc. The shells are cylindrical, conical, and curved panels of circular, elliptical, cycloidal, parabolical, and catenary shapes.

Cantilever Cylindrical Shell

Tabulated solutions for cantilevered cylindrical shells under various loading conditions are presented in Tables 8-3 and 8-4. The shell loaded torsionally (Table 8-3) is also presented.

TABLE 8-3 Cantilevered and Free Cylinders (Ref. 8-2)

	Loading: dead weight (q) $X = 0$ $Y = 0$ $Z = q\cos\theta$	Loading: concentrated load $X = Y = Z = 0$ $V = +1$	Loading: torsion $X = Y = Z = 0$ $N_x = u = 0$ $w = 0$ $M = +1$
N_β	$-qR\cos\theta$	0	0
$N_{x\beta}$	$2q(L-x)\sin\theta = N_{\beta x}$	$\dfrac{\sin\theta}{\pi R}$	$-\dfrac{1}{2\pi R^2}$
N_x	$\dfrac{q}{R}(L-x)^2 \cos\theta$	$\dfrac{(L-x)\cos\theta}{\pi R^2}$	0
Etu	$\dfrac{q x}{R}\left(L^2 - Lx + \dfrac{x^2}{3} + \mu R^2\right)\cos\theta$	$\dfrac{x(2L-x)\cos\theta}{2\pi R^2}$	0
Etv	$\dfrac{q x}{12R^2}\left\{48(1+\mu)LR^2 + 6\left[L^2 - R^2(4+3\mu)\right]x - 4Lx^2 + x^3\right\}\sin\theta$	$\dfrac{x\left[12(1+\mu)R^2 + 3Lx - x^2\right]\sin\theta}{6\pi R^3}$	$\dfrac{1+\mu}{\pi R^2}\left(\dfrac{L}{2} - x\right)$
Etw	$-\dfrac{q}{12R^2}\left\{\left[12R^2(R^2+\mu L^2) + 24(2+\mu)R^2 Lx + 6\left[L^2 - R^2(4+\mu)\right]x^2 - 4Lx^3 + x^4\right]\cos\theta\right\}$	$-\dfrac{\left[6(2+\mu)R^2 x + x^2(3L-x) + 6\mu R^2 L\right]\cos\theta}{6\pi R^3}$	0
$Et\beta$	$-\dfrac{q}{R}\left[R^2 + \mu(L-x)^2\right]\sin\theta$	$-\dfrac{\mu(L-x)\sin\theta}{\pi R^2}$	$\dfrac{1+\mu}{\pi R^3}\left(\dfrac{L}{2} - x\right)$

TABLE 8-4 Cylindrical Shell Loaded by Wind Loading (Ref. 8-2)

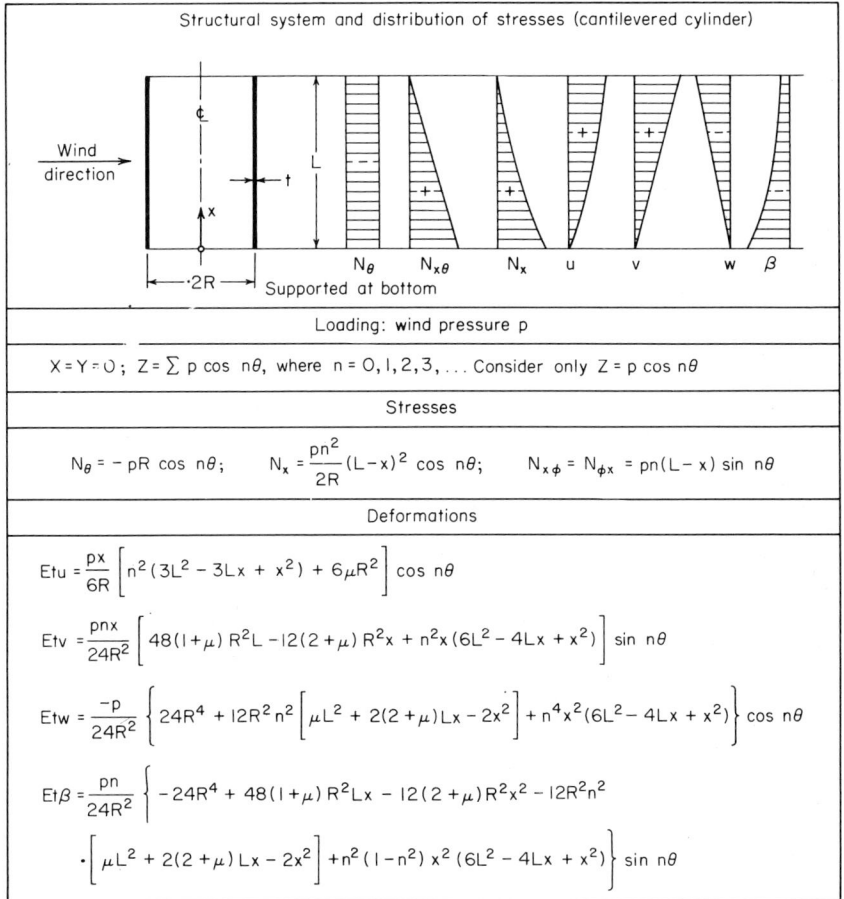

Simple and Fixed-beam Cylindrical Shell

Table 8-5 presents the solutions for cylindrical shells with simple supports and fixed supports under various loading conditions (Refs. 8-1 and 8-2).

TABLE 8-5 Cylindrical Shell Loaded as a Beam (Ref. 8-2)

	Loading: dead weight q $X = 0$ $Y = q\sin\theta$ $Z = q\cos\theta$	Loading: end moment $X = Y = Z = 0$ $M_B = +1$	Hydrostatic loading: $X = Y = 0$ $Z = -\rho(h - R\cos\theta)$ ρ = specific weight of liquid
N_θ	$-qR\cos\theta$	0	$\rho R(h - R\cos\theta)$
$N_{x\theta}$	$q(L-2x)\sin\theta = N_{\theta x}$	$-\dfrac{\sin\theta}{\pi RL} = N_{\theta x}$	$\dfrac{\rho R}{2}(L-2x)\sin\theta = N_{\theta x}$
N_x	$-\dfrac{qx}{R}(L-x)\cos\theta$	$-\dfrac{(L-x)\cos\theta}{\pi R^2}$	$-\dfrac{\rho x}{2}(L-x)\cos\theta$
Etu	$\dfrac{q(L-2x)}{12R}\left[L^2 + 2x(L-x) - 6\mu R^2\right]\cos\theta$	$\dfrac{\left[2L^2 + 12(1+\mu)R^2 - 3x(2L-x)\right]\cos\theta}{6\pi R^2 L}$	$\dfrac{\rho(L-2x)}{24}\left\{\left[L^2 + 2x(L-x) - 12\mu R^2\right]\cos\theta + 12\mu Rh\right\}$
Etv	$\dfrac{qx(L-x)}{12R^2}\left[L^2 + x(L-x) + 6(4+3\mu)R^2\right]\sin\theta$	$\dfrac{x(L-x)(2L-x)\sin\theta}{6\pi R^3 L}$	$\dfrac{\rho x}{24R}(L-x)\left[L^2 + x(L-x) + 12(2+\mu)R^2\right]\sin\theta$
Etw	$-\dfrac{q}{12R^2}\left\{12R^4 + x(L-x)\left[L^2 + x(L-x) + 6(4+\mu)R^2\right]\right\}\cos\theta$	$-\dfrac{(L-x)\left[x(2L-x) - 6R^2\mu\right]\cos\theta}{6\pi R^3 L}$	$\rho R^2 h - \dfrac{\rho}{24R}(L-x)\left\{24R^4 + x(L-x)\left[L^2 + x(L-x) + 24R^2\right]\right\}\cos\theta$
$Et\beta$	$-\dfrac{q}{R}\left[R^2 - \mu x(L-x)\right]\sin\theta$	$\dfrac{\mu(L-x)\sin\theta}{\pi R^2 L}$	$-\dfrac{\rho}{2}\left[2R^2 - \mu x(L-x)\right]\sin\theta$

TABLE 8-5 Continued

Shell	Loading	N_x	N_θ	$N_{x\theta}$
(simply supported shell, load p)	$Y = -p\cos\theta$ $Z = p\sin\theta$	$-p\dfrac{x}{R}(R-x)\sin\theta$	$-pR\sin\theta$	$-p(L-2x)\cos\theta$
(fixed-fixed shell, load p)		$p\left[\dfrac{L^2}{6R} - \mu R - \dfrac{x}{R}(L-x)\right]\sin\theta$ μ = Poisson's ratio	$-pR\sin\theta$	$-p(L-2x)\cos\theta$
(fixed-fixed shell with liquid of depth h)	$Z = -\rho(h - R\sin\theta)$ ρ = specific weight of liquid	$-\rho\left\{\left[\dfrac{L^2}{12} - \mu R^2 - \dfrac{x}{2}(L-x)\right]\sin\theta - \mu Rh\right\}$ μ = Poisson's ratio	$\rho R^2\left(\dfrac{h}{R} - \sin\theta\right)$	$\rho R\left(\dfrac{L}{2} - x\right)\cos\theta$

Continuous Cylindrical Shell under Deadweight (Ref. 8-2)

Fig. 8-1 shows a cylindrical shell under deadweight loading. The system is symmetrical and externally statically indeterminate. For the

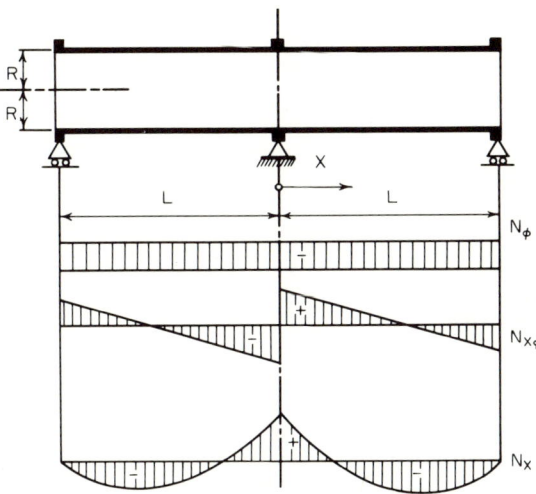

figure 8-1 *Continuous cylindrical shell loaded with deadweight.*

statically indeterminate case, select the moment X_1 above the middle support. For the solution, combine the case of a simple beam loaded by deadweight (case 1) and a simple beam loaded with the end moment (case 2). For reference, see Table 8-5.

For $X_1 = 0$ (case 1)

$$Etu_0 = \frac{qL}{12R}(L^2 - 6\mu R^2)\cos\theta$$

For $X_1 = +1$ (case 2)

$$Etu_1 = \frac{1}{3R^2\pi L}[L^2 + 6(1+\mu)R^2]\cos\theta$$

where u is deflection. Because of symmetry at $x = 0$, it follows that

$$u = u_0 + X_1 u_1 = 0$$

Consequently,

$$X_1 = -\frac{u_0}{u_1} = -\frac{qR\pi L^2}{4}\frac{L^2 - 6\mu R^2}{L^2 + 6(1+\mu)R^2}$$

If $p = q\,2\pi R$ = weight for 1 inch of cylinder is introduced and $\mu = 0$, finally,

$$X_1 = -\frac{pL^2}{8}\frac{1}{1 + 6(R/L)^2}$$

for

$$\frac{R}{L} \to 0 \qquad X_1 = -p\frac{L^2}{8}$$

and for

$$\frac{R}{L} \to \infty \qquad X_1 = 0$$

The nullpoint of shear is located at

$$x_0 = \frac{L}{8}\frac{5L^2 + 6(4+3\mu)R^2}{L^2 + 6(1+\mu)R^2}$$

For the longitudinal stress N_x, the nullpoint is located at

$$x_1 = \frac{L}{4}\frac{L^2 - 6\mu R^2}{L^2 + 6(1+\mu)R^2}$$

Other continuous systems under different loading conditions can be solved in a similar manner.

8-4 Curved Panels (Barrel Vaults)

This section presents a collection of various solutions for curved panels of simple beam systems. The geometry of the curved panels is circular, elliptical, cycloidal, parabolical, catenary, and special shapes. The solutions for various loadings are given in Tables 8-6 to 8-8. The shells under consideration are thin, and linear membrane theory was the basis for the derived formulas.

TABLE 8-6 Curved Circular Panels—Barrel Vaults (Ref. 8-2)

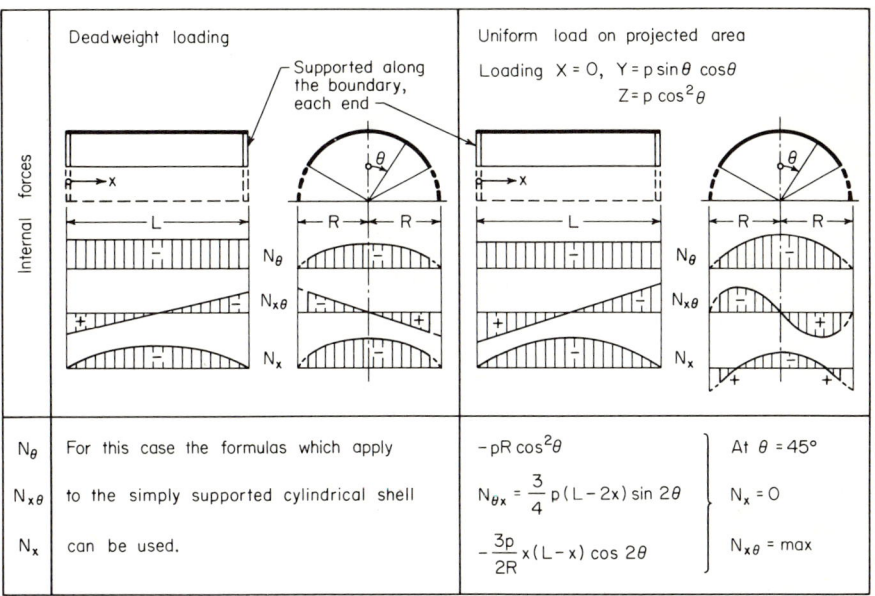

	Deadweight loading	Uniform load on projected area
N_θ	For this case the formulas which apply to the simply supported cylindrical shell can be used.	$-pR\cos^2\theta$
$N_{x\theta}$		$N_{\theta x} = \dfrac{3}{4}p(L-2x)\sin 2\theta$
N_x		$-\dfrac{3p}{2R}x(L-x)\cos 2\theta$

At $\theta = 45°$
$N_x = 0$
$N_{x\theta} = $ max

TABLE 8-7 Barrel Vaults (Ref. 8-1)

N_x, N_θ — In-plane normal stress resultants
$N_{\theta x}$ — In-plane shear stress resultants
X, Y, Z — Surface load components
Boundary conditions: $N_x = 0$ for $x = 0$ and $x = L$

Shell	Loading	N_x	N_θ	$N_{\theta x}$
Circle	$Y = -p\cos\theta$; $Z = p\sin\theta$	$-p\dfrac{x}{R}(L-x)\sin\theta$	$-pR\sin\theta$	$-p(L-2x)\cos\theta$
Circle	$Z = p_w\cos\theta$	$-p_w\dfrac{x}{2R}(L-x)\cos\theta$	$-p_w R\cos\theta$	$p_w\left(\dfrac{L}{2}-x\right)\sin\theta$
Parabola	$Y = -p\sin\theta\cos\theta$; $Z = p\sin^2\theta$	$p\dfrac{x}{2r_0}(L-x)\sin^4\theta$	$-p\dfrac{r_0}{\sin^2\theta}$	$p\left(\dfrac{L}{2}-x\right)\cos\theta$
Parabola	$Y = -p\sin\theta\cos\theta$; $Z = p\sin^2\theta$	0	$-p\dfrac{r_0}{\sin\theta}$	0
Parabola	$Z = p_w\cos\theta$	$p_w\dfrac{x}{2r_0}(L-x)(3+2\sin^2\theta)\sin\theta\cos\theta$	$-p_w r_0\dfrac{\cos\theta}{\sin^3\theta}$	$p_w\left(\dfrac{L}{2}-x\right)\dfrac{1+2\cos^2\theta}{\sin\theta}$
Cycloid	$Y = -p\sin\theta\cos\theta$; $Z = p\sin^2\theta$	$p\dfrac{2x}{r_0}(L-x)\dfrac{1-2\sin^2\theta}{\sin\theta}$	$-p r_0\sin^3\theta$	$-4p\left(\dfrac{L}{2}-x\right)\sin\theta\cos\theta$
Cycloid	$Z = p_w\cos\theta$	$-p_w\dfrac{x}{r_0}(L-x)(1-\cos\theta)\dfrac{\cos\theta}{\sin^3\theta}$	$-p_w r_0\sin\theta\cos\theta$	$-p_w\left(\dfrac{L}{2}-x\right)\dfrac{1-2\sin^2\theta}{\sin\theta}$

TABLE 8-7 Continued

Shell	Loading	N_x	N_θ	$N_{x\theta}$
Ellipse $\eta = a^2\cos^2\theta + b^2\sin^2\theta$	$Y = -p\sin\theta\cos\theta$ $Z = p\sin^2\theta$	$-p\dfrac{3x(L-x)}{2a^2b^2\eta^{1/2}}\left[b^2(a^2\cos^2\theta -b^2\sin^2\theta)+2\eta(\sin^2\theta-\cos^2\theta)\right]$	$-pa^2b^2\dfrac{\sin^2\theta}{\eta^{3/2}}$	$3p\left(\dfrac{L}{2}-x\right)\dfrac{\sin\theta\cos\theta}{\eta}(b^2-2\eta)$
	$Z = p_w\cos\theta$	$-p_w\dfrac{x(L-x)\cos\theta}{2a^2b^2\eta^{1/2}}\left[\eta^2+3\eta(b^2-a^2)-(1-3\sin^2\theta)-6(b^2-a)^2\sin^2\theta\cos^2\theta\right]$	$-p_w a^2b^2\dfrac{\cos\theta}{\eta^{3/2}}$	$3p_w\left(\dfrac{L}{2}-x\right)\dfrac{b^2-a^2}{\eta}(1+\cos^2\theta)\sin\theta$
Catenary	$Y = -p\cos\theta$ $Z = p\sin\theta$	0	$-p\dfrac{r_0}{\sin\theta}$	0
	$Y = -p\sin\theta\cos\theta$ $Z = p\sin^2\theta$	$p\dfrac{x}{2r_0}(L-x)\sin^2\theta(1-2\sin^2\theta)$	$-pr_0$	$-p\left(\dfrac{L}{2}-x\right)\dfrac{1+\cos^2\theta}{\sin\theta}$
	$Z = p_w\cos\theta$	$p_w\dfrac{x}{2r_0}(L-x)(2+\sin^2\theta)\cos\theta$	$-p_w r_0\dfrac{\cos\theta}{\sin^2\theta}$	$p_w\left(\dfrac{L}{2}-x\right)\dfrac{1+\cos^2\theta}{\sin\theta}$
$y = \dfrac{1}{2}\dfrac{\pi b/2-a}{a-b}$	$Y = -p\sin\theta\cos\theta$ $Z = p\sin\theta$	$p\dfrac{x(L-x)(y+\pi/4)}{2a(y+\sin\theta)^3}\left[\gamma(1-6\sin^2\theta)-(2y^2+3\sin^2\theta)\sin\theta\right]$	$-p\dfrac{a}{y+\pi/4}(y+\sin\theta)\sin\theta$	$p\left(x-\dfrac{L}{2}\right)\dfrac{(2y+3\sin\theta)\cos\theta}{y+\sin\theta}$
	$Y = -p\sin\theta\cos\theta$ $Z = p\sin^2\theta$	$p\dfrac{x(L-x)(y+\pi/4)}{2a(y+\sin\theta)^3}\left[\gamma(8-15\sin^2\theta)\sin\theta+(3y^2+4\sin^2\theta)(1-2\sin^2\theta)\right]$	$p\dfrac{a}{y+\pi/4}(y+\sin\theta)\sin^2\theta$	$p\left(x-\dfrac{L}{2}\right)\dfrac{(3y+4\sin\theta)\sin\theta\cos\theta}{y+\sin\theta}$
$r = \dfrac{a}{y+\pi/4}(y+\sin\theta)$	$Z = p_w\cos\theta$	$-p_w\dfrac{x(L-x)(y+\pi/4)}{2a(y+\sin\theta)^3}\cos\theta(1+y^2+\sin^2\theta+4y\sin\theta)$	$-p_w\dfrac{a}{y+\pi/4}(y+\sin\theta)\cos\theta$	$p_w\left(x-\dfrac{L}{2}\right)\dfrac{1-2\sin^2\theta-y\sin\theta}{y+\sin\theta}$

TABLE 8-8 Curved Elliptical and Cycloidal Panels—Barrel Vaults (Ref. 8-2)

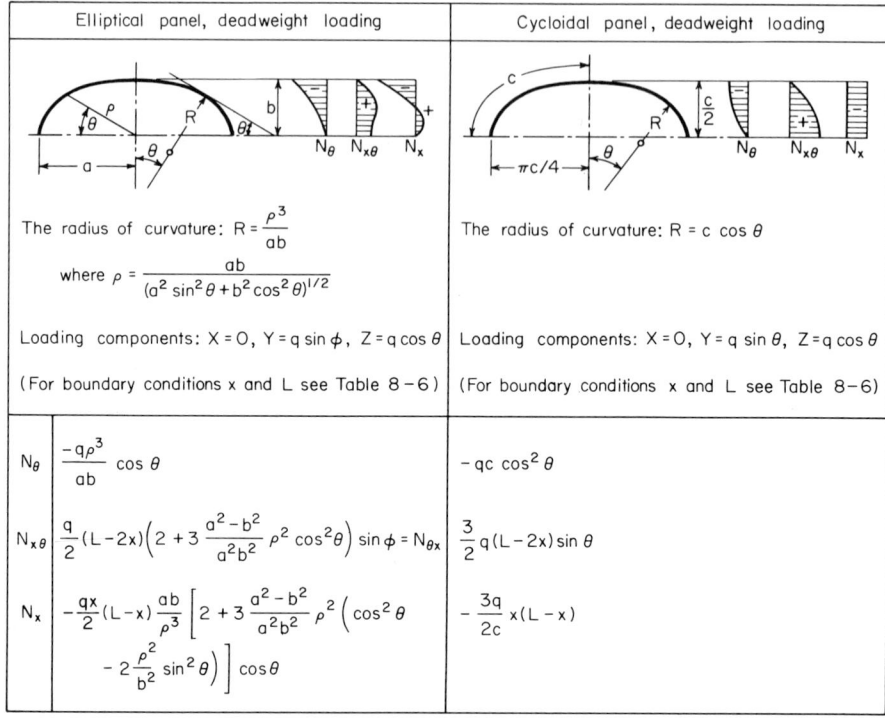

Elliptical panel, deadweight loading	Cycloidal panel, deadweight loading
The radius of curvature: $R = \dfrac{\rho^3}{ab}$ where $\rho = \dfrac{ab}{(a^2 \sin^2\theta + b^2 \cos^2\theta)^{1/2}}$	The radius of curvature: $R = c \cos\theta$
Loading components: $X = 0$, $Y = q\sin\phi$, $Z = q\cos\theta$	Loading components: $X = 0$, $Y = q\sin\theta$, $Z = q\cos\theta$
(For boundary conditions x and L see Table 8–6)	(For boundary conditions x and L see Table 8–6)
N_θ: $\dfrac{-q\rho^3}{ab}\cos\theta$	$-qc\cos^2\theta$
$N_{x\theta}$: $\dfrac{q}{2}(L-2x)\left(2 + 3\dfrac{a^2-b^2}{a^2b^2}\rho^2\cos^2\theta\right)\sin\phi = N_{\theta x}$	$\dfrac{3}{2}q(L-2x)\sin\theta$
N_x: $-\dfrac{qx}{2}(L-x)\dfrac{ab}{\rho^3}\left[2 + 3\dfrac{a^2-b^2}{a^2b^2}\rho^2\left(\cos^2\theta - 2\dfrac{\rho^2}{b^2}\sin^2\theta\right)\right]\cos\theta$	$-\dfrac{3q}{2c}x(L-x)$

REFERENCES

8-1. Pflüger, Alf.: *Elementary Statics of Shells*, 2d ed., McGraw-Hill Book Company, New York, 1961.

8-2. Worch, G.: *Elastische Schalen Beton-Kalendar*, (in German), Wilhelm Ernst & Sohn KG, Berlin, 1958.

Chapter 9

ALLOWABLE STRESSES AND MARGIN OF SAFETY

9-1 Introduction

After the stress distributions are determined in a structural analysis, the questions arise: are the stresses acceptable and does the structure have an adequate factor of safety under the applied loads?

The material of which a shell is made fails if the multiaxial state of stress reaches a certain level. The problem is complicated by the fact that the vast majority of property data for materials are determined by uniaxial tests, whereas the state of stress in a shell is normally biaxial. The multiaxial failing stress for a material cannot be determined in general at the present time because of technical difficulties. The problem is further complicated by the fact that the analyses by which the stress distributions in shells are usually determined assume that the material is in the elastic range. Therefore, considerable effort is being devoted by investigators to developing elastoplastic shell-analysis techiques to determine the stress distributions in shells. These techniques usually employ digital-computer programs for solution and are beyond the scope of this book.

The method discussed here utilizes the results of previous chapters to determine the state of stress, assuming the shell material to remain elastic, and accounts for the biaxial state of stress in predicting the margin of safety. These methods are generally conservative for shells designed by material rupture. The shell failure mode of buckling is discussed in subsequent chapters.

9-2 Allowable Stress

In the uniaxial case, the allowable stress is the limiting stress, such that the maximum actual stress is less than the allowable stress.

The allowable stress (Refs. 9-1, 9-6) is widely employed in civil and mechanical engineering. In other branches of engineering, such as aerospace, a margin of safety is used (Ref. 9-7). The next section will describe the definition of margin of safety.

9-3 Margin of Safety

Instead of the allowable stress, the margin of safety is used in many instances. It is related to the allowable stress as is shown below. When the allowable stress is established, the margin of safety has to be larger than or equal to zero. If the margin of safety is negative, the structure must be revised in order not to exceed the prescribed limit, unless there is additional justification.

$$\text{Margin of safety} = \frac{\text{allowable load}}{\text{applied load}} - 1$$

If the stresses are proportional to the applied loads,

$$\text{Margin of safety} = \frac{\text{allowable stress}}{\text{applied stress}} - 1$$

9-4 Failure Theories

The vast majority of test data on the failure of structural materials have been obtained by uniaxial tests. Design allowables based on these data have subsequently been reduced to account for scatter and are published in documents such as MIL-HDBK-5 (Ref. 9-5).

Determination of allowable stresses or margins of safety does not present any difficulty for the uniaxial case. It is incorrect, however, to compare the stresses for multiaxial cases with uniaxial allowables. If such a comparison is to be made, certain combinations (indicated by σ_i) rather than the stresses themselves may be compared with the uniaxial tension allowables.

Allowable Stresses and Margin of Safety

In order to predict failure under a multiaxial stress state from uniaxial mechanical properties, a number of different approaches have been taken over the years. Various theories of failure have been proposed which have had the objective of describing the onset of failure of a material in a combined-stress state from uniaxial data. Here failure signifies yielding or actual rupture, whichever is more critical.

While some of the theories have had limited success in predicting failure, others do not agree well with test results. Some theories are better suited for brittle materials, while others, strictly speaking, predict the onset of plasticity better than the ultimate failure or material rupture. Table 9-1 summarizes some of the historically proposed theories. Each of the theories presented in Table 9-1 predicts a stress parameter σ_i which must be equal or less than the uniaxial allowable stress, σ_0.

$$\sigma_i \leqslant \sigma_0$$

TABLE 9-1 Failure Theories.

General remarks: Yield stress for tension and compression must be equal, except as noted.

$$\sigma_{1,2} = \frac{\sigma_x + \sigma_y}{2} \pm \sqrt{\left(\frac{\sigma_x - \sigma_y}{2}\right)^2 + \tau_{xy}^2}$$

$\sigma_i \leq \sigma_0$ = allowable stress for uniaxial case

No.	Name	General principle	σ_i	Notes
1	Max. stress theory (Ref. 9-1)	Two states of stress are equivalent if their maximum principal stresses are equal.	σ_{max}	Often used in civil engineering. Does not agree very well with test results, consequently, not recommended.
2	Max. strain theory (suggested by Mariotte) (Ref. 9-1)	Two states of stress are equivalent if their maximum linear strains are equal.	$\sigma_1 - \mu(\sigma_2 + \sigma_3)$	Does not agree very well with test results. Not recommended.
3	Max. shear theory (suggested by Coulomb) (Ref. 9-1)	Two states of stress are equivalent if their maximum shear stresses are equal.	$\sigma_1 - \sigma_3$	$\tau_{max} = \dfrac{\sigma_1 - \sigma_3}{2}$ This theory used in machine design for ductile materials because of simplicity, despite existence of a more accurate theory.
4	Max. strain energy theory (suggested by Beltrami) (Ref. 9-1)	Two states of stress are equivalent if the strain energies are equal.	$\sqrt{\sigma_1^2 + \sigma_2^2 + \sigma_3^2 - 2\mu(\sigma_1\sigma_2 + \sigma_2\sigma_3 + \sigma_3\sigma_1)}$ For biaxial case: $\sqrt{\sigma_1^2 + \sigma_2^2 - 2\mu\sigma_1\sigma_2}$	The results do not agree very well with test results.

TABLE 9-1 Continued

5	Distortion energy theory (Refs. 9-1 and 9-2)	Two states of stress are equivalent if the distortion energies are equal.	$\sqrt{\sigma_1^2 + \sigma_2^2 + \sigma_3^2 - \sigma_1\sigma_2 - \sigma_2\sigma_3 - \sigma_3\sigma_1}$ $= \dfrac{1}{\sqrt{2}}\sqrt{(\sigma_1-\sigma_2)^2 + (\sigma_2-\sigma_3)^2 + (\sigma_3-\sigma_1)^2}$ For non-principal stresses: $\dfrac{1}{\sqrt{2}}\sqrt{(\sigma_x-\sigma_y)^2 + (\sigma_y-\sigma_z)^2 + (\sigma_z-\sigma_x)^2 + 6(\tau_{yz}^2 + \tau_{zx}^2 + \tau_{xy}^2)}$ For biaxial case: $\sqrt{\sigma_1^2 - \sigma_1\sigma_2 + \sigma_2^2}$	The formula represents an ellipsoid. The results agree well with test results. Recommended for ductile materials. (Introduced by Huber, corrected by Hencky/Von Mises) Well justified by tests. Recommended for ductile materials. Equation represents ellipse.
6	Tresca criterion (Ref. 9-4)	Extended by Tresca maximum shear theory. Yielding will occur when any one of the 6 conditions (next column) is reached. For biaxial case:	$\sigma_1 - \sigma_2 = \pm\sigma_0$ $\sigma_2 - \sigma_3 = \pm\sigma_0$ $\sigma_3 - \sigma_1 = \pm\sigma_0$ For biaxial case: $\sigma_1 - \sigma_2 = \sigma_0 \quad \text{if } \sigma_1 > 0,\, \sigma_2 < 0$ $\sigma_1 - \sigma_2 = -\sigma_0 \quad \text{if } \sigma_1 < 0,\, \sigma_2 > 0$ $\sigma_2 = \sigma_0 \quad \text{if } \sigma_2 > \sigma_1 > 0$ $\sigma_1 = \sigma_0 \quad \text{if } \sigma_1 > \sigma_2 > 0$ $\sigma_1 = -\sigma_0 \quad \text{if } \sigma_1 < \sigma_2 < 0$ $\sigma_2 = -\sigma_0 \quad \text{if } \sigma_2 < \sigma_1 < 0$ For pure shear: $\tau = \dfrac{1}{2}\sigma_0$	Well justified by test results. Well justified by test results. Recommended for ductile materials. $\sigma_1 = -\sigma_2 = \tau,\ \sigma_3 = 0$
7	Mohr's theory (Ref. 9-1)	Different states of stresses are equivalent if corresponding Mohr's circles have the same envelope.	$\sigma_1 - m\sigma_3 \leq \sigma$ (allowable uniaxial tensile stress) For non-principal stresses: $\dfrac{1-m}{2}(\sigma_x + \sigma_y) + \dfrac{1+m}{2}\sqrt{(\sigma_x - \sigma_y)^2 + 4\tau_{xy}^2}$ If $m = 1$: $\sigma_1 - \sigma_3 \leq \sigma$	Agree well with test results. Yield $\sigma_t \leq$ Yield σ_c σ_c = compressive stress, σ_t = tensile stress $m = \dfrac{\text{allowable } \sigma_t}{\text{allowable } \sigma_c}$ Only biaxial stresses are considered. Introduction of third stress gives only small difference. Recommended for brittle materials.

213

9-5 Practical Applications

An analysis is performed using the design (actual) loading or ultimate loading. The flow of computations is as shown to design a shell:

1. Determine design loading (maximum applied loading).
2. Analyze shell for above loading and compute σ_i.
3. From corresponding specifications or other requirements select $\sigma_{\text{allowable}}$.
4. If $\sigma_i \leqslant \sigma_{\text{allowable}}$, the structure is safe. If $\sigma_i > \sigma_{\text{allowable}}$, the structure must be redesigned.

Besides the limitation imposed by $\sigma_{\text{allowable}}$, in many cases additional limitations imposed by deflections exist. The actual deflection due to maximum load should not exceed a permissible deflection. The principle of design is analogous to the above procedure.

A slightly different approach is used in aerospace engineering, where analysis is performed using ultimate loading. As an example, if a shell is a part of an aircraft structure, the following steps may be taken:

1. Determine the design load (maximum applied), which may be called P_{limit}.
2. Compute $P_{\text{ult}} = (\text{factor of safety}) \times (P_{\text{limit}})$.
3. Compute σ_1 and σ_2 (ultimate stresses) from P_{ult}.
4. Compute σ_i (distortion energy).
5. Determine ultimate tensile stress F_{tu} from references such as Ref. 9-5.
6. Determine margin of safety

$$\text{M.S.} = \frac{F_{tu}}{\sigma_i} - 1$$

7. If M.S. $\geqslant 0$, the structure is safe. If M.S. < 0, the structure must be redesigned.

In conclusion, it is noted that σ_i always represents a certain combination of stresses at the point of interest, in the elastic range. There is no accurate and practical way for finding the stresses in the plastic range (except as discussed in the introduction). One of the more significant difficulties is that the stresses in the plastic range are not unique but depend upon the loading history.

9-6 Stress Ratios and Interaction Curves (Ref. 9-7)

A means of predicting structural failure under combined loading without determining principal stresses is known as the interaction method. The

Allowable Stresses and Margin of Safety 215

principles of this method are also described in Ref. 9-3. The basis for this method is as follows:

1. The strength under each simple loading condition (tension, shear, bending, buckling, etc.) is determined by test or theory.
2. The combined loading condition is represented by either load or stress ratios R where

$$R = \frac{\text{applied load or stress}}{\text{failing load or stress}}$$

Failing can mean yield, rupture, buckling, etc.

The effect of one loading R_1 on another simultaneous loading R_2 is represented by an equation or interaction curve involving R_1 and R_2. The equation or curve may have been determined by theory, by test, or by a combination of both.

A schematic interaction curve is shown in Fig. 9-1. This curve represents all the possible combinations of R_1 and R_2 that will cause failure.

Using the curve:

1. Let the value of R_1 and R_2 locate point a.
2. R_1 and R_2 can increase proportionately until failure occurs at point b.
3. If R_1 remains constant, R_2 can increase until failure occurs at point c.
4. If R_2 remains constant, R_1 can increase until failure occurs at point d.
5. The factor of safety for (2) is F.S. $= (ob/oa)$ or (oh/oe) or (og/of), and the factor of safety for (3) is F.S. $= (fc/fa)$.

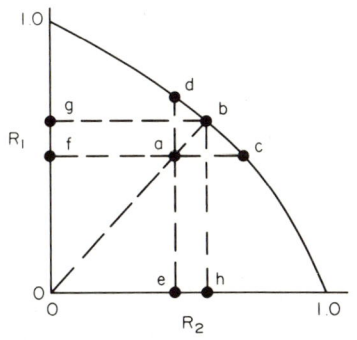

figure 9-1 *Schematic interaction curve.*

In general, the formula for the factor of safety stated analytically for interaction equations where the exponents are only 1 or 2 (one term may be missing) is as follows:

$$\text{F.S.} = \frac{2}{R' + \sqrt{(R')^2 + (R'')^2}}$$

where R' designates the sum of all first-power ratios
R'' designates the sum of all second-power ratios

Table 9-2 summarizes some typical interaction relations.

TABLE 9-2 Interaction Relations

Structure	Loading combination	Figure	Interaction equation	Eq. for factor of safety	Remarks
Compact	Biaxial tension or biaxial compression		$R_x = \frac{f_x}{F}$; $R = \frac{f_y}{F}$	$\frac{1}{R_{max}}$	Use R_x or R_y, whichever is greater
Compact and round tubes (a)	Axial and bending stresses	9-3	$R_s + R_b = 1$	$\frac{1}{R_a + R_b}$	
	Normal and shear stresses	9-3	$R_f^2 + R_s^2 = 1$ $R_f = R_a + R_b$	$\frac{1}{\sqrt{R_f^2 + R_s^2}}$	(b) for $0.5 < \frac{F_s}{F} < 0.75$ For all other values use max. stress equations or Mohr's circle
Round tubes	Bending, torsion and compression		$R_b^2 + R_{st}^2 = (1 - R_c)^2$	$\frac{1}{R_c + \sqrt{R_b^2 + R_{st}^2}}$	
Streamline tubes	Bending and torsion	9-3	$R_b + R_{st} = 1$	$\frac{1}{R_b + R_{st}}$	
Bolts	Tension and shear	9-3	$R_t^2 + R_s^3 = 1$		

NOTE: Care must be exercised in determining whether to check factor of safety for limit or ultimate loads.

Interaction relations are presented graphically on Fig. 9-2.

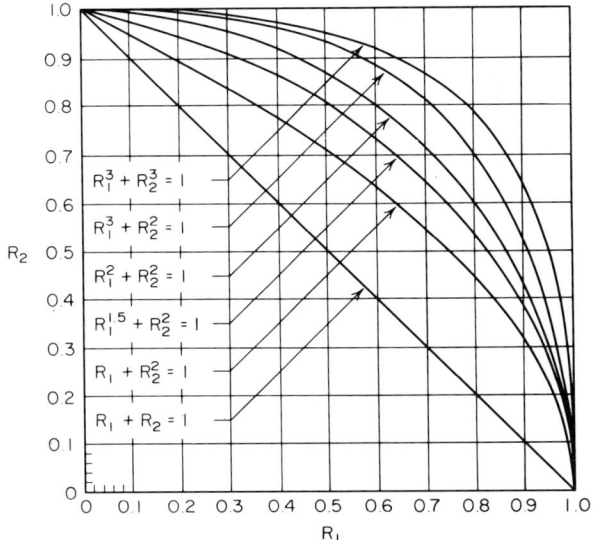

figure 9-2 *Interaction curves.*

9-7 A Theoretical Approach to Interaction (Ref. 9-7)

For combining normal and shear stresses, the principal-stress equations are convenient to use.

Let F and F_s be defined as the failing stress, such as yielding or rupture.
Let $k = F_s/F$; tests of most materials will show this ratio to vary from 0.50 to 0.75.

$$R_f = \frac{f}{F} \qquad R_s = \frac{f_s}{F_s}$$

Maximum Normal Stress Theory:

$$f_{max} = \frac{f}{2} + \sqrt{\left(\frac{f}{2}\right)^2 + f_s^2}$$

Divide by F; replace f_s by $R_s F_s$, f/F by R_f, and F_s/F by k. The resulting equation when $f_{max} = F$ is

$$1 = \frac{R_f}{2} + \sqrt{\left(\frac{R_f}{2}\right)^2 + (kR_s)^2} \tag{9-1}$$

A plot of this equation for $k = 0.50$ and $k = 0.70$ is shown in Fig. 9-3.

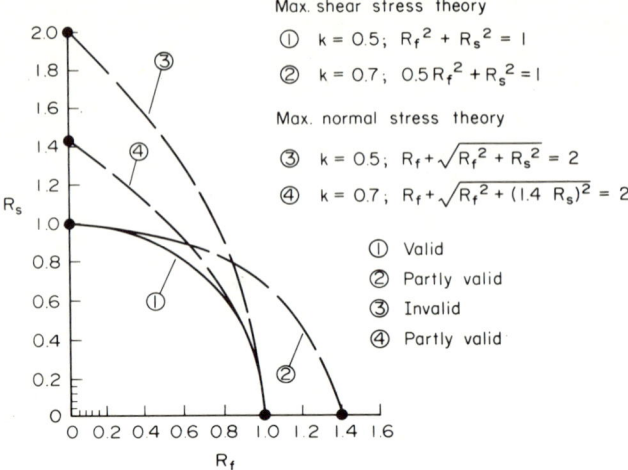

figure 9-3 *Plot of equations for $k = 0.50$ and 0.70.*

Maximum Shear Stress Theory:

$$f_{s_{\max}} = \sqrt{\left(\frac{f}{2}\right)^2 + f_s^2}$$

Divide by F_s; replace f by $R_f F$, f_s/F_s by R_s, and F/F_s by $1/k$. The resulting equation when $f_{s_{\max}} = F_s$ is

$$1 = \sqrt{\left(\frac{R_f}{2k}\right)^2 + R_s^2} \qquad (9\text{-}2)$$

A plot of this equation for $k = 0.50$ and $k = 0.70$ is shown in Fig. 9-3.

Conclusion

From the foregoing analysis, only Eq. 9-2 with $k = 0.5$ is valid for all values of R_f and R_s. It is conservatively safe to use the resulting Eq. 9-3 for values of k ranging from 0.5 to 0.7, since all values within curve ① must also be within the other curves. The use of other curves of Fig. 9-3 may lead to unconservative results.

Based upon the stress ratio curve

$$R_f^2 + R_s^2 = 1 \qquad (9\text{-}3)$$

the factor of safety is

$$\text{F.S.} = \frac{1}{\sqrt{R_f^2 + R_s^2}} \qquad (9\text{-}4)$$

For the graphical solution for factor of safety, the curve $R_1^2 + R_2^2 = 1$ of Fig. 9-2 may be used.

REFERENCES

9-1. Levshin, V. A.: *Strength of Materials* (in Russian), GOSTEHIZDAT, 1961.
9-2. Iljushin, A. A., and V.S. Lenskij: *Strength of Materials* (in Russian), FIZMATGIZ, 1959.
9-3. Shanley, F. R.: *Strength of Materials*, McGraw-Hill Book Company, New York, 1957.
9-4. Mendelson, A.: *Plasticity—Theory and Application*, The Macmillan Company, New York, 1968.
9-5. MIL-HDBK-5: *Strength of Metal Aircraft Elements*, U. S. Government Printing Office, February, 1966.
9-6. Pippard, A., and J. Baker: *The Analysis of Engineering Structures*, Edward Arnold (Publishers) Ltd., London, 1950.
9-7. *Astronautic Structures Manual*, George C. Marshall Space Flight Center, Huntsville, Ala., September, 1961.

Chapter 10

STABILITY OF UNSTIFFENED SHELLS

10-1 General

If a shell structure is subjected to a given compression load and an infinitesimal increase in the load results in a large change in the equilibrium configuration of the shell, the applied load is defined as the buckling load. The change in equilibrium configuration is usually a large increase in the deflections of the shell, which may or may not be accompanied by a change in the basic shape of the shell from the prebuckled shape. Occasionally, the given definition of buckling is difficult to apply to an actual structure. The change in the configuration of the shell may be gradual, and the actual buckling point is rather arbitrary. However, for most types of shells and loading conditions, the buckling load is quite pronounced and easy to identify.

The load-carrying capability of the shell may or may not decrease after buckling. This depends on the type of loading, the geometry of the shell, the stress levels of the buckled shell, etc. Only the buckling load will be discussed in this chapter, because the information available on collapse loads is quite limited. In general, the buckling load and collapse load are nearly the same, and if they are different, the deformations prior to collapse are often very large.

For columns and flat plates, the classical small-deflection theory predicts the buckling load quite well, and in general, the theoretical buckling load is used as the design-allowable buckling load. Therefore, the structure will usually buckle at approximately the design buckling load. In general, this method of design analysis cannot be used for shell structures. The buckling load for some types of shells and loadings may be much less than the load predicted by classical small-deflection theory, and in addition, the scatter of the test data may be quite large. For example, if a set of 10 nominally identical thin-walled cylinders of the same geometry were fabricated from a particular metal, none of the cylinders would fail at the same axial compressive load. In fact, the scatter of results may range to 500 percent and the average buckling load may be one-eighth of the theoretical buckling load. One of the primary sources of this error is the dependence of the buckling load of cylindrical shells on small deviations from the nominal circular cylindrical shape of the structure. Another source of discrepancy is the dependence of buckling loads of cylindrical shells on local edge conditions. Current methods of establishing design data tend to treat both initial imperfections and edge conditions as random effects. Results from all available tests are lumped without regard to specimen construction or method of testing and are analyzed to yield the lower bound or statistical correction factors to be applied to simplified versions of the theoretical results. When a lower-bound correction factor is used, data that do not seem typical are often left out.

When a statistical correction factor is used, a best-fit curve is determined for a given set of data, and the standard deviation of the test data is established. Using this information and small-sample theory, a design curve is obtained at a certain probability level. For instance, if a 90% probability level is chosen to obtain statistical design-allowable curves, the chances are about 9 out of 10 that a shell subjected to the design-allowable buckling load will not buckle. The load at which the shell may be expected to buckle is the load which corresponds to the best-fit curve. Best-fit curves have not been presented in this chapter because, in design analysis, the load at which the shell will not buckle is the primary interest, and approximately half the shells would buckle at loads less than the load corresponding to the best-fit curve.

Because of the lack of data for some types of shells and loading as well as the question of the extent of the range over which the theory is applicable, the recommended design-allowable buckling loads may be too conservative for some cases. Further theoretical and experimental investigations are necessary to justify raising the design curves.

Most analysis procedures presented in this chapter are for shells with simply supported edges. For most applications, simply supported edges should be assumed unless test results are obtained which indicate the effects of the actual boundary condition of the design. The edge of a shell is assumed to be simply supported if at the edge the radial and circumferential displacements are zero and there is no restraint against translation or rotation in the axial direction. For clamped edges, the rotation of the edge is zero.

An attempt has been made to simplify the analysis procedure so that the design-allowable buckling loads may be obtained from hand computations and graphs. In many cases more sophisticated approaches are available, but computer programs are necessary to obtain results. It is not within the scope of this chapter to present an analysis method which requires a computer solution. The references that discuss the more complicated analysis procedure should be obtained if a more detailed investigation is warranted.

In conclusion it is recommended that the analyst attempt to keep abreast of changes in the state of the art of shell buckling, because significant changes may result from recent theoretical and experimental investigations.

10-2 Curved Cylindrical Panels

Introduction

The design-allowable curves presented in this chapter for cylindrical panels are from Ref. 10-1. The buckling curves presented are 90 percent probability level design curves which were obtained statistically from available test data. The theoretical buckling loads and test data for curved panels subjected to various loadings are summarized in Ref. 10-2.

Axial Compression, Curved Panels

Unpressurized: The design-allowable buckling stress for unpressurized curved panels subjected to axial compression is given by

$$\frac{\sigma_{cr}}{\eta} = K_c \frac{\pi^2 E}{12(1-\mu^2)} \left(\frac{t}{b}\right)^2$$

in which b is the width of the panel in the circumferential direction, t is the thickness of the plate, η is the plasticity correction factor, E is Young's modulus of the material, and μ is Poisson's ratio of the material, for elastic buckling $\eta = 1$. For inelastic buckling, the critical stress σ_{cr} may be found using. Sec. 10-7. Design values of the buckling-stress coefficient K_c are given in Fig. 10-1. For simply supported curved panels having a curvature parameter $Z > 30$ and for fixed-edge curved panels

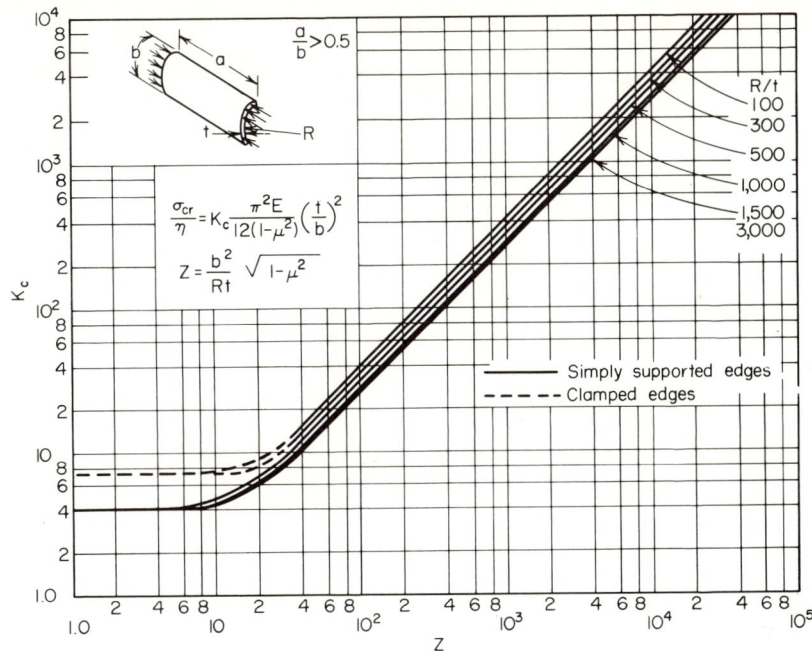

figure 10-1 *Buckling-stress coefficient K_c for unpressurized curved panels subjected to axial compression.*

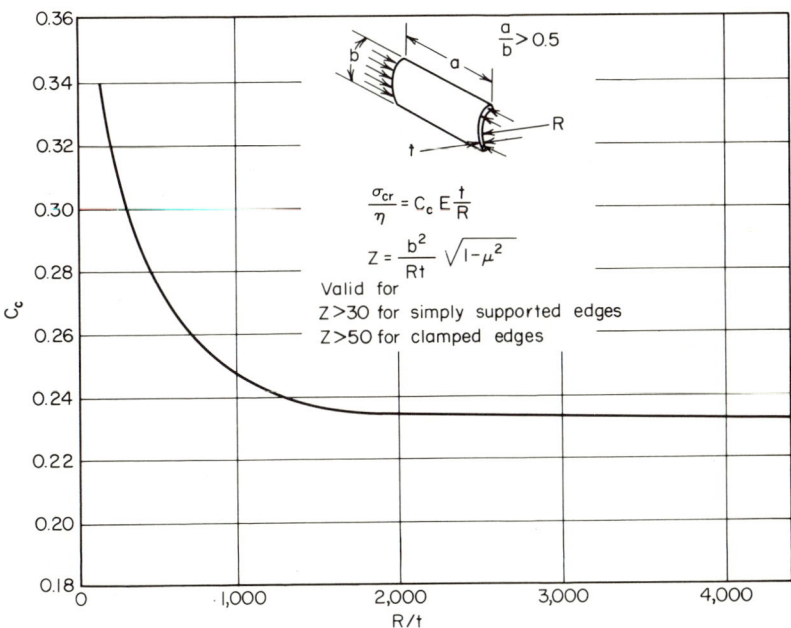

figure 10-2 *Buckling-stress coefficient C_c for unpressurized curved panels subjected to axial compression.*

having a $Z > 50$, Fig. 10-2 may be used instead of Fig. 10-1 to compute the critical stress. The design-allowable buckling stress is then given by

$$\frac{\sigma_{cr}}{\eta} = C_c \frac{Et}{R}$$

in which design values of C_c are given in Fig. 10-2. For elastic buckling, $\eta = 1.0$. For inelastic buckling, the critical stress σ_{cr} may be found by using curves E_1 in Sec. 10-7 on plasticity correction. Note that the design curves in Fig. 10-1 or 10-2 are valid only for $a/b > 0.5$.

Pressurized: The design-allowable buckling stress of curved panels subjected to internal pressure and axial compression may be determined by using Fig. 10-3 together with Fig. 10-2. A curve is presented in Fig. 10-3 that allows the calculation of the increase in

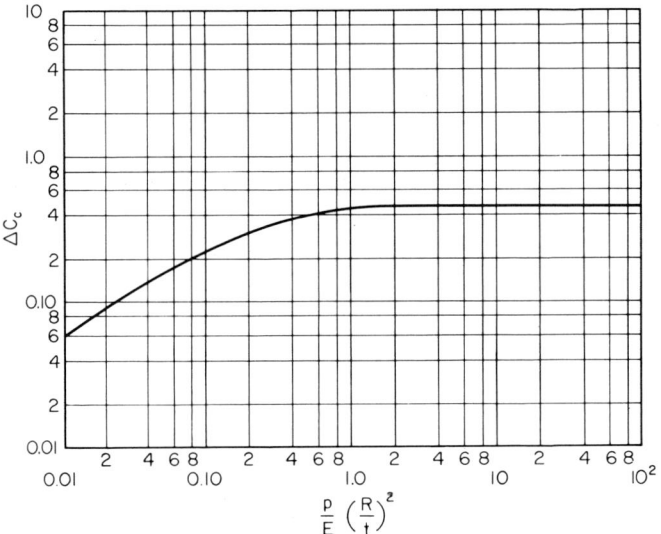

figure 10-3 *Increase in axial-compressive buckling-stress coefficient for curved panels due to internal pressure.*

buckling stress as a function of pressure and geometry only. To calculate the axial compressive buckling stress of a pressurized curved panel, the unpressurized critical stress must first be computed from the design curves in Fig. 10-2 Then, the incremental buckling stress caused by internal pressure is computed by using Fig. 10-3, and this stress is added to the unpressurized value. The design allowable stress is

$$\frac{\sigma_{cr}}{\eta} = (C_c + \Delta C_c) \frac{Et}{R}$$

Stability of Unstiffened Shells

The pressurized curved panel is capable of resisting a total axial compressive load which is the sum of the unpressurized buckling load, the incremental buckling load caused by internal pressure, and an external load sufficient to balance the longitudinal internal-pressure tensile load in the skin. Note that the design curves in Fig. 10-3 are valid only for $a/b > 0.5$. In addition, the curved panels must fall in the domain defined by the straight-line portion of the design curves shown in Fig. 10-1. For inelastic buckling, the critical stress may be found by using curves E_1 of Sec. 10-7. The total stress field should be considered when the platicity correction is determined.

Shear, Curved Panels

Unpressurized: The design-allowable buckling stress for unpressurized, rectangular curved plates subjected to shear is

$$\frac{\tau_{cr}}{\eta} = K_s \frac{\pi^2 E}{12(1-\mu^2)} \left(\frac{t}{b}\right)^2$$

in which the buckling-stress coefficient K_s is given in Figs. 10-4 and 10-5. For elastic buckling, $\eta = 1$. For inelastic buckling, the relation between τ_{cr}/η and τ_{cr} may be determined from the σ_{cr}/η vs. σ_{cr} curves in

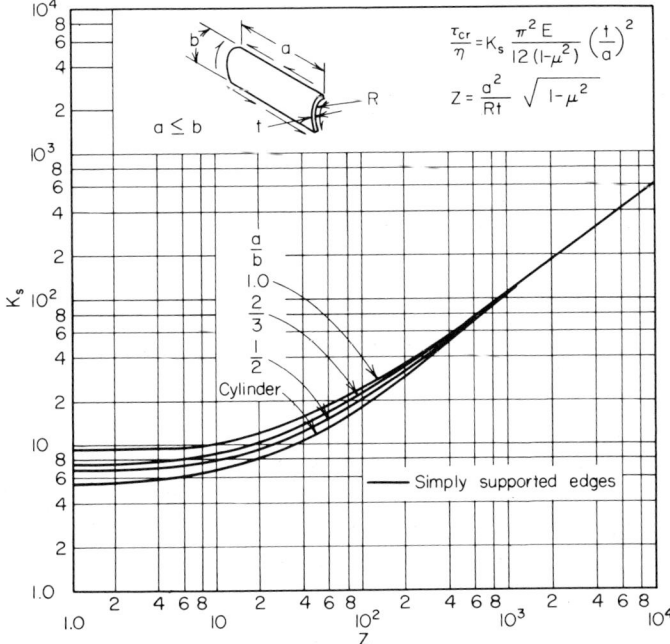

figure 10-4 *Buckling-stress coefficient K_s for unpressurized curved panels subjected to shear.*

Sec. 10-7 on plasticity corrections by using one of the theories for failure of materials (see Chap. 9). There is evidence that the shearing

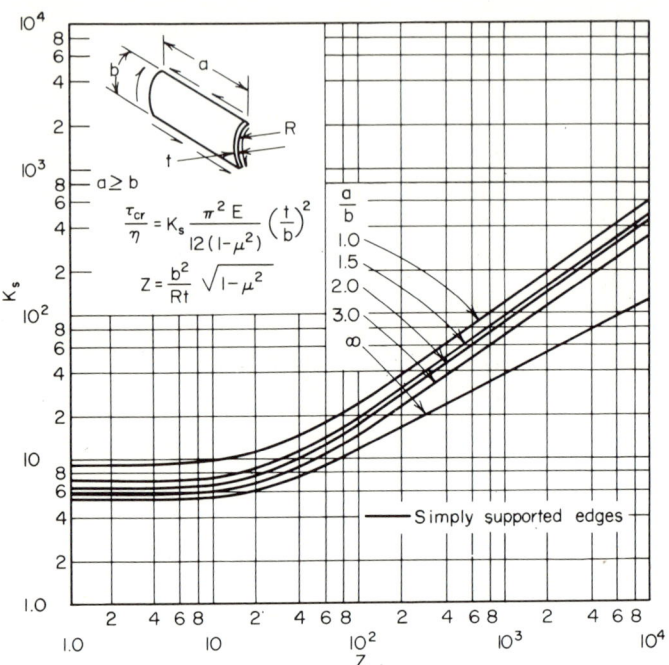

figure 10-5 *Buckling-stress coefficient K_s for unpressurized curved panels subjected to shear.*

stress at which inelastic action occurs in pure shear in ductile materials is related to the analogous tensile stress in pure tension by the equation

$$\tau_{xy} = \frac{1}{\sqrt{3}} \sigma_y$$

The calculated τ_{cr}/η may then be converted to the corresponding σ_{cr}/η by multiplying by $\sqrt{3}$, σ_{cr} may be read from the σ_{cr}/η vs. σ_{cr} curve, and σ_{cr} may be converted back to τ_{cr} by dividing by $\sqrt{3}$. For curved plates in shear, curve A in Sec. 10-7 is suggested.

Pressurized: The design-allowable buckling shear stress for pressurized curved panels may be determined by using Fig. 10-6 with Fig. 10-4 or 10-5. The curves in Fig. 10-6 allow the calculation of the increase in shear buckling stress as a function of pressure and geometry only. To calculate the shear buckling stress of a pressurized curved panel, two quantities must be computed. The unpressurized buckling stress must first be computed from the design curves in Fig. 10-4 or

10-5. Then, the incremental buckling stress caused by internal pressure is computed and added to the unpressurized value.

$$\frac{\tau_{cr}}{\eta} = K_s \frac{\pi^2 E}{12(1-\mu^2)} \left(\frac{t}{b}\right)^2 + \Delta C_s \frac{Et}{R}$$

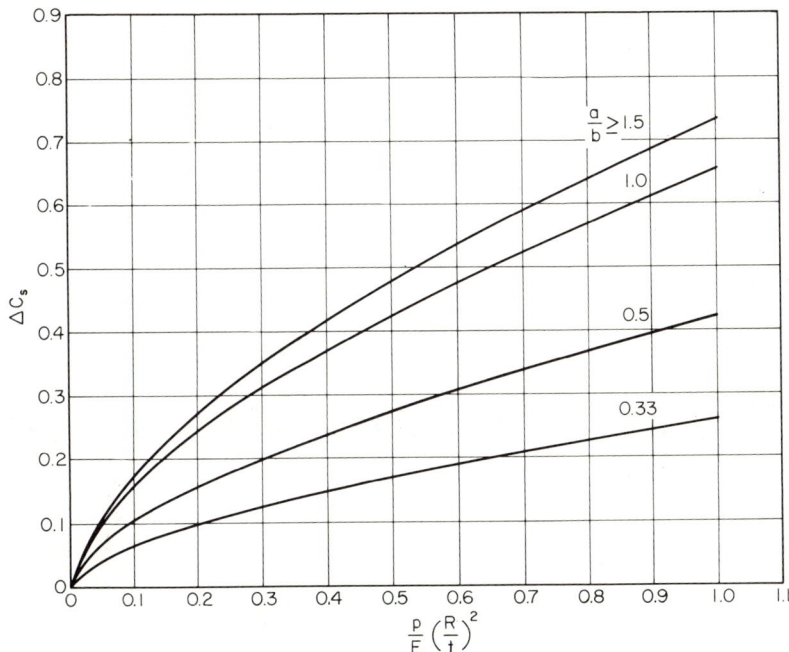

figure 10-6 *Increase in torsional buckling-stress coefficient for curved panels due to internal pressure*

$$\frac{\Delta \tau_{cr}}{\eta} = \Delta C_s E \frac{t}{R}.$$

The design curves should be used for the loading condition in which the axial tensile load caused by internal pressure is not balanced. For inelastic buckling, the critical stress may be found by the procedure recommended for unpressurized curved panels subjected to shear. The total stress field should be considered when the plasticity correction is determined.

Bending, Curved Panels

Test data are not available in sufficiently satisfactory number on the allowable buckling stress of curved plates in bending. However, at low values of the curvature parameter Z, the buckling coefficient for a long, curved plate should approach that for a long, flat plate in bending, and at high values of Z, it should approach that for a long cylinder in bending (Sec. 10-3). These extremes are plotted in Fig. 10-7 with smooth curves faired between. The coefficients are to be used with the equation

$$\frac{\sigma_{cr}}{\eta} = K_b \frac{\pi^2 E}{12(1-\mu^2)} \left(\frac{t}{b}\right)^2$$

For elastic buckling, $\eta = 1$. For inelastic stresses, use the correction suggested for curved panels subjected to axial compression.

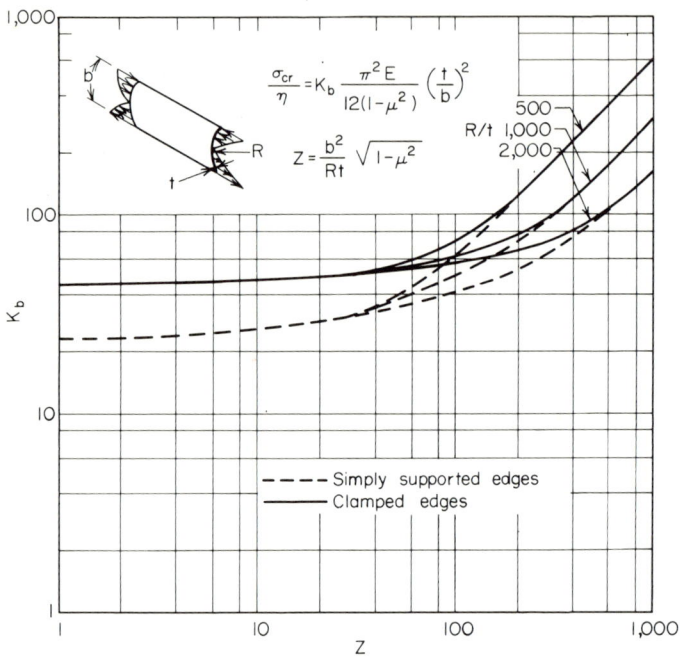

figure 10-7 *Critical buckling-stress coefficients for long, curved panels subjected to bending.*

External Pressure, Curved Panels

Little information is available on the buckling of rectangular plates with single curvature subjected to external pressure. In thin-walled cylinders, external lateral pressure causes buckling by producing a circumferential compressive stress. It is probable that a curved panel which is not shallow may be designed by assuming that it will buckle at a circumferential compressive stress equal to the critical circumferential stress of a thin-walled cylinder of the same proportions. The design-allowable buckling pressure for cylinders subjected to only lateral pressure is given in Sec. 10-3. While edge stiffeners will have generally a stabilizing effect, the panel may be less stable than a geometrically similar cylinder if the stiffeners are torsionally weak and the circumferential load in the skin is not applied to the stiffeners near their shear center.

Combined Loading, Curved Panels

An interaction curve for buckling of rectangular curved plates under combined compression and shear is shown in Fig. 10-8. σ_{cr} is found from Fig. 10-1 or 10-2 and τ_{cr} from Fig. 10-4 or 10-5. To use the curve given in Fig. 10-8, a straight line is drawn through the origin with slope

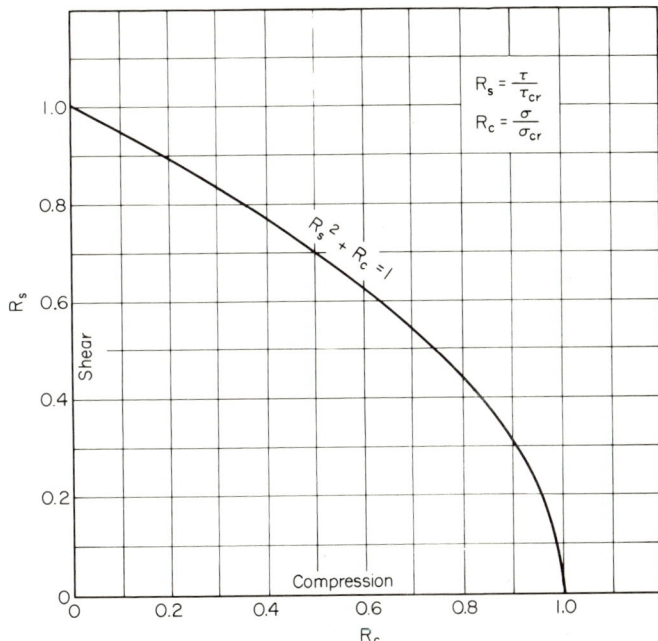

figure 10-8 *Buckling-stress interaction curve for rectangular curved plates under combined shear and compression.*

R_s/R_c, and R_s or R_c is read at the intersection of this line with the given curve.

10-3 Cylinders

Axial Compression, Thin-walled Cylinders

Unpressurized: The design-allowable buckling stress for an unpressurized thin-walled circular cylinder subjected to axial compression is given by the equation

$$\frac{\sigma_{cr}}{\eta} = K_c \frac{\pi^2 E}{12(1-\mu^2)} \left(\frac{t}{L}\right)^2$$

For short cylinders ($\gamma Z < \pi^2 K_{c0}/2\sqrt{3}$), the buckling coefficient may be expressed approximately by the equation

$$K_c = K_{c0} + \frac{12}{\pi^4} \frac{\gamma^2 Z^2}{K_{c0}}$$

and for moderately long cylinders ($\gamma Z > \pi^2 K_{c0}/2\sqrt{3}$) by

$$K_c = \frac{4\sqrt{3}}{\pi^2} \gamma Z$$

where

$$Z = \frac{L^2}{Rt} \sqrt{1 - \mu^2}$$

$K_{c0} = 1$ for simply supported edges. $K_{c0} = 4$ for clamped edges. γ may be obtained from Fig. 10-9.

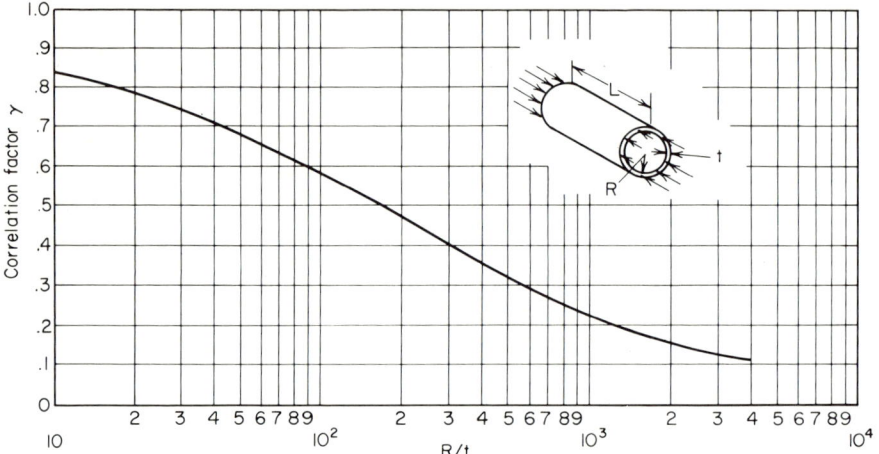

figure 10-9 *Correlation factors for unstiffened unpressurized circular cylinders subjected to axial compression.*

The factor γ has been included to account for the difference between theoretical and experimental results. The classical theoretical value for the buckling coefficient is $\gamma = 1$. The majority of the test data available for cylinders subjected to axial compression are presented in Refs. 10-3 and 10-4. Very few data are available in the short-cylinder range.

The formula for the buckling stress of moderately long cylinders may be rewritten in a more useful form

$$\frac{\sigma_{cr}}{\eta} = \gamma C_c \frac{Et}{R}$$

where

$$C_c = \frac{1}{[3(1-\mu^2)]^{1/2}} \approx 0.6$$

For elastic buckling the value of $\eta = 1$ used. For inelastic buckling of a moderately long cylinder, the critical stress σ_{cr} may be found from curve E_1 in Sec. 10-7.

If the cylinder is very short ($Z\gamma \leq 1$), curve G in Sec. 10-7 should be used. For a cylinder with a γZ that falls between the ranges of very short and moderately long, a linear interpolation between the E_1 and G curves with the Z parameter is sufficiently accurate.

Very long cylinders must be checked for buckling as an Euler column, by the equation

$$\frac{\sigma_{cr}}{\eta} = \frac{\pi^2}{2} cE \left(\frac{R}{L}\right)^2$$

where c = column-fixity coefficient ($c = 1$ for simple support). If the buckling is inelastic, curve G in Sec. 10-7 should be used.

Pressurized: The buckling stress of moderately long cylinders subjected to internal pressure and axial compression may be determined by using Fig. 10-10 in conjunction with Fig. 10-9. Figure 10-10 presents

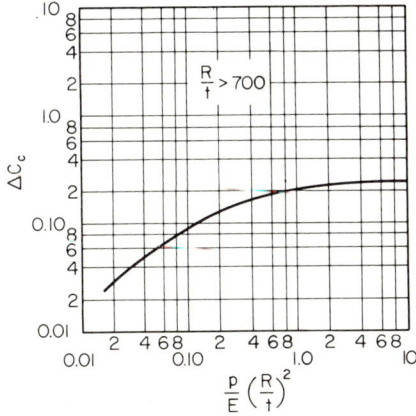

figure 10-10 *Increase in axial-compressive buckling-stress coefficient for cylinders due to internal pressure.*

a curve that allows the calculation of the increase in buckling stress as a function of pressure and geometry only.

The design-allowable buckling stress is

$$\frac{\sigma_{cr}}{\eta} = (\gamma C_c + \Delta C_c)\frac{Et}{R}$$

where γ is obtained from Fig. 10-9 and ΔC_c is obtained from Fig. 10-10. For inelastic buckling, the critical stress may be found by using curves E_1 of Sec. 10-7. The total stress field should be considered when determining the plasticity correction. The pressurized cylinder is capable of resisting a total compressive load P_{cr}, which may be obtained from the equation

$$P_{cr} = 2\pi R t \sigma_{cr} + \pi R^2 p$$

It should be noted that the pressurized design curve in Fig. 10-10 is valid only for moderately long cylinders, and $R/t > 700$. Very long cylinders must be checked for buckling as Euler columns.

Shear or Torsion, Unstiffened Cylinders

Unpressurized: The design-allowable buckling stress of thin-walled circular cylinders of moderate length $100 < Z < 78(R/t)^2(1 - \mu^2)$, subjected to torsion, is given by

$$\frac{\tau_{cr}}{\eta} = C_s \frac{Et}{RZ^{1/4}}$$

For long cylinders, $Z > 78(R/t)^2(1 - \mu^2)$, the design-allowable buckling stress is given by

$$\frac{\tau_{cr}}{\eta} = \frac{0.261 C_s E}{(1 - \mu^2)^{3/4}} \left(\frac{t}{R}\right)^{3/2}$$

The coefficient C_s is given in Fig. 10-11. The classical theoretical value for C_s is 0.735 as derived in Ref. 10-5 for moderate-length cylinders and in Ref. 10-6 for long cylinders. The majority of the test data available for cylinders subjected to torsion are presented in Ref. 10-5.

For elastic buckling, the plasticity correction term $\eta = 1.0$ is used. For inelastic buckling, the critical shear stress τ_{cr} may be found by the procedure outlined in Sec. 10-3.

Pressurized: The shear buckling stress of long thin-walled cylinders subjected to internal pressure and torsion may be determined by using Fig. 10-12 in conjunction with Fig. 10-11. Figure 10-12 presents curves that allow the calculation of the increase in buckling stress as a function of pressure and geometry only.

The design-allowable shear buckling stress is given by

$$\frac{\tau_{cr}}{\eta} = C_s \frac{Et}{RZ^{1/4}} + \Delta C_s \frac{Et}{R}$$

where C_s is obtained from Fig. 10-11 and ΔC_s is obtained from Fig. 10-12.

. Two curves are presented in Fig. 10-12 for calculating the increment in critical stress caused by pressurization. One curve, labeled "No external axial load," should be used for calculating the critical stress of a cylinder

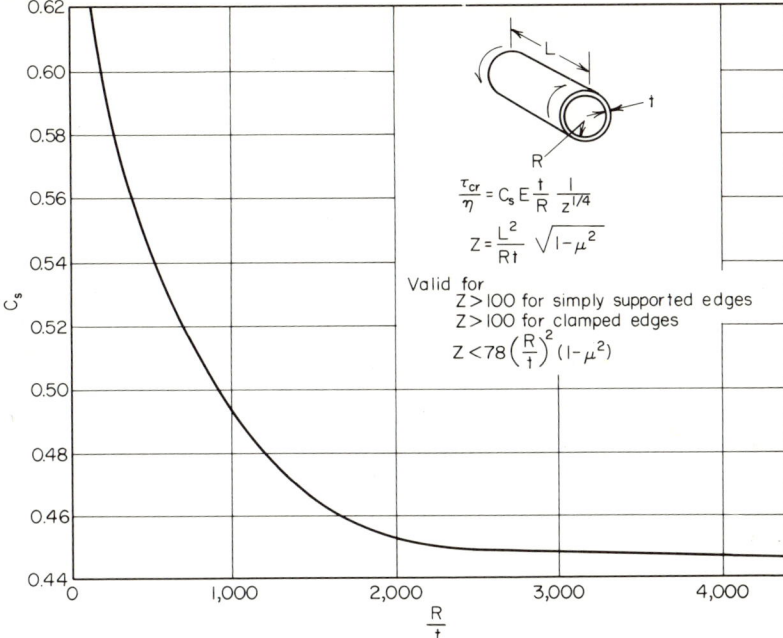

figure 10-11 *Buckling-stress coefficient C_s for unstiffened, unpressurized circular cylinders subjected to torsion.*

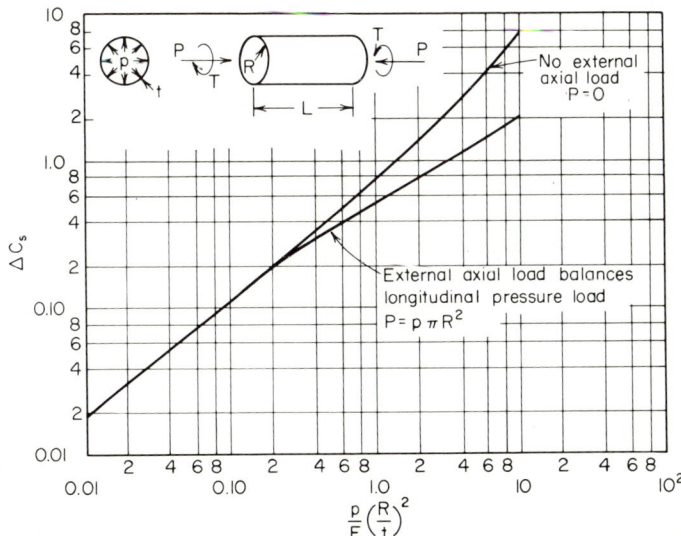

figure 10-12 *Increase in torsional buckling-stress coefficient for cylinders due to internal pressure.*

subjected to torsion and internal pressure only. The second curve, labeled "External axial load balances longitudinal pressure load," should be used to calculate the critical stress of a cylinder subjected to torsion and internal pressure plus an external axial compression load equal to the internal pressure load $\pi R^2 p$, acting on the heads of the cylinder. It should be noted that the pressurized design curves of Fig. 10-12 are valid only for moderately long cylinders. For inelastic buckling the critical shear stress may be obtained by following the procedure outlined in Sec. 10-3. The total stress field should be taken into consideration when the plasticity correction is determined.

Bending, Unstiffened Cylinders

Unpressurized: The design-allowable buckling stress for a thin-walled circular cylinder subjected to bending may be obtained from the equations presented for axial compression if γ is obtained from Fig. 10-13. The classical theoretical buckling stress is

$$\sigma_c = \frac{Et}{R[3(1 - \mu^2)]^{1/2}}$$

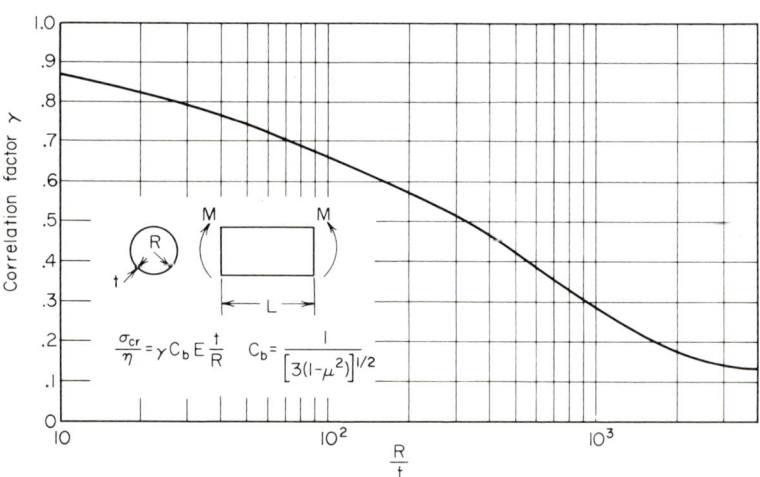

figure 10-13 *Correlation factors for unstiffened unpressurized circular cylinders subjected to bending.*

The majority of the test data available for cylinders subjected to bending are presented in Refs. 10-2 and 10-3. Very few data are available in the short-cylinder range.

Stability of Unstiffened Shells

The formula for the buckling stress of moderately long cylinders subjected to bending may be written in the more useful form

$$\frac{\sigma_{cr}}{\eta} = \gamma C_b \frac{Et}{R}$$

where

$$C_b = \frac{1}{[3(1 - \mu^2)]^{1/2}}$$

σ_{cr} is the maximum stress due to the bending moment (e.g., the outer fiber stress). For elastic buckling, the plasticity correction term $\eta = 1.0$ is used. For inelastic buckling, the critical stress σ_{cr} may be found by using the plasticity corrections suggested in axial compression.

If the stresses are elastic, the allowable moment is

$$M_{cr} = \pi R^2 \sigma_{cr} t$$

Pressurized: The buckling stress of moderately long cylinders subjected to internal pressure and bending may be determined by using Fig. 10-14 in conjunction with Fig. 10-13. Figure 10-14 presents curves

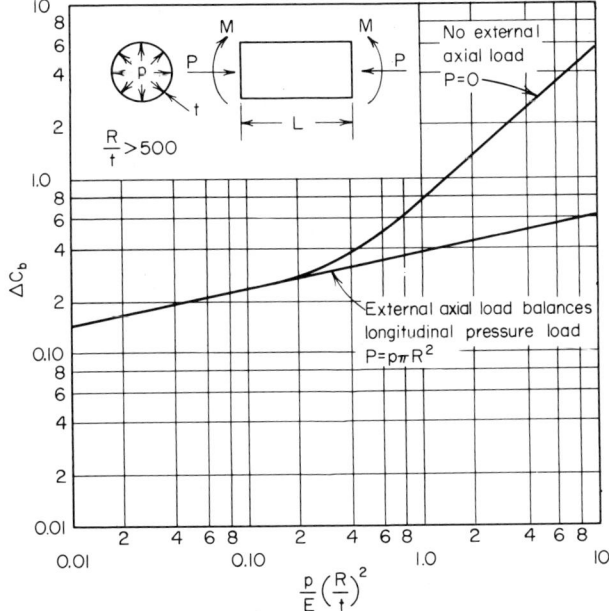

figure 10-14 *Increase in bending buckling-stress coefficient for cylinders due to internal pressure.*

that allow the calculation of the increase in critical stress as a function of pressure and geometry only. The design-allowable buckling stress is

$$\frac{\sigma_{cr}}{\eta} = (\gamma C_b + \Delta C_b) \frac{Et}{R}$$

where γ is obtained from Fig. 10-13 and ΔC_b is obtained from Fig. 10-14.

Two curves for calculating the increment in critical stress caused by pressurization are presented in Fig. 10-14. The curve labeled "No external axial load" should be used to calculate the critical stress of a cylinder subjected to bending and internal pressure only. The curve labeled "External axial load balances longitudinal pressure load" should be used to calculate the critical stress of a cylinder subjected to bending and internal pressure plus an external axial compression load equal to the internal pressure load $\pi R^2 p$ acting on the heads of the cylinder.

If the curve for no axial load is used and the stresses are elastic, the design-allowable moment is

$$M_{cr} = \pi R^2 \left(\sigma_{cr} t + \frac{pR}{2} \right)$$

It should be noted that the pressurized design curves in Fig. 10-14 are valid only for moderately long cylinders. For inelastic buckling, the critical stress σ_{cr} may be found by using curves E_1 in Sec. 10-7. The total stress field should be taken into consideration when determining the plasticity correction.

External Pressure, Unstiffened Cylinders

If a cylindrical shell with simply supported edges is subjected to uniform external pressure p, the design-allowable buckling stress in the circumferential direction is

$$\frac{\sigma_{cr}}{\eta} = K_p \frac{\pi^2 E}{12(1 - \mu^2)} \left(\frac{t}{L} \right)^2$$

The buckling coefficient K_p and a definition of the geometrical parameters are given in Fig. 10-15. For elastic buckling, $\eta = 1$ is used. For moderate-length cylinders ($100 < Z < 11R^2/t^2$) in the inelastic range, Ref. 10-7 recommends

$$\eta = \frac{E_s}{E} \sqrt{\left(\frac{E_t}{E_s} \right)^{1/2} \left(\frac{1}{4} + \frac{3}{4} \frac{E_t}{E_s} \right)}$$

where E_s = secant modulus
E_t = tangent modulus

Stability of Unstiffened Shells   237

For inelastic stresses, σ_{cr} may be obtained for the E_1 curves of Sec. 10-7 because, for most materials, the value of η for the E_1 curves doesn't vary

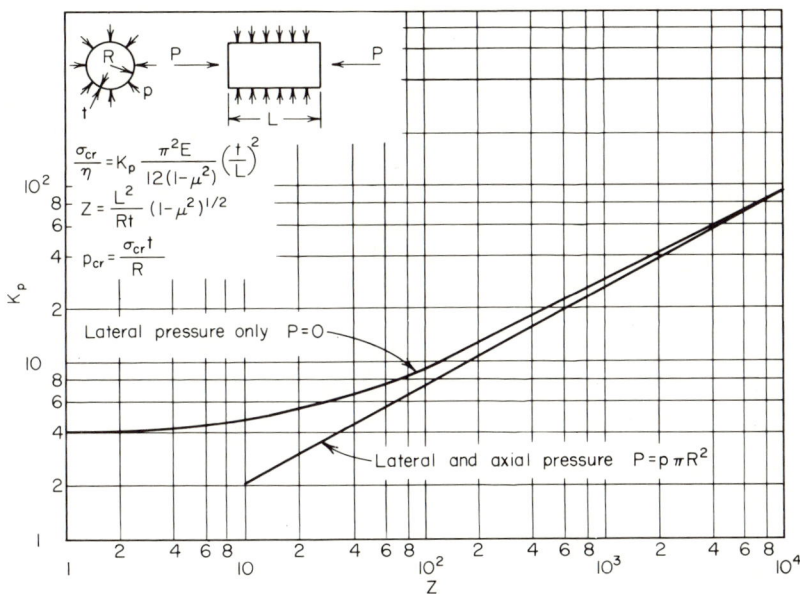

figure 10-15 *Buckling coefficients for circular cylinders subjected to external pressure.*

appreciably from the value obtained from the preceding formula. For short cylinders ($Z < 10$), the C curves of Sec. 10-7 should be used. For $10 < Z < 100$, a linear interpolation between the E_1 and C curves with the Z parameter is probably sufficiently accurate. The theoretical buckling coefficient for $Z > 100$ is given in Ref. 10-5 as $K_p = 1.04 \sqrt{Z}$ for lateral pressure. The theoretical coefficient for lateral and axial pressure is also $K_p = 1.04 \sqrt{Z}$ if $Z > 10^3$ and only slightly different if $10^2 < Z < 10^3$. The majority of test data for cylinders subjected to external pressure are presented in Refs. 10-2, 10-3, and 10-47. For long cylinders (e.g., $L^2/R^2 > 11 R/t$), the design-allowable buckling stress is

$$\frac{\sigma_{cr}}{\eta} = \frac{\gamma E}{4(1-\mu^2)} \left(\frac{t}{R}\right)^2$$

The factor γ was introduced to reduce the theory to a design value. Reference 10-7 recommends $\gamma = 0.9$. For inelastic buckling, Ref. 10-2 recommends

$$\eta = \frac{E_s}{E}\left(\frac{1}{4} + \frac{3}{4}\frac{E_t}{E_s}\right)$$

238   Structural Analysis of Shells

Sufficiently accurate values of σ_{cr} may be obtained by using the E curves of Sec. 10-7.

The design-allowable pressure for any length of cylinder may be obtained from the formula

$$p_{cr} = \frac{\sigma_{cr} t}{R}$$

The pressure p_{cr} is the design-allowable pressure for complete buckling of the shell (e.g., when buckles have formed all the way around the cylinder). For some values of the parameters (large R/t and/or large initial imperfections), single buckles will occur at pressures less than p_{cr}, but complete buckling will occur at higher pressures. Therefore, for some applications these results should be used with caution.

The plasticity correction factors recommended in this chapter were obtained primarily for the case of lateral pressure, but they are probably sufficiently accurate for the case of lateral and axial pressure.

Combined Loading, Unstiffened Cylinders

The criterion for structural failure of a member under combined loading is frequently expressed in terms of a stress-ratio equation $R_1^x + R_2^y + R_3^z = 1$. In general, the stress ratio R is the ratio of the allowable value of the stress caused by a particular kind of load in a combined-loading condition to the allowable stress for the same kind of load when it is acting alone. The subscripts denote the stress due to a particular kind of loading (compression, shear, etc.), and the exponents (usually empirical) express the general relationship of the quantities for failure of the member. The stress ratio is most easily understood if it is defined first for a particular loading condition. In combined compression and torsion loading ($R_c^x + R_s = 1$), the stress ratio R_c is defined as the ratio of compressive stress at which buckling occurs under the combined loading to the compressive stress at which buckling occurs under compression alone. A curve drawn from such a stress-ratio equation is termed a stress-ratio interaction curve. In simple loadings, the term "stress ratio" is used to denote the ratio of applied to allowable stress.

Combined Torsion and Axial Loading: A semiempirical interaction curve for circular cylinders under combined torsion and axial loading is given in Fig. 10-16. σ_{cr} is found using Fig. 10-9, and τ_{cr} is found using Fig. 10-11. In Fig. 10-16, the curves for R/t ratios of 600, 800, and 1,000 were determined by test data. Curves for R/t of 1,500 and 2,000 were drawn by extrapolation.

Bending and Torsion: Test results shown in Ref. 10-2 indicate that a conservative estimate of the interaction for cylinders under combined

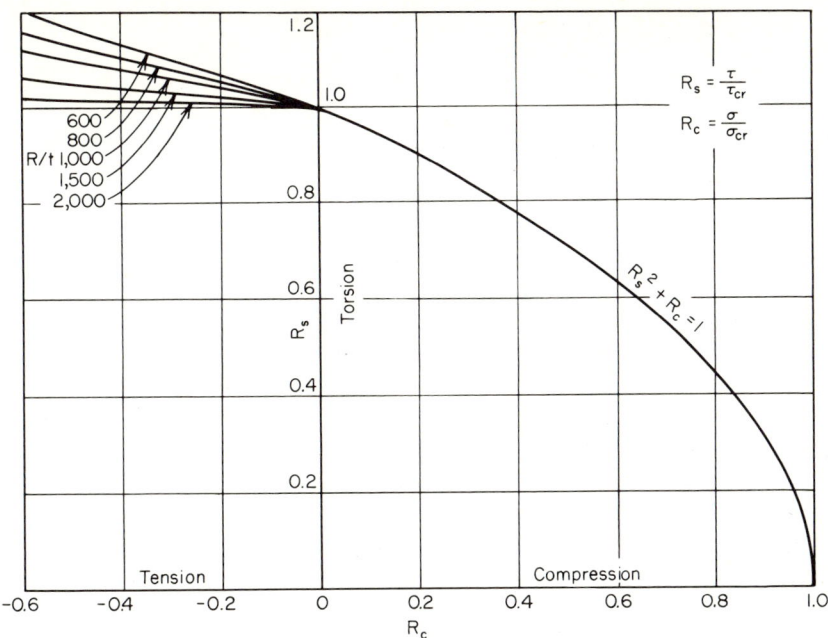

figure 10-16 *Buckling-stress interaction curve for unstiffened circular cylinders under combined torsion and axial loading.*

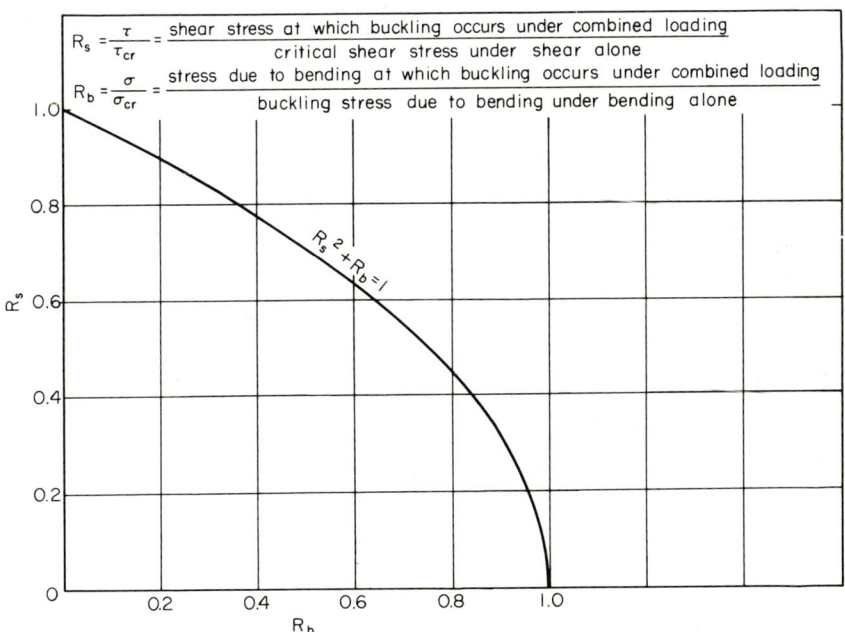

figure 10-17 *Buckling-stress interaction curve for unstiffened circular cylinders under combined bending and torsion.*

240 *Structural Analysis of Shells*

bending and torsion may be obtained from Fig. 10-17; σ_{cr} is found from Fig. 10-13 and τ_{cr} from Fig. 10-11.

Axial Compression and Bending: The test data presented in Refs. 10-2 and 10-3 indicate that the linear interaction for the case of cylinders under combined axial compression and bending shown in Fig. 10-18

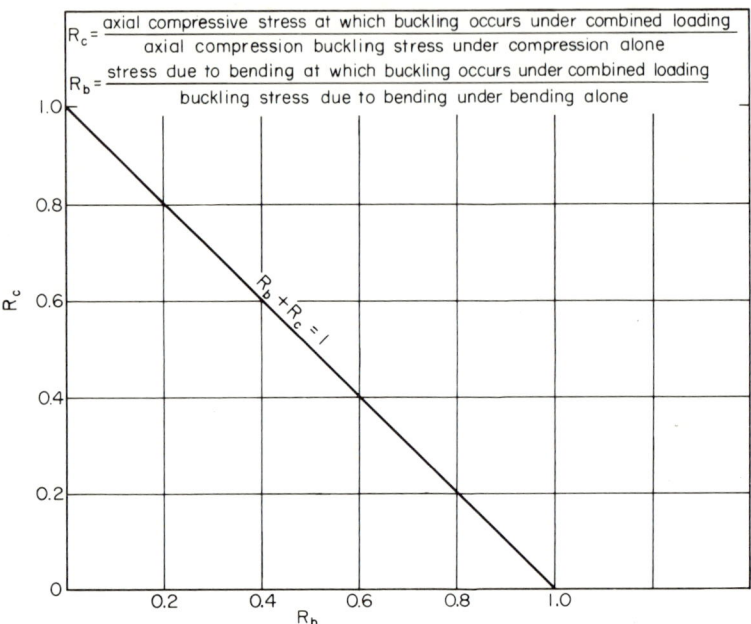

figure 10-18 *Buckling-stress interaction curve for unstiffened circular cylinders under combined axial compression and bending.*

may be used. The buckling stress due to bending alone may be found from Fig. 10-13, and the buckling stress under axial compression alone may be found using Fig. 10-9.

Axial Compression and External Pressure: The limited test data from Ref. 10-3 for cylinders subjected to axial compression and external lateral and axial pressure indicate that the linear interaction curve presented in Fig. 10-19 may be used for design. σ_{cr} is found using Fig. 10-9 and p_{cr} using Fig. 10-15.

Stability of Unstiffened Shells 241

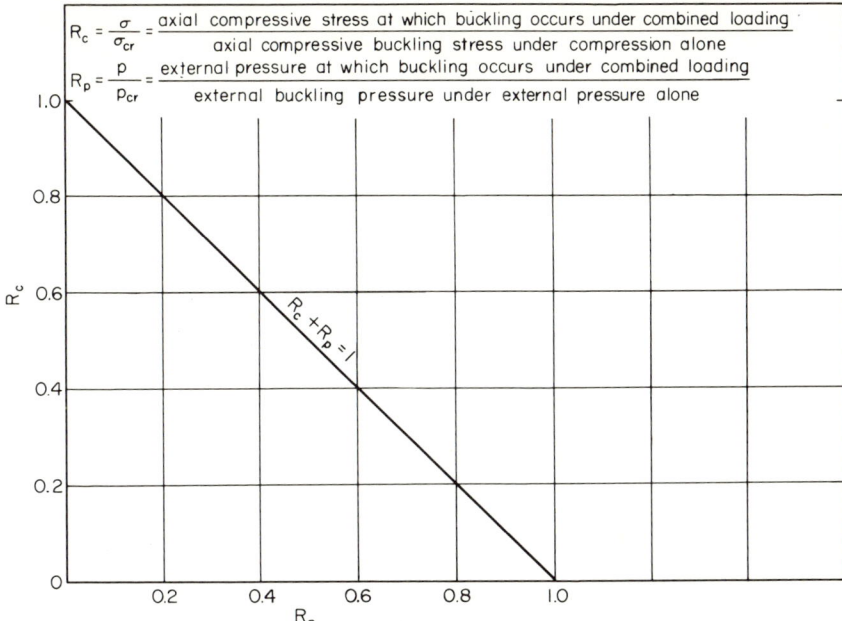

figure 10-19 *Buckling-stress interaction curve for unstiffened circular cylinders under combined external pressure and axial compression.*

Cylinders with an Elastic Core, General

The term "cylinder with an elastic core" defines a thin cylindrical shell enclosing an elastic material that can either be solid or have a hole in its center. This type of shell closely approximates a solid-propellant-filled missile structure. The propellant is generally of a viscoelastic material and therefore is strain-rate-sensitive. The core modulus should be obtained from tension or compression tests of the core material simulating its expected strain rate.

Although there are some analytical data for orthotropic shells (Ref. 10-8), design curves are given only for isotropic unstiffened shells and cores. The inverse problem of a core or cushion on the outside of the cylindrical shell is analyzed in Ref. 10-9. Not enough data are available, however, to recommend design curves for this problem.

Axial Compression, Cylinders with Elastic Core

The buckling behavior of cylindrical shells with a solid elastic core in axial compression is given in Ref. 10-10. Analytical results obtained from this reference are shown graphically in Fig. 10-20. For small values of ϕ_1,

$$\frac{\sigma_p}{\sigma_c} \approx 1 + \phi_1$$

where
$$\sigma_c = \frac{\gamma E}{\sqrt{3(1-\mu^2)}} \frac{t}{R}$$

$$\phi_1 = \frac{\sqrt[4]{12(1-\mu^2)}}{4(1-\mu_c^2)} \frac{E_c}{E} \left(\frac{R}{t}\right)^{3/2}$$

E_c = Young's modulus of core

μ_c = Poisson's ratio of core

This approximation is accurate for ϕ_1 less than $1/2$. For larger values of ϕ_1, say ϕ_1 greater than 3,

$$\frac{\sigma_p}{\sigma_c} \approx \frac{3}{2}(\phi_1)^{2/3}$$

The experimental data tabulated in Ref. 10-10 suggest that the value of γ to be used in calculating σ_c can be taken as that for isotropic cylinders in compression. Then

$$\gamma = 1 - 0.901(1 - e^{-\phi})$$

where

$$\phi = \frac{1}{16}\sqrt{\frac{R}{t}}$$

The plasticity correlation factors given for isotropic cylinders in axial compression may be applied also to the cylinder with an elastic core.

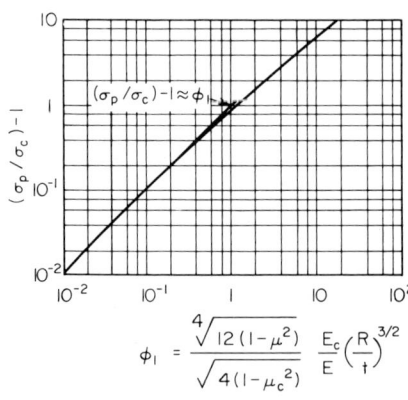

figure 10-20 *Variation of compressive buckling stress with core-stiffness parameter.*

External Pressure, Cylinders with Elastic Core

Analytical curves for the lateral-pressure case are presented in Ref. 10-10. A plot of k_{pc} against $\pi R/L$ for $R/t = 100$, 200, 500, or 1,000 is shown graphically in Fig. 10-21. The parameter k_{pc} is expressed by

$$k_{pc} = \frac{pR^3}{D} \qquad D = \frac{Et^3}{12(1-\mu^2)}$$

where p is the external buckling pressure.

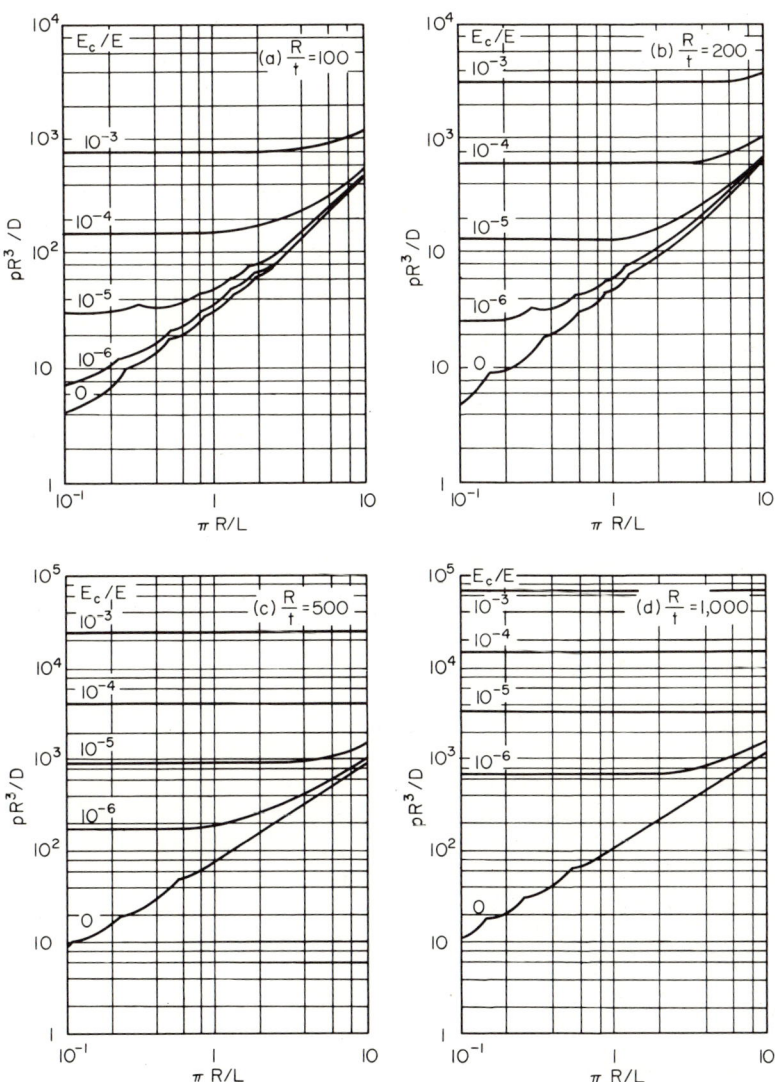

figure 10-21 Variation of buckling-pressure coefficient with length and modulus ratio ($\mu = 0.3$, $\mu_c = 0.5$).

These curves are to be used for finite cylinders loaded by lateral pressure.

Some cylinders are long enough for the critical pressure to be independent of length (Fig. 10-21); the single curve shown in Fig. 10-22

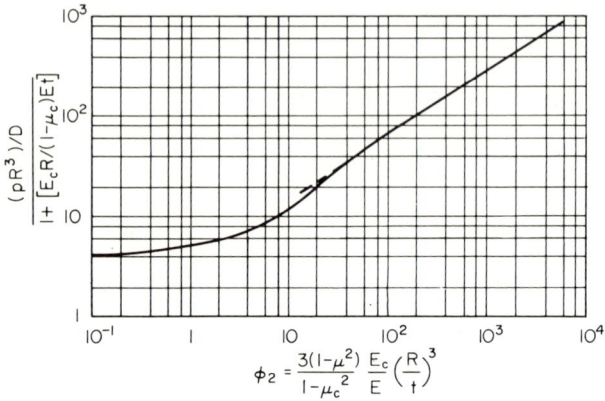

figure 10-22 *Buckling-pressure coefficients for long cylinder with a solid core.*

can then be used. The straight-line portion of the curve can be approximated by the equation

$$\frac{k_{pc}}{1 + E_c R/[Et(1 - \mu_c)]} = 3\phi_2^{2/3}$$

where

$$\phi_2 = \frac{3(1 - \mu^2)}{1 - \mu_c^2} \frac{E_c}{E} \left(\frac{R}{t}\right)^3$$

The few experimental data points available indicate good agreement between analysis and experiment, but one test point falls 4 percent below theory. Hence, a correlation factor of 0.90 is recommended for use in conjuction with the curves in Figs. 10-21 and 10-22; reinvestigation of this factor may be warranted as more data become available.

The plasticity factors given for isotropic cylinders subjected to external pressure may be used also for cylinders filled with an elastic core.

Torsion, Cylinders with Elastic Core

The buckling behavior of cylindrical shells with an elastic core is analytically described in Ref. 10-11 and is shown graphically in Fig. 10-23.

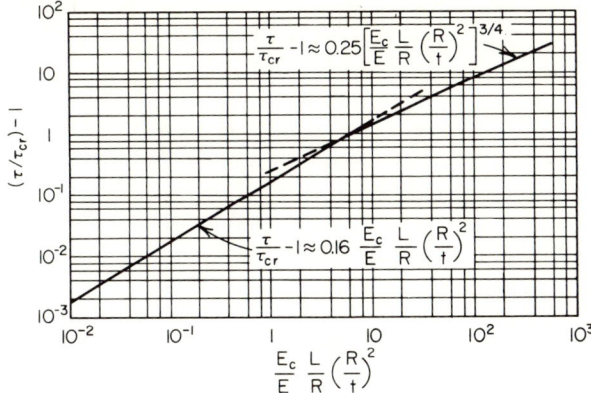

figure 10-23 *Torsional-buckling coefficients for cylinders with an elastic core.*

For small values of ϕ_3 ($\phi_3 < 7$) the analytical results can be approximated by

$$\frac{\tau}{\tau_{cr}} = 1 + 0.16\phi_3$$

where

$$\phi_3 = \frac{E_c}{E}\left(\frac{L}{R}\right)\left(\frac{R}{t}\right)^2$$

τ_{cr} is the torsional buckling stress for a cylinder without the core, and τ is the torsional buckling stress with the core. When ϕ_3 is greater than 10, the analytical results follow the curve

$$\frac{\tau}{\tau_{cr}} = 1 + 0.25\phi_3^{3/4}$$

Experimental data are not available for this loading condition. The experimental points obtained for cylinders with an elastic core for axial compression and external pressure, however, show better correlation with theory than the corresponding experimental results for the unfilled cylinder.

Combined Axial Compression and Lateral Pressure

Interaction curves for cylinders with an elastic core subjected to combined axial compression and lateral pressure are shown in Fig. 10-24. These curves were obtained analytically in Ref. 10-10 and indicate that for a sufficiently stiff core, the critical lateral pressure is insensitive to axial compression. Until more experimental data become available, the use of a straight-line interaction curve is recommended for conservative design.

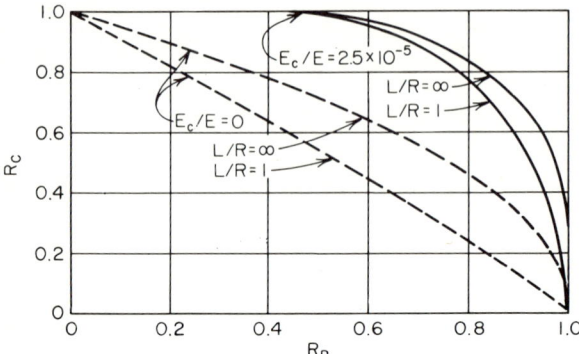

figure 10-24 *Interaction curves for cylinders ($R/t = 300$) with an elastic core.*

10-4 Cones

Axial Compression, Unstiffened Cones

Unpressurized: The equivalent-cylinder approach recommended in Ref. 10-3 will be used for determining the buckling stress for a right circular cone subjected to axial compression. The statistical reduction of experimental cone data presented in Ref. 10-12 indicates that the equivalent-cylinder approach may be conservative for large radius-to-thickness ratios, but the results of Ref. 10-12 would be unconservative for small cone angles (e.g., cones that are almost cylinders). The design-allowable buckling stress may be obtained from the formula

$$\frac{\sigma_{cr}}{\eta} = \gamma C_c \frac{Et}{R_e}$$

where

$$C_c = \frac{1}{[3(1 - \mu^2)]^{1/2}}$$

σ_{cr} is the stress at the small end of the cone. The coefficient γ and a definition of the geometrical parameters are given in Fig. 10-25, as are estimates of the limitations of the buckling equation. The curve of γ vs. R_e/t given in Fig. 10-25 for cones is the same curve given in Fig. 10-9 for cylinders. Test data for cones subjected to axial compression are presented in Refs. 10-12 and 10-13. The theoretical value for σ_{cr} is obtained by setting $\gamma = 1$.

Stability of Unstiffened Shells

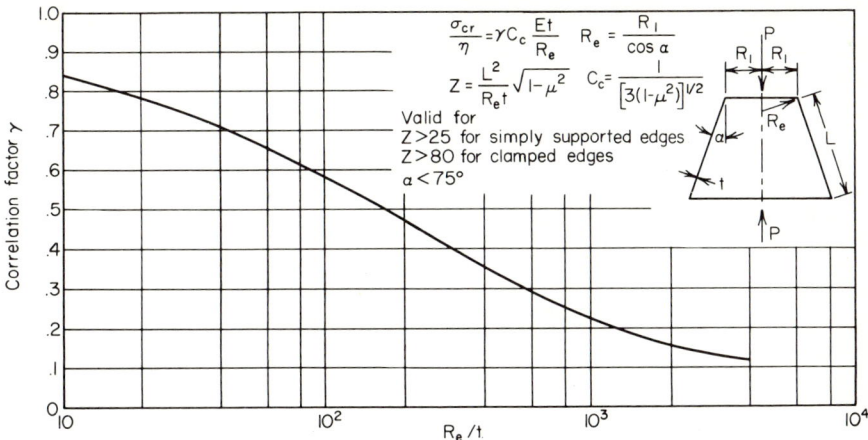

figure 10-25 *Correlation factors for unstiffened, unpressurized circular cones subjected to axial compression.*

For elastic buckling, $\eta = 1$ is used. In the inelastic range, the critical stress σ_{cr} may be found by using curves E_1 in Sec. 10-7. The plasticity correction is based on the largest membrane stress in the cone and may be very conservative for some cases. The design-allowable total compressive load P_{cr} may be obtained from the equation

$$P_{cr} = 2\pi R_e \sigma_{cr} t \cos^2 \alpha$$

Pressurized: The design-allowable buckling stress for cones subjected to internal pressure and axial compression may be determined by using Fig. 10-25 in conjunction with Fig. 10-26, which was obtained from Fig. 10-10. Fig. 10-26 presents a curve that allows the calculation of the increase in buckling stress as a function of pressure and geometry only. The design-allowable buckling stress may be obtained from the formula

$$\frac{\sigma_{cr}}{\eta} = (\gamma C_c + \Delta C_c) \frac{Et}{R_e}$$

where γ is obtained from Fig. 10-25 and ΔC_c is obtained from Fig. 10-26. For elastic buckling, $\eta = 1$ is used. In the inelastic range, the critical stress σ_{cr} may be found by using curves E_1 in Sec. 10-7. The stress in the hoop and axial direction should be taken into consideration when determining the plasticity correction. The pressurized cone is capable of resisting a total compressive load P_{cr}, which may be obtained from the equation

$$P_{cr} = 2\pi R_e \sigma_{cr} t \cos^2 \alpha + \pi R_e^2 p \cos^2 \alpha$$

The P_{cr} found for pressurized cones subjected to axial compression may be conservative for certain values of the parameters, because it has been shown both theoretically and experimentally that internal pressure increases the buckling load of cones more than it increases the buckling load of cylinders. Test data for pressurized cones, however, are too limited to determine an empirical design curve based on parameters from the theoretical buckling analysis of pressurized cones.

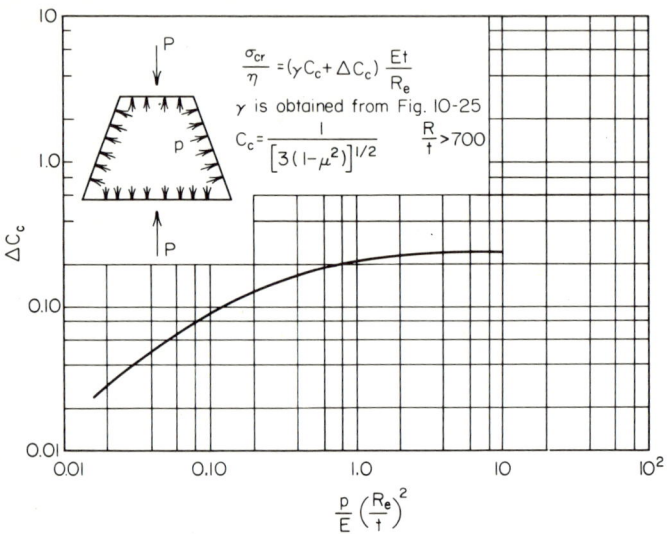

figure 10-26 *Increase in axial compressive buckling-stress coefficient of cones due to internal pressure.*

Shear or Torsion, Unstiffened Cones

Unpressurized: The equivalent-cylinder approach recommended in Ref. 10-14 will be used to determine the buckling stress for a right circular cone subjected to torsion. The design-allowable buckling stress is

$$\frac{\tau_{cr}}{\eta} = \frac{R_e^2}{R_1^2} C_s \frac{Et}{R_e Z^{1/4}}$$

τ_{cr} is the shear stress at the small end of the cone. The buckling-stress coefficient C_s and a definition of the geometrical parameters are given in Fig. 10-27, as are the limitations of the buckling equation. The curve of C_s vs. R_e/t given in Fig. 10-27 for cones is the same curve given in Fig. 10-11 for cylinders. Test data for cones subjected to torsion are presented in Refs. 10-13 and 10-15. For elastic buckling, the plasticity correction term $\eta = 1.0$ is used. For inelastic buckling, the critical shear stress τ_{cr} may be found by using the procedure recommended for curved panels subjected to torsion. The plasticity correction is based

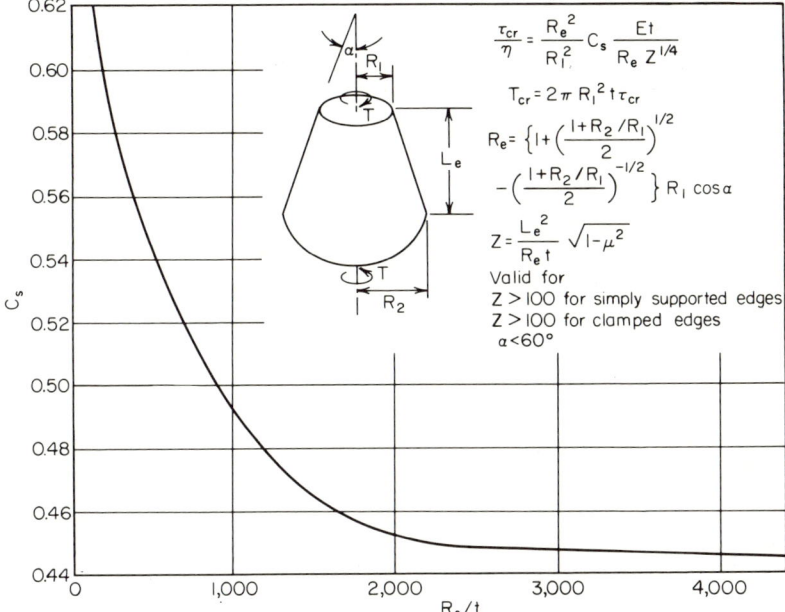

figure 10-27 *Buckling-stress coefficient C_s for unstiffened, unpressurized circular cones subjected to torsion.*

on the largest membrane stress in the cone and may be very conservative for some cases. The design-allowable torque T_{cr} may be obtained from the equation

$$T_{cr} = 2\pi R_1^2 t \tau_{cr}$$

Pressurized: The theoretical results and the test results of Ref. 10-15 show that internal pressure will increase the torsional buckling load of cones, but simple design formulas for computing this increase are not available.

Bending, Unstiffened Cones

Unpressurized: The equivalent-cylinder approach recommended in Ref. 10-3 will be used to determine the buckling stress of a right circular cone subjected to bending. The design-allowable buckling stress is

$$\frac{\sigma_{cr}}{\eta} = \gamma C_b \frac{Et}{R_e}$$

where

$$C_b = \frac{1}{[3(1-\mu^2)]^{1/2}}$$

and σ_{cr} is the maximum stress at the small end of the cone. The coefficient γ and a definition of the geometrical parameters are given in Fig. 10-28, as are estimates of the limitations of the buckling equations. The curve of γ vs. R_e/t given in Fig. 10-28 for cones is the same curve as

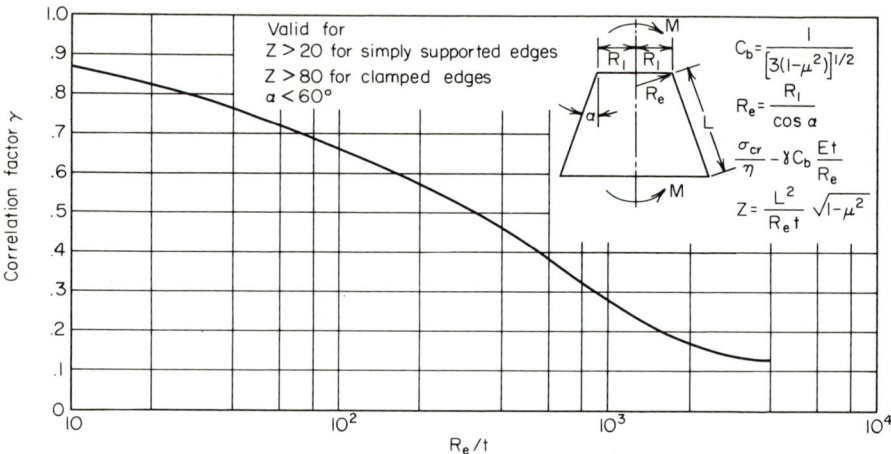

figure 10-28 *Correlation factors for unstiffened, unpressurized circular cones subjected to bending.*

that given in Fig. 10-13 for cylinders. Test data for cones subjected to bending are available in Refs. 10-3 and 10-13. For elastic buckling, $\eta = 1$ is used. In the inelastic range, the critical stress σ_{cr} may be found by using curves E_1 in Sec. 10-7. The plasticity correction is based on the largest membrane stress in the cone and may be very conservative for some cases. If the stresses are elastic, the allowable moment may be obtained from the formula

$$M_{cr} = \pi R_1^2 \sigma_{cr} t \cos \alpha$$

Pressurized: An estimate of the design-allowable buckling stress for a cone under internal pressure and axial compression may be determined by using Fig. 10-28 in conjunction with Fig. 10-29. Fig. 10-29 presents a curve that allows the calculation of the increase in buckling stress as a function of pressure and geometry only. The design-allowable buckling stress may be obtained from the formula

$$\frac{\sigma_{cr}}{\eta} = (\gamma C_b + \Delta C_b) \frac{Et}{R_e}$$

where γ is obtained from Fig. 10-28 and ΔC_b is obtained from Fig. 10-29. For elastic buckling, $\eta = 1$ is used. In the inelastic range, the critical stress σ_{cr} may be found by using the E_1 curves in Sec. 10-7. The stress due

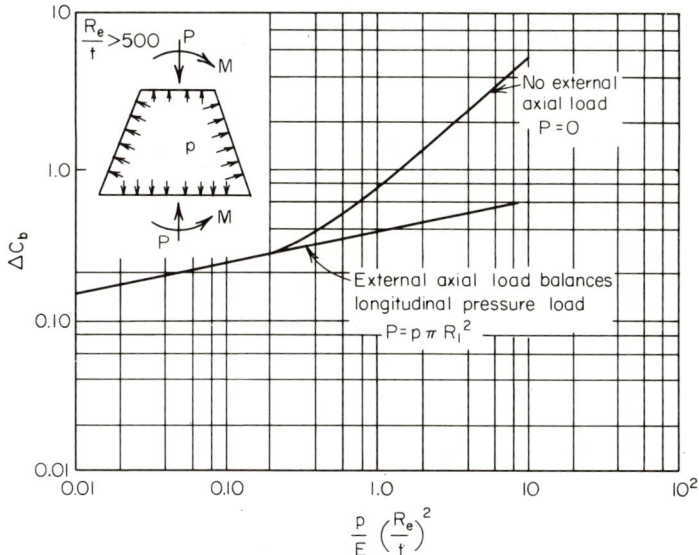

figure 10-29 *Increase in bending buckling-stress coefficient for cones due to internal pressure.*

to bending and pressure should be taken into consideration when the plasticity correction is determined. If the stresses are elastic and no external axial load is applied, the allowable moment may be obtained from the formula

$$M = \pi R_1^2 t \sigma_{cr} \cos \alpha + \pi p \frac{R_1^3}{2}$$

Lateral and Axial External Pressure, Unstiffened Cones

The equivalent cylinder recommended in Ref. 10-3 will be used to determine the design-allowable buckling stress for a circular right cone subjected to lateral and axial external pressure. The design-allowable buckling stress may be obtained from the formula

$$\frac{\sigma_{cr}}{\eta} = K_p \frac{\pi^2 E}{12(1 - \mu^2)} \left(\frac{t}{L}\right)^2 \frac{R_2}{R_e \cos \alpha}$$

σ_{cr} is the circumferential membrane stress at the large end of the cone due to an external pressure p_{cr}. The buckling-stress coefficient K_p and a definition of the geometrical parameters are given in Fig. 10-30. Fig. 10-30 is for simply supported edges and will be conservative for fixed edges. Test data are presented in Refs. 10-3 and 10-13. For elastic buckling, $\eta = 1$ is used. In the inelastic range, the critical stress σ_{cr} may be found by using the method discussed for cylinders subjected to

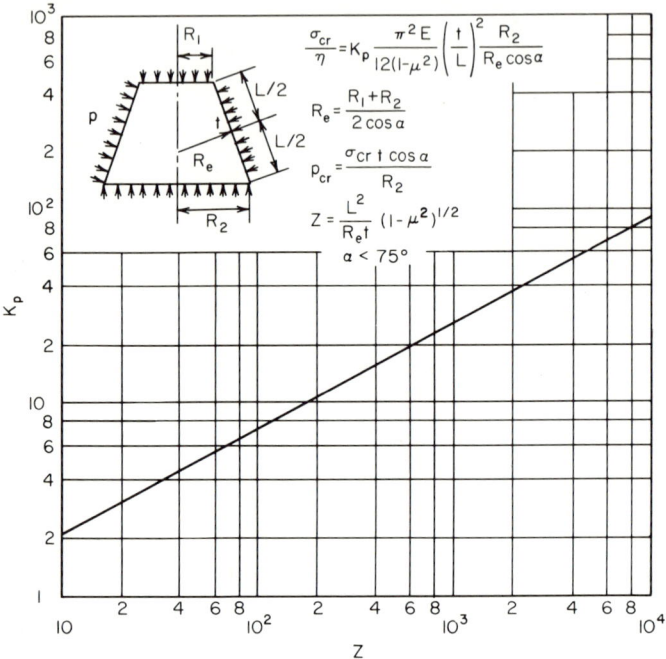

figure 10-30 *Buckling-pressure coefficients for circular cones subjected to external radial and axial pressure.*

external pressure. The plasticity correction is based on the largest membrane stress in the cone and may be very conservative for some cases. The design-allowable external pressure may be obtained from the formula

$$p_{cr} = \frac{\sigma_{cr} t \cos \alpha}{R_2}$$

The pressure p_{cr} is the design-allowable pressure for complete buckling of the shell (e.g., when buckles have formed all the way around the cone). For some values of the parameters (such as large R_e/t or large initial imperfections), single buckles will occur at lower pressures. Therefore, for some applications, these results should be used with caution.

It has been shown in the literature that the critical pressure is a function of the quantity $(1 - R_1/R_2)$, but the effect is generally small and the available information has not been reduced for design purposes.

Combined Loading, Unstiffened Cones

The concept of stress-ratio interaction curves as described for cylinders will also be used for cones.

Axial Compression and Torsion: Reference 10-14 has shown that for unstiffened right conical shells, the curve given in Fig. 10-16 may be used for predicting the interaction between axial compression and torsion. σ_{cr} is found from Fig. 10-25 and τ_{cr} from Fig. 10-27.

Axial Compression and Bending: The very limited test data in Ref. 10-3 indicate that the linear interaction equation shown in Fig. 10-18 may be used for right circular cones subjected to combined axial compression and bending. The buckling stress due to bending alone may be found from Fig. 10-28, and the buckling stress under axial compression alone may be found from Fig. 10-25.

Axial Compression and External Pressure: The limited test data from Ref. 10-3 indicated that the curve given in Fig. 10-19 may be used for right circular cones subjected to axial compression and external lateral and axial pressure. σ_{cr} may be obtained from Fig. 10-25 and p_{cr} from Fig. 10-30.

10-5 Spherical Shells

Uniform External Pressure, Spherical Caps

The buckling of a spherical cap under uniform external pressure (Fig. 10-31) has been treated extensively. Theoretical results are presented in Refs. 10-18 and 10-19 for axisymmetric snap-through of shallow spherical shells with edges that are restrained against translation but are either free to rotate or are clamped. Results for asymmetric buckling are given in Refs. 10-20 and 10-21 for the same boundary conditions. The results reported in these references are presented as the ratio of the buckling pressure p_{cr} for the spherical cap and the classical buckling pressure p_{cl} for a complete spherical shell as a function of a geometry parameter λ:

$$\frac{p_{cr}}{p_{cl}} = f(\lambda)$$

with

$$p_{cl} = \frac{2}{[3(1-\mu^2)]^{1/2}} E \left(\frac{t}{R}\right)^2$$

$$\lambda = [12(1-\mu^2)]^{1/4} \left(\frac{R}{t}\right)^{1/2} 2 \sin\frac{\phi}{2}$$

where ϕ is half the included angle of the spherical cap (Fig. 10-31).

The function $f(\lambda)$ depends on the boundary conditions imposed on the shell.

Most of the available test data apply to spherical shells, and values are lower than theoretically predicted buckling pressures. The discrepancy between theory and experiment can be largely attributed to initial deviations from the ideal spherical shape (Refs. 10-19, 10-22, and 10-23) and to differences between actual and assumed edge conditions (Refs. 10-24 and 10-25). Most of the available data are summarized in Ref. 10-26; some other test results are given in Refs. 10-22 and 10-27. A lower bound to the data for clamped shells is given by

$$\frac{p_{cr}}{p_{cl}} = 0.14 + \frac{3.2}{\lambda^2} \qquad \lambda > 2$$

which is plotted in Fig. 10-32. While the λ parameter is used in shallow-shell analysis, Fig. 10-32 may be applied to deep shells as well as to shallow shells.

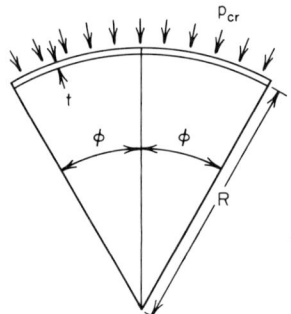

figure 10-31 *Geometry of spherical cap under uniform external pressure.*

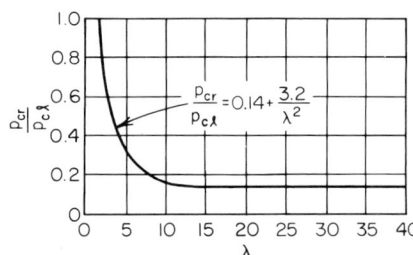

figure 10-32 *Recommended design buckling pressure of spherical caps.*

Concentrated Load at the Apex, Spherical Caps

Spherical caps under concentrated load at the apex (Fig. 10-33) will buckle under certain conditions. Theoretical results for edges that are free to rotate and to expand in the direction normal to the axis of revolution and for clamped edges are given in Ref. 10-28 for axisymmetric snap-through and in Refs. 10-17 and 10-29 for asymmetric buckling. Experimental results for loads which approximate concentrated loading are described in Refs. 10-30 to 10-34.

For shells with unrestrained edges, buckling will not occur if λ is less than about 3.8. In this range of shell geometry, deformation will increase with increasing load until collapse resulting from plasticity effects occurs. For shells with values of λ greater than 3.8, theoretical and experimental results are in good agreement for axisymmetric snap-through but disagree when theory indicates that asymmetric buckling should occur first. In this case, buckling and collapse are apparently not synonymous, and only collapse loads have been measured. A lower-bound relationship between the collapse-load parameter and the geometry parameter for the data of Refs. 10-17, 10-30, and 10-31 for shells with unrestrained edges is given by

$$\frac{p_{cr}R}{Et^3} = \frac{1}{24}\lambda^2 \qquad 4 \leqslant \lambda \leqslant 18$$

For spherical caps with clamped edges, theory indicates that buckling will not occur if λ is less than about 8. For values of λ between 8 and 9, axisymmetric snap-through will occur, with the shell continuing to carry increasing load. For larger values of λ, asymmetrical buckling will occur first, but the shell will continue to carry load. Although imperfections influence the initiation of symmetric or asymmetric buckling, few measurements have been made of the load at which symmetric or asymmetric deformations first occur. Experimental results indicate that the collapse loads of clamped spherical caps loaded over a small area are conservatively estimated by the loads calculated in Ref. 10-17 and shown in Fig. 10-34. When the area of loading becomes large, local buckling may occur at a lower load.

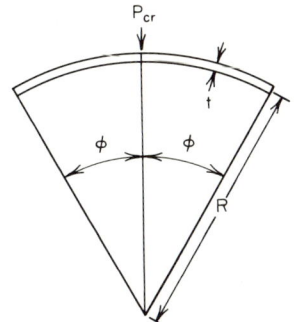

figure 10-33 *Geometry of spherical cap under concentrated load at the apex.*

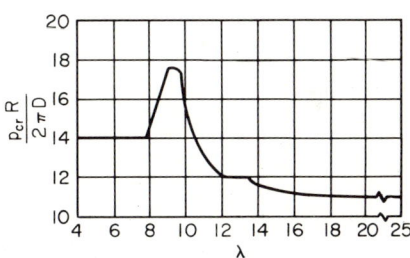

figure 10-34 *Theoretical buckling loads for clamped spherical cap under concentrated load.*

Uniform External Pressure and Concentrated Load at the Apex, Spherical Caps

Clamped spherical caps subjected to combinations of uniform external pressure and concentrated load at the apex are discussed in Ref. 10-35. The experimental and theoretical data given there are insufficient, however, to yield conclusive results. A straight-line interaction curve is recommended:

$$\frac{P}{P_{cr}} + \frac{p}{p_{cr}} = 1$$

where P is the applied concentrated load, p the applied uniform pressure, P_{cr} the critical concentrated load without external pressure, and p_{cr} the critical uniform external pressure without a concentrated load.

10-6 Other Shapes

Uniform External Pressure, Complete Ellipsoidal Shells

Ellipsoidal shells of revolution subjected to uniform external pressure are shown in Fig. 10-35. Calculated theoretical buckling pressures for prolate spheroids are shown in Figs. 10-36 and 10-37. Experimental results given in Ref. 10-36 for prolate spherical shells with $4 > A/B > 1.5$ are in reasonably close agreement with the theoretical results. For $A/B \geqslant 1.5$, the theoretical pressure should be multiplied by the factor 0.75 to provide a lower bound to the data. Results given in Ref. 10-37 for half of a prolate spheroidal shell $(A/B = 3)$ closed by an end plate are in good agreement with those for the complete shell.

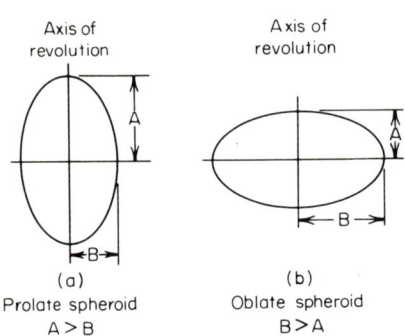

figure 10-35 *Geometry of ellipsoidal shells.*

The analysis indicates that theoretical results for thin, oblate spheroidal shells are similar to those for a sphere of radius

$$R_A = \frac{B^2}{A}$$

The data of Ref. 10-38 show that experimental results are similar, as well. Thus, the external buckling pressure for a thin, oblate spheroid may be approximated by the relationship

$$\frac{\sqrt{3(1-\mu^2)}}{2}\left(\frac{R_A}{t}\right)^2 \frac{p}{E} = 0.14$$

which is the limit for a sphere as λ becomes large.

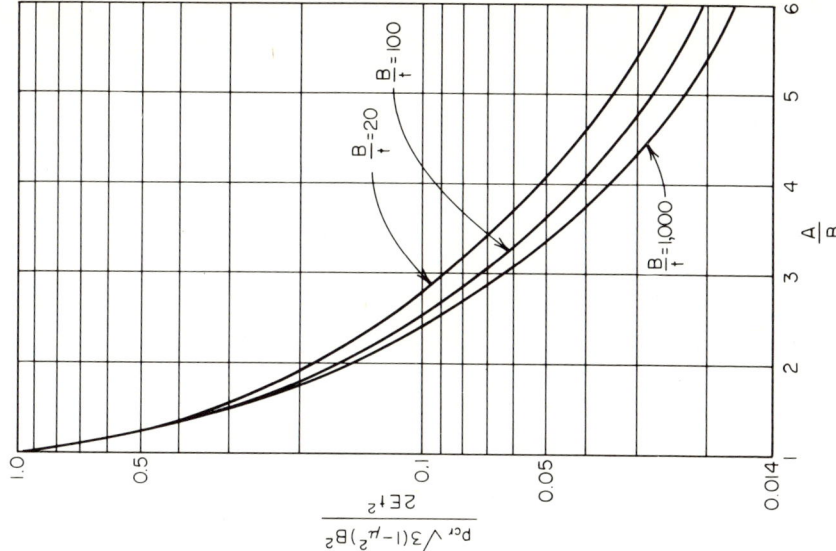

figure 10-37 *Theoretical external buckling pressures of prolate spheroids ($\mu = 0.3$).*

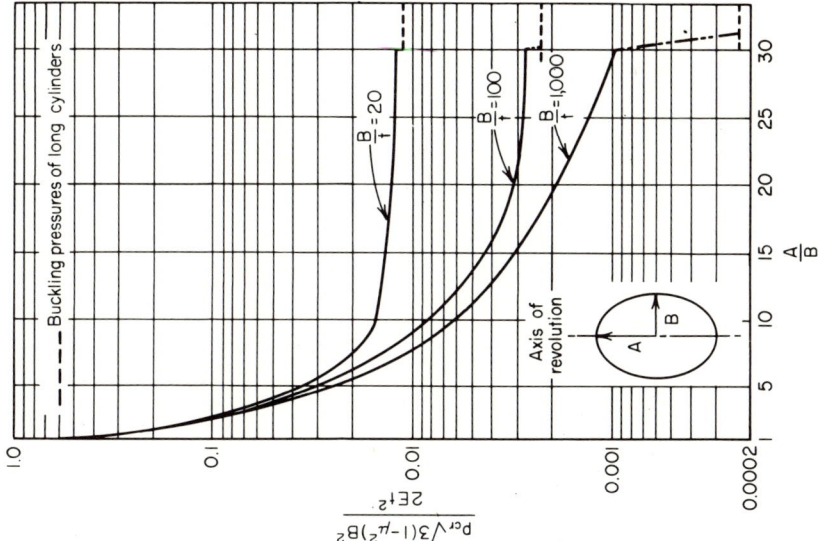

figure 10-36 *Theoretical external buckling pressures of prolate spheroids ($\mu = 0.3$).*

257

Uniform Internal Pressure, Complete Oblate Spheroidal Shells

When the radius ratio A/B of an oblate spheroid is less than $\sqrt{2}/2$, internal pressure produces compressive stresses in the shell and hence allows instability to occur. Theoretical values of the critical internal pressures are shown in Fig. 10-38. No experimental results are available,

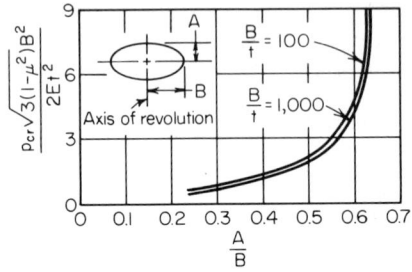

figure 10-38 *Theoretical buckling pressures of oblate spheroids under internal pressure* ($\mu = 0.3$).

but the study of imperfection sensitivity indicates that there should be good agreement between theory and experiment for shells with $0.5 < A/B < 0.7$.

Internal Pressure, Ellipsoidal and Toroidal Bulkheads

Clamped oblate spheroidal (ellipsoidal) bulkheads (Fig. 10-39) may have the ratio of length of minor and major axes A/B less than $\sqrt{2}/2$ without buckling under internal pressure, provided that the thickness exceeds a certain critical value. This problem is investigated in Ref. 10-39. Nonlinear bending theory is used to determine the prebuckling stress distribution. The regions of stability are shown in Fig. 10-40; the calculated variation of buckling pressure with thickness

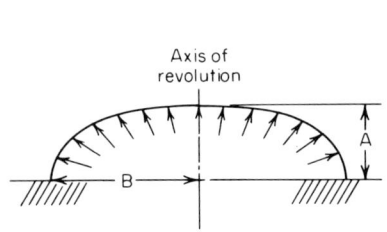

figure 10-39 *Clamped ellipsoidal bulkhead under internal pressure.*

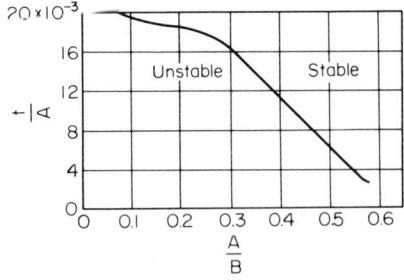

figure 10-40 *Region of stability for ellipsoidal closures subjected to internal pressure* ($\mu = 0.3$).

is shown in Fig. 10-41. The theory has not been verified by experimental results, however, and should be used with caution.

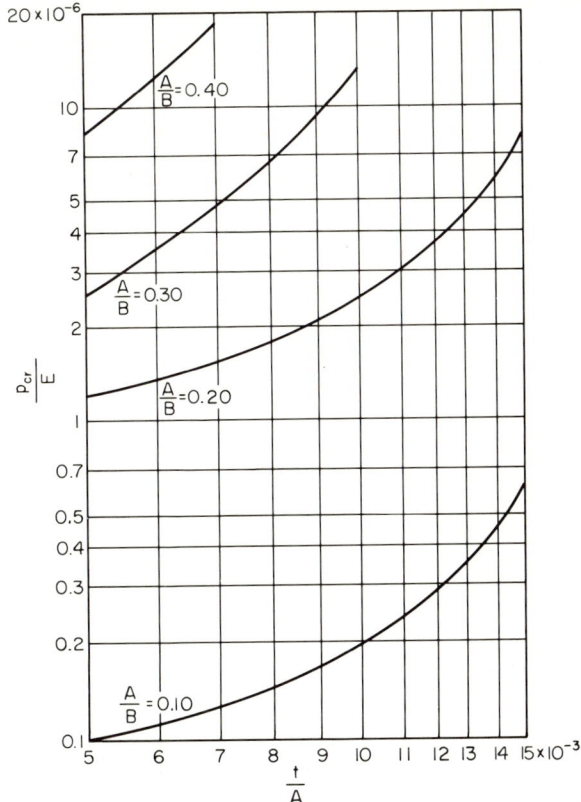

figure 10-41 *Theoretical results for clamped ellipsoidal bulkheads subjected to uniform internal pressure* ($\mu = 0.3$).

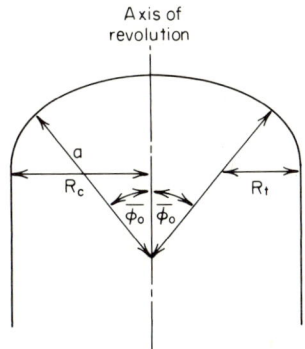

figure 10-42 *Geometry of torispherical closure.*

Torispherical end closures, shown in Fig. 10-42, are also investigated in Ref. 10-39. Calculations are made for the prebuckling stress distribution in these bulkheads for ends restrained by cylindrical shells and for buckling pressures for torispherical bulkheads with clamped edge conditions after buckling. The results are shown in Fig. 10-43. The experimental results of Ref. 10-40 indicate that the theoretically predicted buckling pressures should be multiplied by a correlation factor γ equal to 0.7.

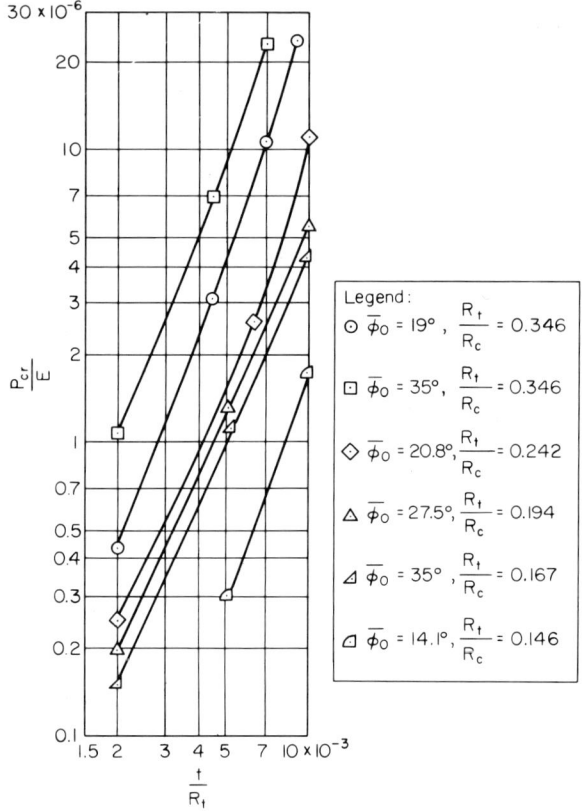

figure 10-43 *Theoretical results for torispherical closures subjected to uniform internal pressure ($\mu = 0.3$).*

Uniform External Pressure, Complete Circular Toroidal Shells

The complete circular toroidal shell under uniform external pressure (Fig. 10-44) has been investigated and is described in Ref. 10-41; the theoretical results obtained are shown in Fig. 10-45.

Experimental results are given in Ref. 10-41 for values of b/a of 6.3 and 8, and indicate good agreement with theory. For values of b/a equal to or greater than 6.3, the theoretical buckling pressure should be

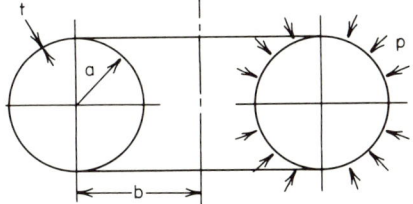

figure 10-44 *Geometry of a toroidal shell under uniform external pressure.*

multiplied by a factor of 0.9 to yield design values. This correction factor has been recommended in Ref. 10-7 for long cylindrical shells, which correspond to a value of b/a of ∞. For values of b/a less than 6.3, the buckling pressure should be verified by test.

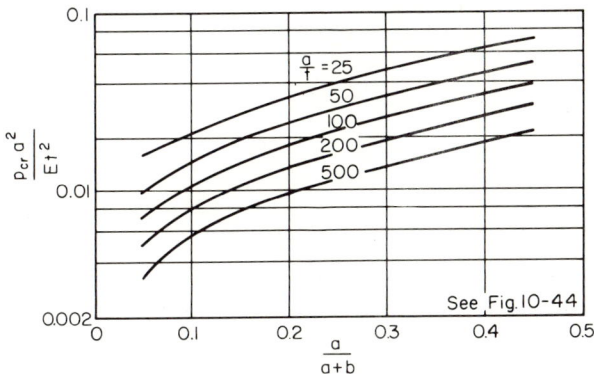

figure 10-45 *Theoretical buckling coefficients for toroidal shells under uniform external pressure.*

Axial Loading, Shallow Bowed-out Toroidal Segments

A bowed-out equatorial toroidal segment under axial tension (Fig. 10-46) will undergo compressive circumferential stress and will thus be susceptible to buckling. An analysis for simply supported shallow segments is given in Ref. 10-42 and yields the relationship

$$\frac{Nl^2}{\pi^2 D} = \frac{1}{(r/a)\beta^2 - 1}\left\{(1+\beta^2)^2 + 12\frac{\gamma^2 Z^2}{\pi^4}\left(\frac{1+(r/a)\beta^2}{1+\beta^2}\right)^2\right\}$$

262 *Structural Analysis of Shells*

where

$$N = \text{running load per unit length}$$

$$Z = \frac{l^2(1-\mu^2)^{1/2}}{rt}$$

$$D = \frac{Et^2}{12(1-\mu^2)}$$

a, r, l are defined in Fig. 10-49

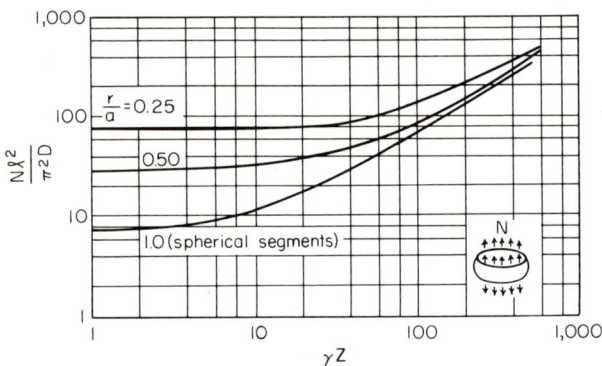

figure 10-46 *Buckling of bowed-out toroidal segments under axial tension.*

The correlation coefficient γ has been inserted to account for discrepancies between theory and experiment. The values obtained by minimizing the equation with respect to β, which is the buckle wavelength parameter, are shown in Fig. 10-46. The straight-line portion of the curves is represented by the relationship

$$\frac{Nl^2}{\pi^2 D} = \frac{4\sqrt{3}}{\pi^2} \gamma Z$$

A similar analytical investigation described in Ref. 10-43 for clamped truncated hemispheres in axial tension yields results in close agreement with those for the curve of Fig. 10-46 for $r/a = 1$.

Experimental results for the truncated hemisphere given in Ref. 10-43 indicate that the correlation coefficient for the curve for r/a equal to 1 is $\gamma = 0.35$. The same value of the correlation coefficient may be used for other values of r/a.

Some results for bowed-out equatorial toroidal segments under axial compression are given in Ref. 10-44; the equatorial spherical shell segment loaded by its own weight is treated in Ref. 10-45.

Uniform External Pressure, Shallow Toroidal Segments

The term "lateral pressure" designates an external pressure which acts only on the curved walls of the shell and not on the ends; "hydrostatic pressure" designates an external pressure that acts on both the curved walls and the ends of the shell. Expressions for simply supported shallow equatorial toroidal segments subjected to uniform external lateral or hydrostatic pressure, as shown in Figs. 10-47 and 10-48, are given in Ref. 10-46 as

$$\frac{p_{cr}rl^2}{\pi^2 D} = \frac{1}{\beta^2} \left\{ (1+\beta^2)^2 + \frac{12}{\pi^4} \gamma^2 Z^2 \left[\frac{1 \pm (r/a)\beta^2}{1+\beta^2} \right]^2 \right\}$$

for lateral pressure, and as

$$\frac{p_{cr}rl^2}{\pi^2 D} = \frac{1}{\beta^2 \left(1 \mp \frac{1}{2}\frac{r}{a}\right) + \frac{1}{2}} \left\{ (1+\beta^2)^2 + \frac{12}{\pi^4} \gamma^2 Z^2 \left[\frac{1 \pm (r/a)\beta^2}{1+\beta^2} \right]^2 \right\}$$

for hydrostatic pressure. Definitions of *a*, *r*, *l* are given in Fig. 10-49.

$$Z = \frac{l^2(1-\mu^2)^{1/2}}{rt} \qquad D = \frac{Et^3}{12(1-\mu^2)}$$

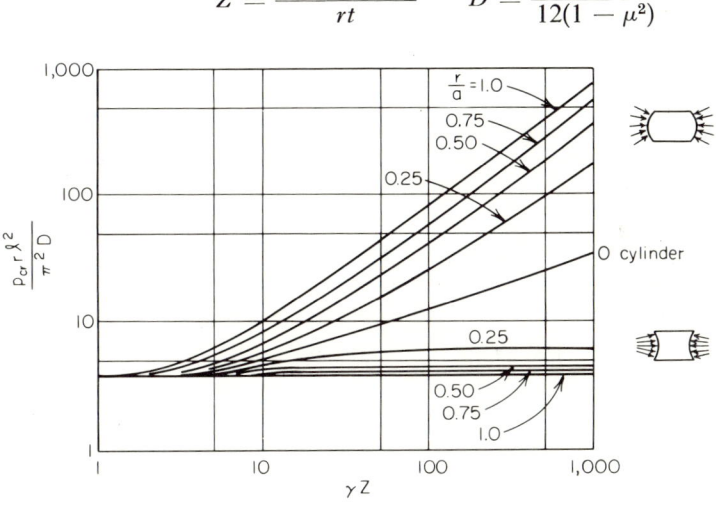

figure 10-47 *Buckling of toroidal segments under uniform lateral pressure.*

The upper sign in these equations refers to segments of type (*a*) of Fig. 10-49, while the lower sign refers to segments of type (*b*) of Fig. 10-49. The correlation coefficient γ has been introduced to account for discrepancies between theory and experiment. The results of minimizing the buckling pressure with respect to the circumferential wavelength parameter β are shown in Figs. 10-47 and 10-48. The straight-line portions of the curve for the shells of type (a) of Fig. 10-49 are represented by the relationships

$$\frac{p_{cr}rl^2}{\pi^2 D} = \frac{4\sqrt{3}}{\pi^2}\frac{r}{a}\gamma Z \qquad \text{lateral pressure}$$

$$\frac{p_{cr}rl^2}{\pi^2 D} = \frac{8\sqrt{3}}{2-r/a}\left(\frac{r}{a}\right)\gamma Z \qquad \text{hydrostatic pressure}$$

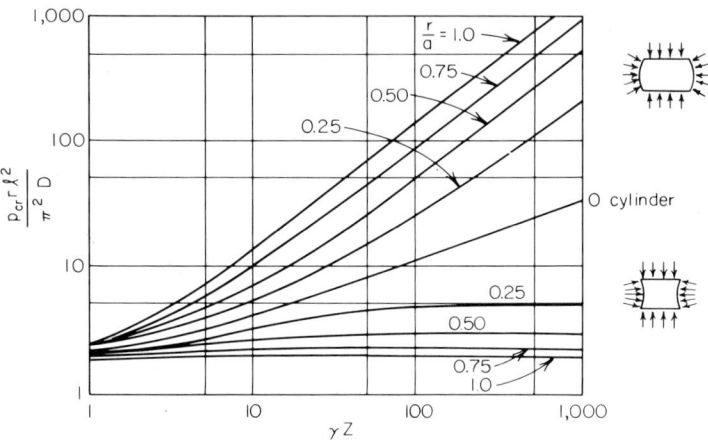

figure 10-48 *Buckling of toroidal segments under uniform external hydrostatic pressure.*

No experimental data are available except for the cylindrical shell for which a correlation factor of $\gamma = 0.56$ was recommended in Ref. 10-7. The same correlation factor can be used for shells with r/a near zero but should be used with caution for shells of type (b) with values of r/a near unity. For shells of type (a) with values of r/a near unity, the shell can be conservatively treated as a sphere, or the buckling pressure should be verified by test.

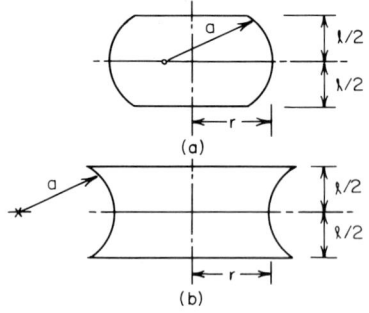

figure 10-49 *Geometry of toroidal segments near equators.*

10-7 Inelastic Buckling

General, Inelastic Buckling

The formulas presented to determine the allowable buckling load are based primarily on theoretical results which have assumed that the compressive modulus of the material is a constant. If the buckling stress is below the proportional limit, this is a reasonable assumption; if the stresses are in the inelastic range, however, the modulus of the material becomes a function of the stresses. The modulus of the material decreases at inelastic stresses; therefore, there is a decrease in the stiffness of the shell and a corresponding decrease in the buckling load.

The Euler formula, which was derived for an elastic column, is used for the case of inelastic buckling of a column. However, the elastic modulus in the formula is replaced by the tangent modulus of the material. The agreement between the predicted buckling stress and test data has been quite good. It is considerably more difficult to include the effects of plasticity for shells. Methods have been developed, but in general, they are quite complicated, and computer programs are needed to obtain results. Plasticity correction factors derived for some types of loadings and shells are briefly discussed in the next section. Until additional information is available, this method is recommended as a simple way to account for the effects of plasticity on the buckling load.

Plasticity Correction Factor

The effect of plasticity on the buckling of shells can be accounted for by the use of the plasticity coefficient η. This coefficient is defined by the ratio

$$\eta = \frac{\sigma_{cr}}{\sigma_e}$$

where σ_{cr} = actual buckling stress
σ_e = elastic buckling stress (the stress at which buckling would occur if the modulus remained constant at any stress level)

The elastic buckling stress, therefore, is given by the equation

$$\sigma_e = \frac{\sigma_{cr}}{\eta}$$

The definition of η depends on σ_{cr}/σ_e, which is a function of the loading, the type of shell, the boundary conditions, and the type of construction. For example, the η recommended in Ref. 10-2 for homogeneous isotropic

266   Structural Analysis of Shells

cylindrical shells with simply supported edges subjected to axial compression is

$$\eta = \frac{\sqrt{E_t E_s}}{E}\left(\frac{1-\mu_e^2}{1-\mu^2}\right)^{1/2}$$

where E_t, E_s, and μ are the tangent modulus, secant modulus, and Poisson's ratio, respectively, at the actual buckling stress, and μ_e is the elastic Poisson's ratio.

For a given material, temperature, and η, a chart may be prepared for σ_{cr}/η vs. σ_{cr}. Then, for a given problem the elastic buckling stress σ_{cr}/η can be calculated using the formulas presented in the preceding sections and the actual buckling stress σ_{cr} can be read from the chart of σ_{cr}/η vs. σ_{cr}. This method eliminates an iterative procedure which would otherwise be necessary.

The formulas for η are, in general, determined theoretically; and the testing performed to evaluate the theoretical η provides only qualitative agreement. In addition, the number of charts of σ_{cr}/η vs. σ_{cr} necessary to cover all combinations of materials, temperatures, loadings, shells, boundary conditions, and types of construction would be excessively large, and in many cases, the curves would be very close to each other.

To reduce the number of σ_{cr}/η vs. σ_{cr} curves, only the η's defined in Refs. 10-2 and 10-48 will be presented, because these curves have already been computed and, in general, cover the range of possible η within the accuracy of which the actual η is known. The curve that gives the best agreement with experimental and theoretical results of shell structures will be recommended whenever possible.

Figures 10-50 through 10-64 present curves of σ_{cr}/η vs. σ_{cr} for materials and temperatures commonly encountered in industry. In many cases, the curves are so close together that they are drawn as one curve.

The η's used to determine each curve are defined in Table 10-1.

The formulas for η for curves A through G were obtained from Ref. 10-48, which is based on Ref. 10-49. However, Ref. 10-49 assumes $\mu = 1/2$. The constants outside the radical for curve B differ from Ref. 10-49 because of a correction that was made. Although the value of μ is a function of the stresses for stresses in excess of the proportional limit, the plasticity curves were obtained assuming the conservative value of $\mu = 1/3$. The difference between using the value of $\mu = 1/3$ and $\mu = 1/2$ is small except for curves E and F.

It is worth noting that for curve A, $\eta = E_s/E$; for curve G, $\eta = E_t/E$; and on the remaining curves, η is a function of both E_t and E_s. It can be seen that curve A and curve G bound the range of η. Curve G is the most conservative, while curve A results in the smallest possible reduction in the buckling load due to plasticity.

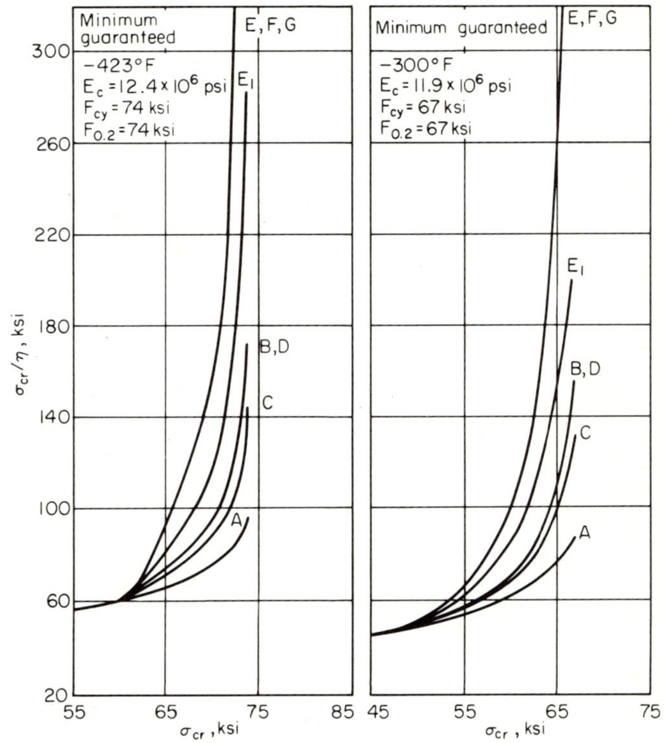

figure 10-50 *2014-T6, -T651 aluminum-alloy sheet and plate. Plasticity correction curves ($-423°F$, $-300°F$).*

TABLE 10-1 Definition of η Factors

Curve	η	Ref.
A	$\dfrac{E_s}{E}$	
B	$\dfrac{E_s}{E}\left[0.330 + 0.670\sqrt{\mu^2 + (1-\mu^2)\dfrac{E_t}{E_s}}\right]$	10-48
C	$\dfrac{E_s}{E}\left[\dfrac{1}{2} + \dfrac{1}{2}\sqrt{\mu^2 + (1-\mu^2)\dfrac{E_t}{E_s}}\right]$	
D	$\dfrac{E_s}{E}\left[0.352 + 0.648\sqrt{\mu^2 + (1-\mu^2)\dfrac{E_t}{E_s}}\right]$	
E	$\mu^2 \dfrac{E_s}{E} + (1-\mu^2)\dfrac{E_t}{E}$	
F	$0.046\dfrac{E_s}{E} + 0.954\dfrac{E_t}{E}, \quad \mu = 0.33$	
G	$\dfrac{E_t}{E}$	
E_i	$\dfrac{\sqrt{E_t E_s}}{E}\left[\dfrac{1-\mu_e^2}{1-\mu^2}\right]^{1/2}$	10-2

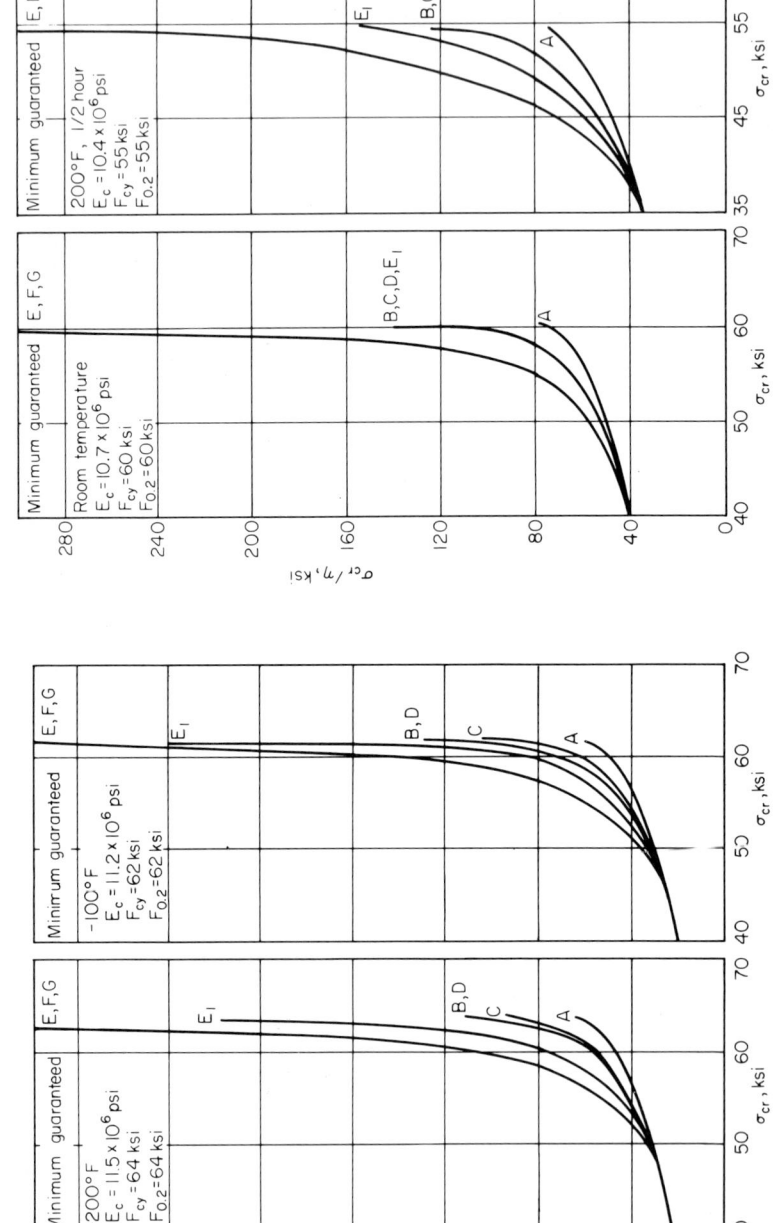

figure 10-51 2014-T6, -T651 aluminum-alloy sheet and plate. Plasticity correction curves ($-200°F$, $-100°F$).

figure 10-52 2014-T6, -T651 aluminum-alloy sheet and plate. Plasticity correction curves (R.T.; 200°F, 1/2 hr).

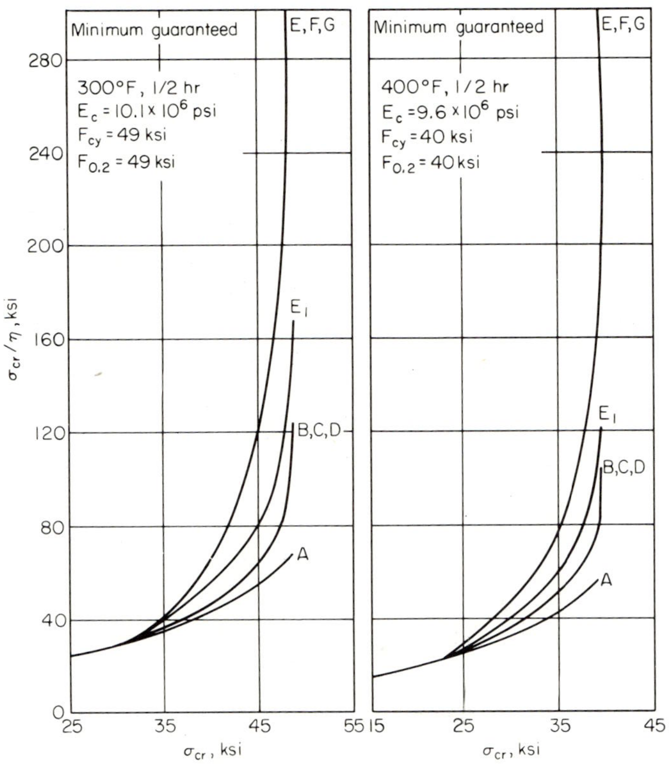

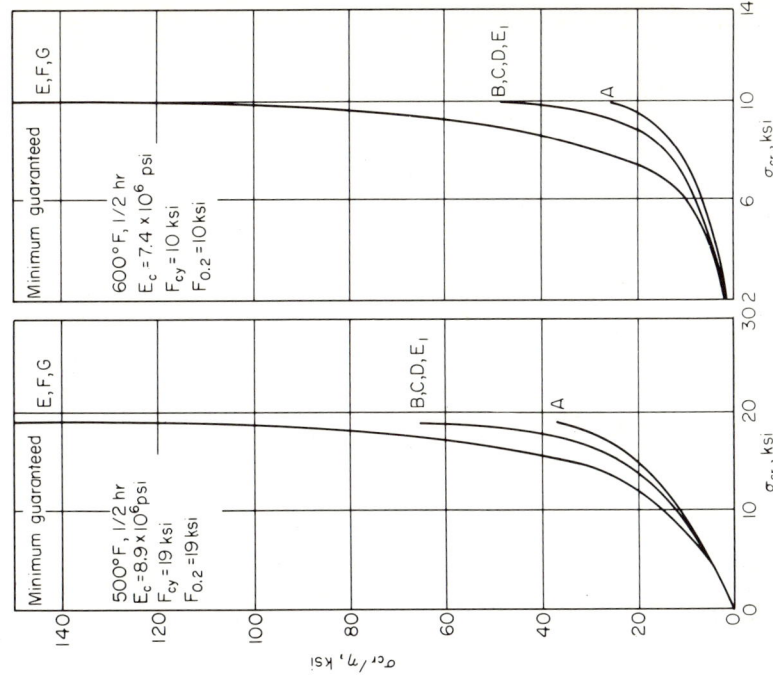

figure 10-53 2014-T6, -T651 aluminum-alloy sheet and plate. Plasticity correction curves (300°F, 1/2 hr; 400°F, 1/2 hr).

figure 10-54 2014-T6, -T651 aluminum-alloy sheet and plate. Plasticity correction curves (500°F, 1/2 hr; 600°F, 1/2 hr).

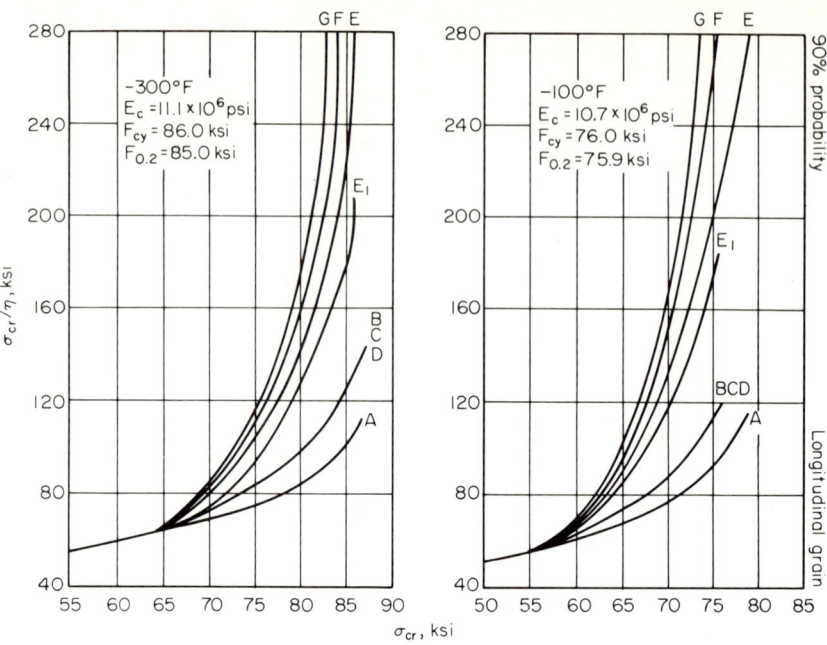

figure 10-55 *Plasticity correction bare 7075-T6 aluminum-alloy sheet ($-300°F$, $-100°F$).*

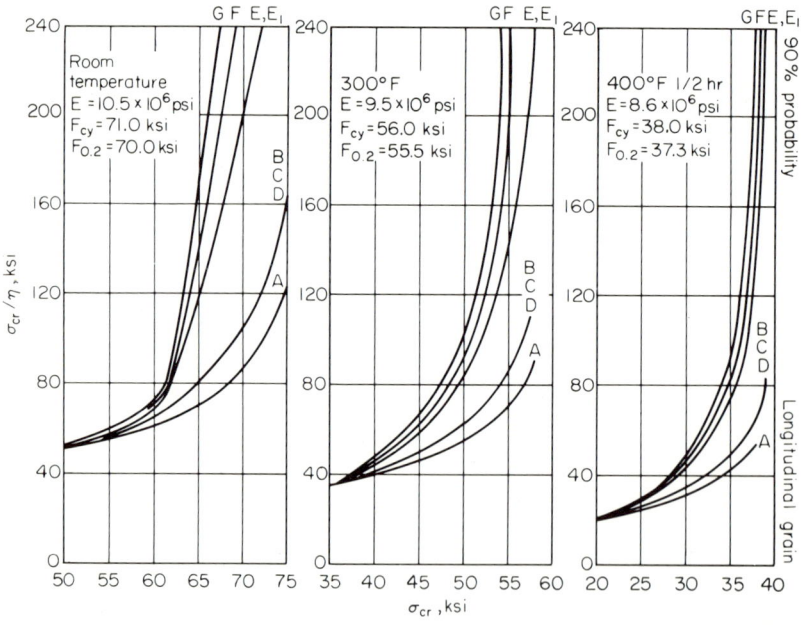

figure 10-56 *Plasticity correction bare 7075-T6 aluminum-alloy sheet (R.T., 300°F, 400°F).*

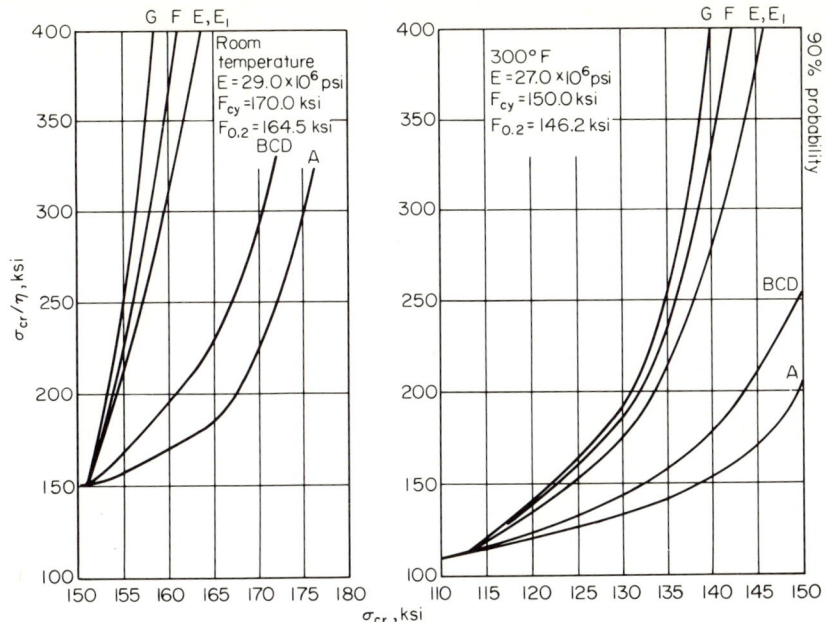

figure 10-57 *Plasticity correction alloy steel 4130, 4140, 4340—H.T. 180,000 psi (R.T., 300°F).*

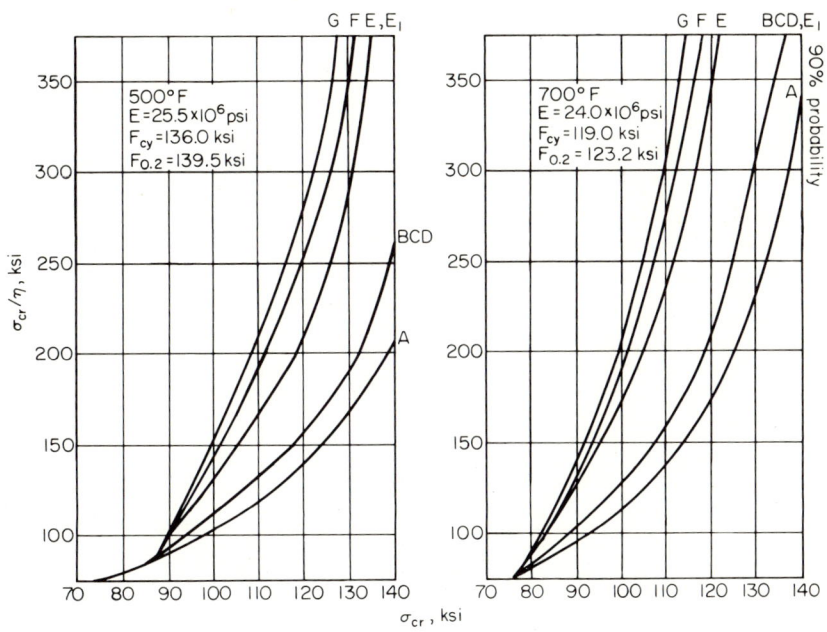

figure 10-58 *Plasticity correction alloy steel 4130, 4140, 4340—H.T. 180,000 psi (500°F, 700°F).*

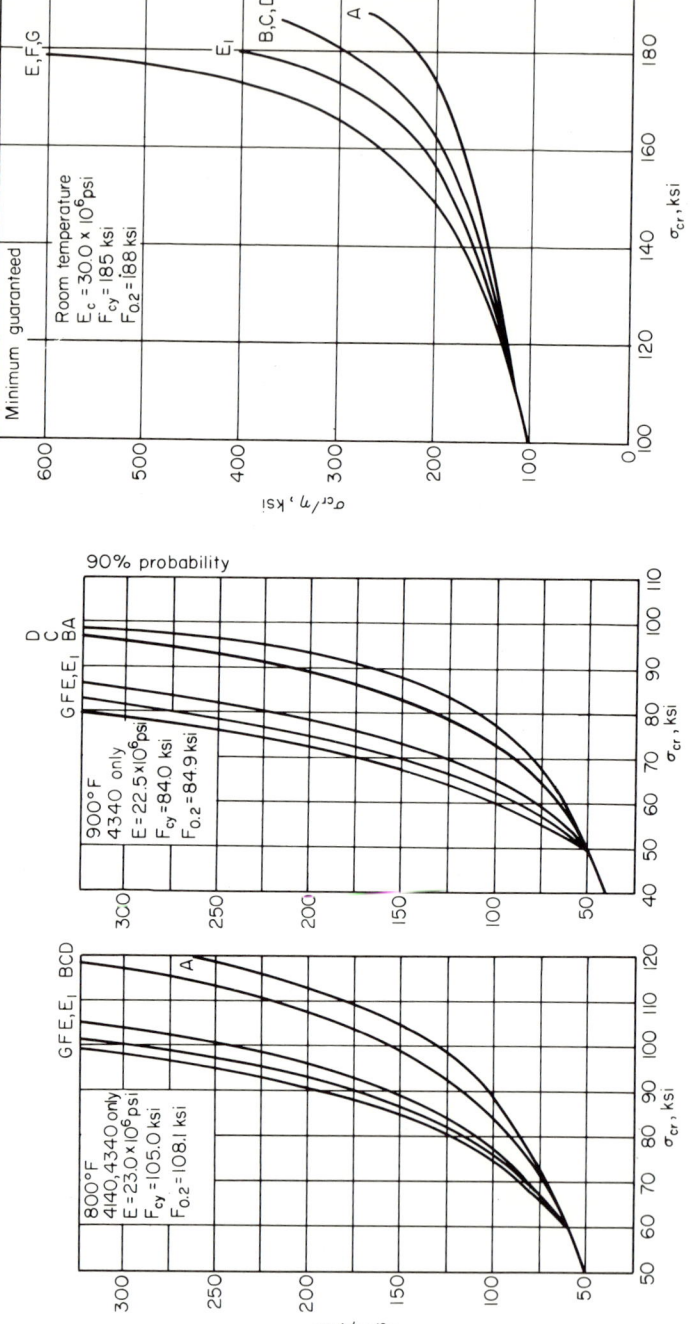

figure 10-60 *PH 15-7 Mo stainless-steel and plate—RH 1050, RH 1075. Plasticity correction curves (R.T.).*

figure 10-59 *Plasticity correction alloy steel 4140, 4340—H.T. 180,000 psi (800°F, 900°F).*

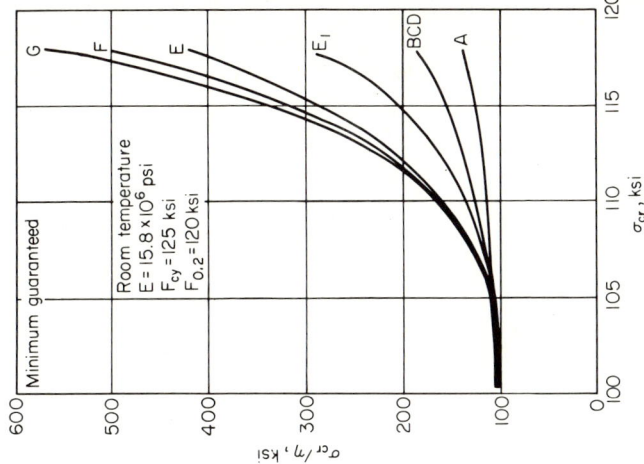

figure 10-62 *Plasticity correction titanium-alloy sheet 6AL-4V annealed LB0170-I13 (R.T.).*

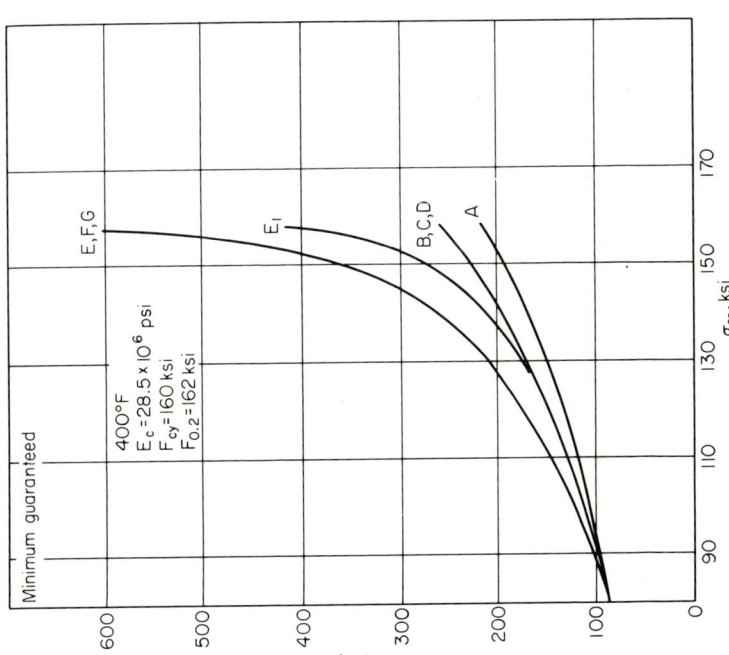

figure 10-61 *PH 15-7 Mo stainless-steel sheet and plate—RH 1050, RH 1075. Plasticity correction curves (400°F).*

274 *Structural Analysis of Shells*

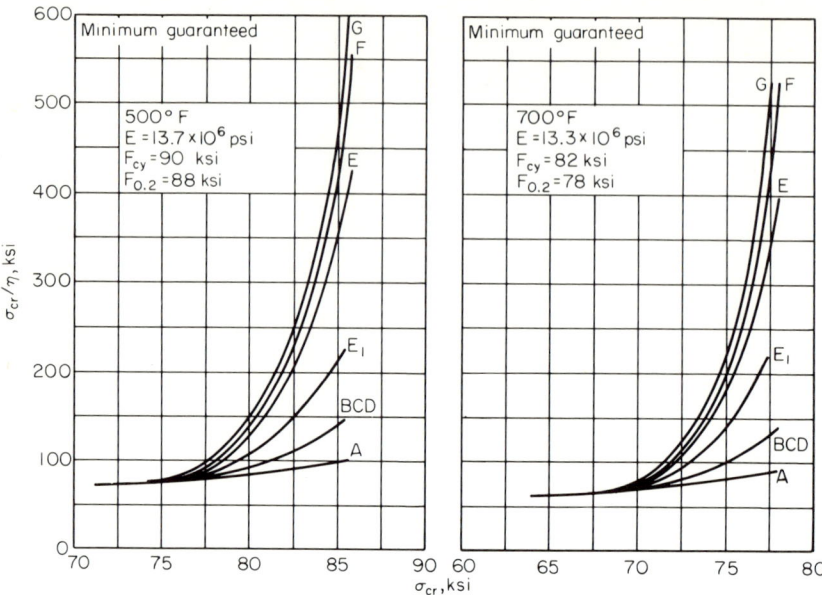

figure 10-63 *Plasticity correction titanium-alloy sheet 6AL-4V annealed (500°F, 700°F).*

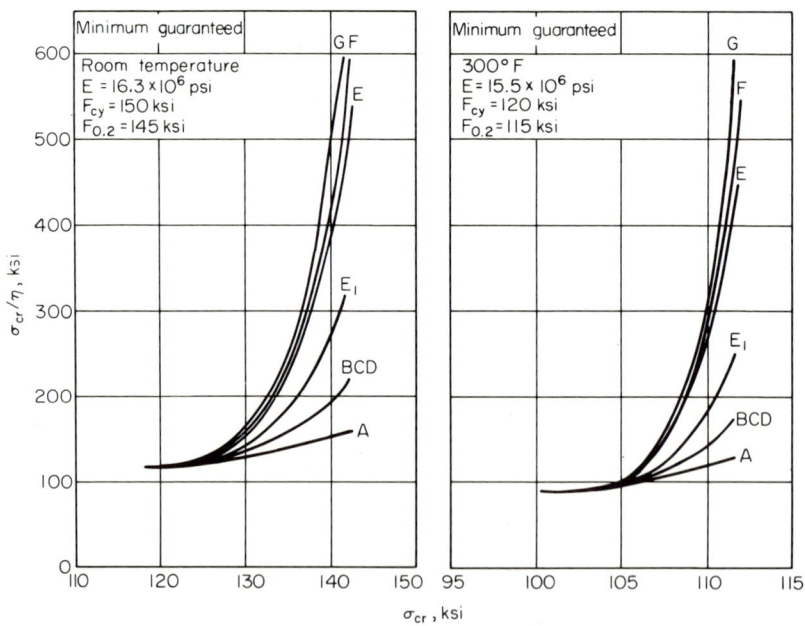

figure 10-64 *Plasticity correction titanium-alloy sheet 6AL-4V condition S.T.A. (R.T., 300°F).*

Combined Loadings, Inelastic Buckling

The information on the inelastic stability analyses of shell structures subjected to combined loadings is limited. Very little theoretical work has been done in this field because of the complexity of the problem, and in general, plasticity correction factors are not available.

Methods of determining whether the stresses are in the inelastic range are discussed in Chap. 9. The method that can be used for ductile materials will be described.

The stress intensity σ_i and strain intensity e_i are obtained from the formulas

$$\sigma_i = \sqrt{\sigma_\phi^2 + \sigma_\theta^2 - \sigma_\phi \sigma_\theta + 3\tau^2}$$

$$e_i = \frac{2}{\sqrt{3}} \sqrt{\epsilon_\phi^2 + \epsilon_\theta^2 + \epsilon_\phi \epsilon_\theta + \frac{\gamma_{\phi\theta}^2}{4}}$$

where σ_ϕ, σ_θ = stresses in the ϕ and θ directions, respectively (for a cylinder or cone, the ϕ direction is the x direction)
τ = in-plane shear stress
ϵ_ϕ, ϵ_θ = strains in the ϕ and θ directions, respectively
$\gamma_{\phi\theta}$ = shear strain

For a ductile material, the σ_i vs. e_i curve for a biaxial stress field is very close to the σ vs. ϵ curve in a uniaxial stress field. Therefore, if σ_i is greater than the uniaxial proportional limits of the material, the stress field is in the inelastic range.

It can be seen from the formula for σ_i that each of the individual stresses may be less than the proportional limits of the material, but σ_i may be in the plastic range.

If the stress is in the inelastic range for a shell subjected to combined loads, an estimate of the inelastic buckling load can be obtained by using the plasticity correction factor associated with σ_i to modify the elastic buckling load. This method is useful if the plasticity correction factor for each of the pure loading cases is approximately the same.

A cylinder subjected to external lateral and axial pressure will be investigated as an example. The stress in the circumferential direction σ_θ is twice as large as the stress in the axial direction σ_x; therefore,

$$\sigma_i = \sqrt{\left(\frac{\sigma_\theta}{2}\right)^2 + \sigma_\theta^2 - \frac{\sigma_\theta}{2} \sigma_\theta}$$

$$= \frac{\sqrt{3}}{2} \sigma_\theta$$

The elastic critical circumferential buckling stress σ_{cr}/η may be obtained from Sec. 10-3 for external lateral and axial pressure for unstiffened cylinders. The elastic stress intensity σ_i/η is therefore

$$\frac{\sigma_i}{\eta} = \frac{\sqrt{3}}{2}\frac{\sigma_{cr}}{\eta}$$

If σ_i/η is less than the proportional limit, $\eta = 1$. If the stresses are inelastic, σ_i may be obtained from preceding figures using σ_i/η and curves E. Then $\sigma_{cr} = 2\sigma_i/\sqrt{3}$, and the design-allowable pressure is $p_{cr} = \sigma_{cr} t/R$.

REFERENCES

10-1. *S & ID Structures Manual*, NAA S & ID, 543-G-11, North American Rockwell, Downey, Calif., 1964.
10-2. Gerard, G., and H. Becker: *Handbook of Structural Stability*, part III, *Buckling of Curved Plates and Shells*, NACA TN 3783, 1957.
10-3. Weingarten, V., E. Morgan, and P. Seide: *Final Report on Development of Design Criteria for Elastic Stability of Thin Shell Structures*, Space Technology Laboratories, Inc., STL/TR-60-0000-19425, 1960.
10-4. Harris, L. A., H. S. Suer, W. T. Skene, and R. J. Benjamin: The Stability of Thin Walled Unstiffened Cylinders under Axial Compression Including the Effects of Internal Pressure, *J. Aerospace Sci.*, August, 1957.
10-5. Batdorf, S. B.: A Simplified Method of Elastic Stability Analysis for Thin Cylindrical Shells, *NACA Rept.* 874, 1947.
10-6. Timoshenko, S.: *Theory of Elastic Stability*, McGraw-Hill Book Company, New York, 1936.
10-7. *Buckling of Thin-walled Circular Cylinders*, NASA SP 8007, 1968.
10-8. Holston, A., Jr.: Stability of Inhomogeneous Anisotropic Cylindrical Shells Containing Elastic Cores, *J. Am. Inst. Aeronautics Astronautics*, vol. 5, no. 6, pp. 1135–1138, June, 1967.
10-9. Brush, D. O., and E. V. Pittner: Influence of Cushion Stiffness on the Stability of Cushion-loaded Cylindrical Shells, *J. Am. Inst. Aeronautics Astronautics*, vol. 3, no. 2, pp. 308–316, February, 1965.
10-10. Seide, P.: The Stability under Axial Compression and Lateral Pressure of a Circular-cylindrical Shell with an Elastic Core, *J. Aeron. Sci.*, vol. 29, no. 7, pp. 851–862, July, 1962.
10-11. Weingarten, V. I.: Stability under Torsion of Circular Cylinder Shells with an Elastic Core, *Am. Rocket Soc. J.*, vol. 82, no. 4, pp. 637–639, April, 1962.
10-12. Hausrath, A., and F. Dittoe: *Development of Design Strength Levels for the Elastic Stability of Monocoque Cones under Axial Compression*, NASA TN D-1510, 1962.
10-13. Stocker, J.: *A Review of the Literature on the Buckling Characteristics of Conical Shells*, Boeing Co., Seattle, Wash., D2-23835-1, 1965.
10-14. Seide, P.: On Buckling of Truncated Conical Shells in Torsion, *J. Appl. Mech.*, vol. 29, pp. 320–238, 1962.
10-15. Weingarten, V.: Experimental Investigation of the Stability of Internally Pressurized Conical Shells under Torsion, *J. Am. Inst. Aeronautics Astronautics*, vol. 2, no. 10, 1964.

10-16. *Buckling of Thin-walled Doubly Curved Shells*, NASA SP-8032, August, 1969.
10-17. Fitch, J. R.: The Buckling and Post-Buckling Behavior of Spherical Caps under Concentrated Load, *Intern. J. Solids Structures*, vol. 4, no. 4, pp. 421–446, April, 1968.
10-18. Weinitschke, H.: On the Stability Problem for Shallow Spherical Shells, *J. Math. Phys.*, vol. 38, no. 4, pp. 209–231, December, 1960.
10-19. Budiansky, B.: Buckling of Clamped Shallow Spherical Shells, *Proc. UITAM Symp. Theory of Thin Elastic Shells*, pp. 64–94, North-Holland Publishing Company, Amsterdam, 1960.
10-20. Huang, N. C.: Unsymmetrical Buckling of Thin Shallow Spherical Shells, *J. Appl. Mech.*, vol. 31, no. 3, pp. 447–457, September, 1964.
10-21. Weinitschke, H.: On Asymmetric Buckling of Shallow Spherical Shells, *J. Math. Phys.*, vol. 44, no. 2, pp. 141–163, June, 1965.
10-22. Thurston, G. A., and F. A. Penning: Effect of Axisymmetric Imperfections on the Buckling of Spherical Caps under Uniform Pressure, *J. Am. Inst. Aeronautics Astronautics*, vol. 4, no. 2, p. 319, February, 1966.
10-23. Bushnell, D.: Nonlinear Axisymmetric Behavior of Shells of Revolution, *J. Am. Inst. Aeronautics Astronautics*, vol. 5, no. 3, pp. 432–439, March, 1967.
10-24. Wang, L. R.-L.: Effects of Edge Restraint on the Stability of Spherical Caps, *J. Am. Inst. Aeronautics Astronautics*, vol. 4, no. 4, pp. 718–719, April, 1966.
10-25. Bushnell, D.: Buckling of Spherical Shells Ring-supported at the Edges, *J. Am. Inst. Aeronautics Astronautics*, vol. 5 no. 11, pp. 2041–2046, November 1967.
10-26. Wang, L. R.-L.: Discrepancy of Experimental Buckling Pressures of Spherical Shells, *J. Am. Inst. Aeronautics Astronautics*, vol. 5, no. 2, pp. 357–359, February, 1967.
10-27. McComb, H. G., Jr., and W. B. Fitcher: *Buckling of a Sphere of Extremely High Radius-thickness Ratio, Collected Papers on Instability of Shell Structures*, NASA TN D-1510, pp. 561–570, 1962.
10-28. Mescall, J. F.: Large Deflections of Spherical Shells under Concentrated Loads, *J. Appl. Mech.*, vol. 32, no. 4, pp. 936–938, December, 1965.
10-29. Bushnell, D.: Bifurcation Phenomena in Spherical Shells under Concentrated and Ring Loads, *J. Am. Inst. Aeronautics Astronautics*, vol. 5, no. 11, pp. 2034–2040, November, 1967.
10-30. Ashwell, D. G.: On the Large Deflection of a Spherical Shell with an Inward Point Load, *Proc. UITAM Symp. Theory of Thin Elastic Shells*, pp. 43–63, North-Holland Publishing Company, Amsterdam, 1960.
10-31. Evan-Iwanowski, R. M., H. S. Cheng, and T. C. Loo: Experimental Investigations and Deformation and Stability of Spherical Shells Subjected to Concentrated Loads at the Apex, *Proc. 4th U.S. Natl. Eng. Appl. Mech.*, 1962, pp. 563–575.
10-32. Penning, F. A., and G. A. Thurston: *The Stability of Shallow Spherical Shells under Concentrated Load*, NASA CR-265, 1965.
10-33. Penning, F. A.: Experimental Buckling Modes of Clamped Shallow Shells under Concentrated Load, *J. Appl. Mech.*, vol. 33, no. 2, pp. 297–304, June, 1966.
10-34. Penning, F. A.: Nonaxisymmetric Behavior of Shallow Shells Loading at the Apex, *J. Appl. Mech.*, vol. 33, no. 3, pp. 699–700, September, 1966.
10-35. Loo, T. C., and R. M. Evan-Iwanowski: Interaction of Critical Concentrated Loads Acting on Shallow Spherical Shells, *J. Appl. Mech.*, vol. 33, no. 3, pp. 612–616, September, 1966.
10-36. Hyman, B. I., and J. J. Healey: Buckling of Prolate Spheroidal Shells under Hydrostatic Pressure, *J. Am. Inst. Aeronautics Astronautics*, vol. 5, no. 8, pp. 1469–1477, August, 1967.

10-37. Nickell, E. H.: Experimental Buckling Tests of Magnesium Monocoque Ellipsoidal Shells Subjected to External Hydrostatic Pressure, Lockheed Missiles & Space Company, *Tech. Rept.* 3-42-61-2, vol. IV, June 30, 1961.

10-38. Meyer, R. R., and R. J. Bellinfante: Fabrication and Experimental Evaluation of Common Domes Having Waffle-like Stiffening, Douglas Aircraft Company, Inc., *Rept.* SM-47742, 1964.

10-39. Thurston, G. A., and A. E. Holston, Jr: *Buckling of Cylindrical Shell End Closures by Internal Pressure*, NASA CR-540, 1966.

10-40. Adachi, J., and M. Benicek: Buckling of Torispherical Shells under Internal Pressure, *Exptl. Mech.*, vol. 4, no. 8, pp. 217–222, August, 1964.

10-41. Sobel, L. H., and W. Flügge: Stability of Toroidal Shells under Uniform External Pressure, *J. Am. Inst. Aeronautics Astronautics*, vol. 5, no. 3, pp. 425–431, March, 1967.

10-42. Hutchinson, J. W.: Initial Post-buckling Behavior of Toroidal Shell Segments, *Intern. J. Solids Structures*, vol. 3, no. 1, pp. 97–115, January, 1967.

10-43. Yao, J. C.: Buckling of a Truncated Hemisphere under Axial Tension, *J. Am. Inst. Aeronautics Astronautics*, vol. 1, no. 10, pp. 2316–2320, October, 1963.

10-44. Babcock, C. D., and E. E. Sechler: *The Effect of Initial Imperfections on the Buckling Stress of Cylindrical Shells, Collected Papers on Instability of Shell Structures*, NASA TN D-1510, 1962, pp. 135–142.

10-45. Blum, R. E., and McComb, H. G., Jr.: *Buckling of an Equatorial Segment of a Spherical Shell Loaded by Its Own Weight*, NASA TN D-4921, 1968.

10-46. Stein, M., and J. A. McElman: Buckling of Segments of Toroidal Shells, *J. Am. Inst. Aeronautics Astronautics*, vol. 3, no. 9, pp. 1704–1709, September 1965.

10-47. Gerard, G.: *Handbook of Structural Stability*, supplement to Part III—*Buckling of Curved Plates and Shells*, NASA TN D-163, September, 1959.

10-48. Silver, P.: *The Effects of Poisson's Ratio on Plasticity Corrections*, NAA S & ID, STR55, North American Rockwell, Space Division, April, 1957.

10-49. Stowell, E.: *A Unified Theory of Plastic Buckling of Columns and Plates*, NACA TN 1556, April, 1943.

Chapter 11

STABILITY OF ORTHOTROPIC COMPOSITE SHELLS

11-1 General

This section deals with shells with stiffness properties which are different in the circumferential and meridional directions. The stiffness properties of the shell wall in the two directions may be completely independent (i.e., bending stiffness, extensional stiffnesses, inplane shear stiffness, and twisting stiffness of the walls of the shell are not necessarily interrelated as they are in a homogeneous isotropic shell or a homogeneous orthotropic shell). An example of this type of shell is a multilayered filament-wound cylinder. Other types of construction, such as integrally stiffened waffle with closely spaced stiffeners, can be idealized as orthotropic by assuming that discrete stiffening elements are evenly distributed per unit width of wall. It is difficult to determine how close the stiffeners must be to treat the shell as an orthotropic shell. For the case of buckling, a buckle must include several stiffeners before orthotropic-shell theory would be a useful tool in predicting the buckling load.

Definitions of the stiffness properties (elastic constants) for an orthotropic-shell wall are given in Sec. 11-2, and approximate formulas are given for computing the stiffness properties for several types of construction.

The buckling formulas presented in this chapter are based on the classical small-deflection theory of orthotropic shells. Experimental and theoretical studies indicate that the discrepancy between test and theory as well as the scatter of the test data may be much smaller for certain types of orthotropic cylinders than it is for homogeneous isotropic cylinders. However, the number of tests conducted to date on orthotropic shells typical of large manufactured parts is limited and covers only a small range of possible parameters. Therefore, the results of unstiffened homogeneous isotropic cylinder tests are used to modify the orthotropic theory until more information is available.

The analysis presented assumes that the centroid planes of the orthotropic wall in the axial direction coincide with the centroid planes of the wall in the circumferential direction. This assumption eliminates coupling between many of the internal stress resultants and simplifies the analysis. Although the effects of coupling can be large for some types of construction, they are usually small for the types of construction presented in this section (multilayered, integrally stiffened, homogeneous orthotropic).

The effects of Poisson's ratio have been included in the analysis procedures that are presented. However, the analysis may be simplified by neglecting Poisson's ratio. For many types of orthotropic construction, the effect of Poisson's ratio is very small.

11-2 Elastic Constants

If transverse shear deformations are neglected and the shell is very thin, the relations between the resultant forces and the centroidal surface strain for an orthotropic shell can be written as

$$N_x = B_x(\epsilon_x + \mu_\theta' \epsilon_\theta)$$
$$N_\theta = B_\theta(\epsilon_\theta + \mu_x' \epsilon_x)$$
$$N_{x\theta} = G_{x\theta} \gamma_{x\theta}$$

and the corresponding relations between the resultant moments and curvature and twist are

$$M_x = D_x(\chi_x + \mu_\theta \chi_\theta)$$
$$M_\theta = D_\theta(\chi_\theta + \mu_x \chi_x)$$
$$M_{x\theta} = \frac{D_{x\theta}}{2} \chi_{x\theta}$$

where B_x, B_θ, $G_{x\theta}$, D_x, D_θ, $D_{x\theta}$, μ_x, μ_θ, μ_x', μ_θ' are the elastic constants (stiffness properties) for the shell wall. $\gamma_{x\theta}$ is the inplane shear strain, $\chi_{x\theta}$ is the twisting of the centroidal surface, and the remaining symbols are defined in Chap. 1 and Chap. 7.

The equations presented for orthotropic shells may be used provided that the preceding six equations are valid for the shell wall.

Approximate formulas are given in this section for computing the elastic constants for several types of construction that may be idealized as orthotropic construction. These constants can be used in Chap. 7 for stress analysis or in this chapter for stability analysis. The formulas for the constants have been derived for flat orthotropic plates but are sufficiently accurate for thin orthotropic shells.

The x direction is the axial direction for cylinders and cones and the meridional direction for spheres. The θ direction is the circumferential direction. The nomenclature for these elastic constants is as follows:

B_x, B_θ = extension stiffness of the shell wall in the x and θ directions, respectively, lb/in.

D_x, D_θ = bending stiffness of the shell wall in the x and θ directions, respectively, in.-lb

$D_{x\theta}$ = twisting stiffness of the shell wall, in.-lb

μ_x, μ_θ = Poisson's ratios associated with bending in the x and θ directions, respectively

μ_x', μ_θ' = Poisson's ratios associated with extension in the x and θ directions, respectively

$G_{x\theta}$ = shear stiffness of the shell wall in the $x\theta$ plane, lb/in.

From the reciprocity theorem, it can be shown that the following useful relationships exist:

$$D_x \mu_\theta = D_\theta \mu_x$$

$$B_x \mu_\theta' = B_\theta \mu_x'$$

If transverse shear deflections are included in the analysis, two additional elastic constants will be necessary, D_{Q_x} and D_{Q_θ}, the transverse shear stiffness of the shell wall in the x and θ directions, respectively. Sandwich-type construction is the only orthotropic construction in which shear deflections are likely to be important; therefore, these constants are presented only for sandwiches. D_{Q_x} and D_{Q_θ} are not needed in the stability analysis of orthotropic shells presented because shear deflections have been neglected in the basic analysis. In general, the data of Chap. 13 should be used for the stability analysis of sandwich shells. However, if Chap. 13 does not include a particular type of sandwich (for instance, facings made from a different material), and if the core is very stiff in transverse shear (Chap. 13 can be used to estimate if the transverse shear stiffness is large), the design buckling load can be estimated using the formulas from Secs. 11-3 and 11-4 and the elastic constants given in this section.

Orthotropic Layered Shells

The elastic constants for orthotropic layered shells were obtained from Ref. 11-1. A typical multilayered cross section is shown in Fig. 11-1.

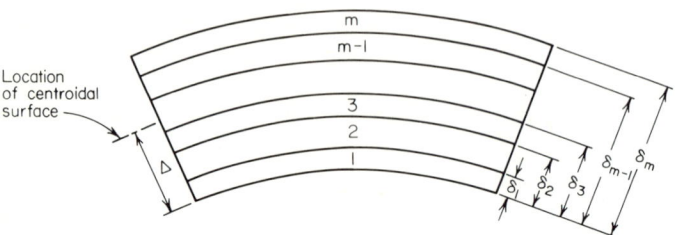

figure 11-1 *Layered construction.*

It can be seen that there are m layers. The inner layer is δ_1 thick; the next layer is $(\delta_2 - \delta_1)$ thick; etc.; and the total thickness of the shell is δ_m. If each layer is homogeneous and orthotropic and the natural axis of the orthotropic layer coincides with the x and θ directions, then Hooke's law for each layer may be written

$$\epsilon_{x_i} = \frac{\sigma_{x_i}}{E_{x_i}} - \mu_{\theta_i}\frac{\sigma_{\theta_i}}{E_{\theta_i}}$$

$$\epsilon_{\theta_i} = \frac{\sigma_{\theta_i}}{E_{\theta_i}} - \mu_{x_i}\frac{\sigma_{x_i}}{E_{x_i}}$$

$$\gamma_{x\theta_i} = \frac{\tau_{x\theta_i}}{G_{x\theta_i}}$$

where E_{x_i}, E_{θ_i} = moduli of elasticity in the x and θ direction, respectively (lb/in.²)
$\mu_{x_i}, \mu_{\theta_i}$ = Poisson's ratios associated with stretching in the x and θ directions, respectively
$G_{x\theta_i}$ = inplane shear modulus in the $x\theta$ plane (lb/in.²)

Subscript i represents the layer to which the material property responds ($i = 1, 2, \ldots, m$).

The stresses may be obtained in terms of the strains and moduli by solving the equations presented above:

$$\sigma_{x_i} = C_{x_i}\epsilon_{x_i} + C_{\mu_i}\epsilon_{\theta_i}$$

$$\sigma_{\theta_i} = C_{\theta_i}\epsilon_{\theta_i} + C_{\mu_i}\epsilon_{x_i}$$

$$\tau_{x\theta_i} = G_{x\theta_i}\gamma_{x\theta_i}$$

where

$$C_{x_i} = \frac{E_{x_i}}{1 - \mu_{x_i}\mu_{\theta_i}} \qquad C_{\theta_i} = \frac{E_{\theta_i}}{1 - \mu_{x_i}\mu_{\theta_i}} \qquad C_{\mu_i} = \frac{\mu_{x_i}E_{\theta_i}}{1 - \mu_{x_i}\mu_{\theta_i}}$$

If a layer is orthotropic but the natural axes α, β of the layer are at some angle Y to the x and θ directions (see Fig. 11-2), the orthotropic-shell analysis which is presented in the following sections cannot be used. However, if the natural axes α, β of one layer of an orthotropic material are at an angle Y to the x, θ axes and the natural axis of another layer of the same material and thickness is at an angle of $-Y$ to the x, θ axes, these two layers in combination can be treated as an orthotropic material with the natural axes oriented in the x, θ direction provided that the total shell wall is greater than about eight layers in thickness. The necessary properties in the x and θ directions of the two layers at $\pm Y$ can be computed from the equations

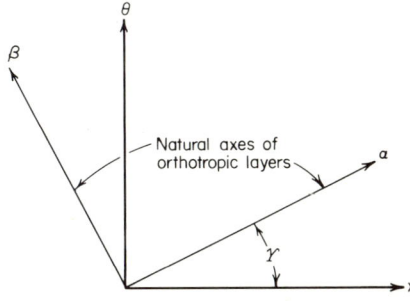

figure 11-2 *Coordinate system.*

$$C_{x_i} = C_\alpha \cos^4 Y + 2(C_{\mu\alpha} + 2G_{\alpha\beta}) \sin^2 Y \cos^2 Y + C_\beta \sin^4 Y$$

$$C_{\theta_i} = C_\alpha \sin^4 Y + 2(C_{\mu\alpha} + 2G_{\alpha\beta}) \sin^2 Y \cos^2 Y + C_\beta \cos^4 Y$$

$$C_{\mu_i} = (C_\alpha + C_\beta - 4G_{\alpha\beta}) \sin^2 Y \cos^2 Y + C_{\mu\alpha}(\sin^4 Y + \cos^4 Y)$$

$$C_\alpha = \frac{E_\alpha}{1 - \mu_\alpha\mu_\beta} \qquad C_\beta = \frac{E_\beta}{1 - \mu_\alpha\mu_\beta} \qquad C_{\mu\alpha} = \frac{\mu_\alpha E_\beta}{1 - \mu_\alpha\mu_\beta}$$

where E_α, E_β = moduli of elasticity in the α and β directions, respectively
μ_α, μ_β = Poisson's ratios associated with stretching in the α and β directions, respectively
$G_{\alpha\beta}$ = shear modulus in $\alpha\beta$ plane

The equations to be presented for a layered orthotropic shell are valid only if the various centroid surfaces of the shell walls coincide. For instance, the neutral axis for bending in the x direction must be the same as the neutral axis for bending in the θ direction, or else there is a coupling between the resultant forces and resultant moments (see

Ref. 11-1). Therefore, the following relationships must be approximately satisfied:

$$\varDelta = \frac{\sum\limits_{i=1}^{m} C_{x_i}(\delta_i{}^2 - \delta_{i-1}^2)}{2\sum\limits_{i=1}^{m} C_{x_i}(\delta_i - \delta_{i-1})} = \frac{\sum\limits_{i=1}^{m} C_{\theta_i}(\delta_i{}^2 - \delta_{i-1}^2)}{2\sum\limits_{i=1}^{m} C_{\theta_i}(\delta_i - \delta_{i-1})}$$

$$= \frac{\sum\limits_{i=1}^{m} C_{\mu_i}(\delta_i{}^2 - \delta_{i-1}^2)}{2\sum\limits_{i=1}^{m} C_{\mu_i}(\delta_i - \delta_{i-1})} = \frac{\sum\limits_{i=1}^{m} G_{x\theta_i}(\delta_i{}^2 - \delta_{i-1}^2)}{2\sum\limits_{i=1}^{m} G_{x\theta_i}(\delta_i - \delta_{i-1})}$$

The parameter $\varDelta$ locates the centroidal surface (see Fig. 11-1).

The elastic constants for a layered shell are given by the following equations:

$$B_x = \sum_{i=1}^{m} C_{x_i}(\delta_i - \delta_{i-1})$$

$$B_\theta = \sum_{i=1}^{m} C_{\theta_i}(\delta_i - \delta_{i-1})$$

$$D_x = \frac{1}{3} \sum_{i=1}^{m} C_{x_i}[(\delta_i{}^3 - \delta_{i-1}^3) - 3\varDelta(\delta_i{}^2 - \delta_{i-1}^2) + 3\varDelta^2(\delta_i - \delta_{i-1})]$$

$$D_\theta = \frac{1}{3} \sum_{i=1}^{m} C_{\theta_i}[(\delta_i{}^3 - \delta_{i-1}^3) - 3\varDelta(\delta_i{}^2 - \delta_{i-1}^2) + 3\varDelta^2(\delta_i - \delta_{i-1})]$$

$$D_{x\theta} = \frac{2}{3} \sum_{i=1}^{m} G_{x\theta_i}[(\delta_i{}^3 - \delta_{i-1}^3) - 3\varDelta(\delta_i{}^2 - \delta_{i-1}^2) + 3\varDelta^2(\delta_i - \delta_{i-1})]$$

$$G_{x\theta} = \sum_{i=1}^{m} G_{x\theta_i}(\delta_i - \delta_{i-1})$$

$$\mu_x = \frac{1}{3D_\theta} \sum_{i=1}^{m} C_{\mu_i}[(\delta_i{}^3 - \delta_{i-1}^3) - 3\varDelta(\delta_i{}^2 - \delta_{i-1}^2) + 3\varDelta^2(\delta_i - \delta_{i-1})]$$

$$\mu_\theta = \frac{\mu_x D_\theta}{D_x}$$

Stability of Orthotropic Composite Shells

$$\mu_x' = \frac{1}{B_\theta} \sum_{i=1}^{m} C_{\mu_i}(\delta_i - \delta_{i-1})$$

$$\mu_\theta' = \frac{\mu_x' B_\theta}{B_x}$$

D_{Q_x} and D_{Q_θ} are effectively infinite for most layered shells with the possible exception of sandwich construction. For simplicity, μ_x could be assumed to equal μ_x'. For a single layer t thick ($\delta_1 = t$), the fomulas reduce to

$$B_x = C_{x_1} t$$

$$B_\theta = C_{\theta_1} t$$

$$D_x = \frac{C_{x_1} t^3}{12}$$

$$D_\theta = \frac{C_{\theta_1} t^3}{12}$$

$$D_{x\theta} = \frac{2 G_{x\theta_1} t^3}{12}$$

$$G_{x\theta} = G_{x\theta_1} t$$

$$\mu_x = \mu_{x_1}' = \mu_{x_1}$$

$$\mu_\theta = \mu_{\theta_1}' = \mu_{\theta_1}$$

If the layer is isotropic with a Young's modulus of E, Poisson's ratio of μ, and shear modulus of G, the constants are

$$B_x = B_\theta = \frac{Et}{1 - \mu^2}$$

$$D_x = D_\theta = \frac{Et^3}{12(1 - \mu^2)}$$

$$D_{x\theta} = \frac{2Gt^3}{12}$$

$$G_{x\theta} = Gt$$

$$\mu = \mu_x = \mu_\theta = \mu_x' = \mu_\theta'$$

Sandwich Shells

The elastic constants presented in this section are for a sandwich construction with a core that resists very little bending or stretching (such as a honeycomb core) and with thin facing sheets relative to the overall thickness of the sandwich. A typical sandwich element and a definition of the geometrical parameters are given in Fig. 11-3.

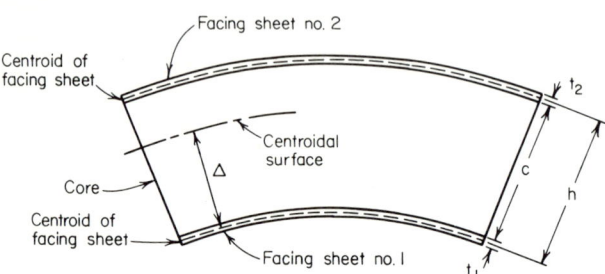

figure 11-3 *Sandwich construction.*

The material properties of a facing sheet are:

E_{x_i}, E_{θ_i} = moduli of elasticity in the x and θ directions, respectively
$\mu_{x_i}, \mu_{\theta_i}$ = Poisson's ratios associated with stretching in the x and θ directions, respectively
$G_{x\theta_i}$ = inplane shear modulus

Subscript i represents the facing sheet to which the material corresponds ($i = 1, 2$).

The only property required for the type of core considered is G_{xz} and $G_{\theta z}$, which are the transverse shear moduli of the core in the x and θ directions, respectively. The formulas for the elastic constants are approximate but sufficiently accurate for engineering purposes.

Sandwich with Orthotropic Facings (same material for both facings):

$$C_{x_i} = \frac{E_{x_i}}{1 - \mu_{x_i}\mu_{\theta_i}}$$

$$C_{\theta_i} = \frac{E_{\theta_i}}{1 - \mu_{x_i}\mu_{\theta_i}}$$

$$C_{\mu_i} = \frac{\mu_{x_i}E_{\theta_i}}{1 - \mu_{x_i}\mu_{\theta_i}}$$

$$B_x = C_{x_1}t_1 + C_{x_2}t_2$$

$$B_\theta = C_{\theta_1}t_1 + C_{\theta_2}t_2$$

Stability of Orthotropic Composite Shells

$$G_{x\theta} = G_{x\theta_1}t_1 + G_{x\theta_2}t_2$$

$$D_{x\theta} = 2\frac{G_{x\theta_1}G_{x\theta_2}t_1t_2h^2}{G_{x\theta_1}t_1 + G_{x\theta_2}t_2}$$

$$D_x = \frac{C_{x_1}C_{x_2}t_1t_2h^2}{C_{x_1}t_1 + C_{x_2}t_2}$$

$$D_\theta = \frac{C_{\theta_1}C_{\theta_2}t_1t_2h^2}{C_{\theta_1}t_1 + C_{\theta_2}t_2}$$

$$\mu_x' = \frac{1}{B_\theta}(C_{\mu_1}t_1 + C_{\mu_2}t_2)$$

$$\mu_\theta' = \frac{\mu_x'B_\theta}{B_x}$$

$$\mu_x = \frac{C_{\mu_1}C_{\mu_2}t_1t_2h^2}{C_{\mu_1}t_1 + C_{\mu_2}t_2}$$

$$\mu_\theta = \frac{\mu_x D_\theta}{D_x}$$

$$D_{Q_x} = \frac{G_{xz}h^2}{c}$$

$$D_{Q_\theta} = \frac{G_{\theta z}h^2}{c}$$

where D_{Q_x} and D_{Q_θ} are the shear stiffnesses per inch of width (lb/in.) in the xz plane and the θz plane, respectively. G_{xz} and $G_{\theta z}$ are the shear moduli (lb/in.^{2}) in the xz plane and the θz plane, respectively.

As discussed in the previous section, the centroidal surfaces should coincide in order to use the formulas which are presented for orthotropic shells. Therefore, the following relationships must be approximately satisfied:

$$\varDelta = \frac{C_{x_2}t_2h}{C_{x_1}t_1 + C_{x_2}t_2} = \frac{C_{\theta_2}t_2h}{C_{\theta_1}t_1 + C_{\theta_2}t_2} = \frac{C_{\mu_2}t_2h}{C_{\mu_1}t_1 + C_{\mu_2}t_2} = \frac{G_{x\theta_2}t_2h}{G_{x\theta_1}t_1 + G_{x\theta_2}t_2}$$

The parameter $\varDelta$ locates the centroidal surface (see Fig. 11-3).
Sandwich with Isotropic Facings (same material for both facings):

$$B_x = B_\theta = \frac{E}{1 - \mu^2}(t_1 + t_2)$$

$$D_x = D_\theta = \frac{Et_1t_2h^2}{(1-\mu^2)(t_1+t_2)}$$

$$G_{x\theta} = G(t_1 + t_2)$$

$$D_{x\theta} = 2G\frac{t_1 t_2 h^2}{t_1 + t_2}$$

$$\mu_x = \mu_\theta = \mu_x' = \mu_\theta' = \mu$$

$$D_{Q_x} = \frac{G_{xz} h^2}{c}$$

$$D_{Q_\theta} = \frac{G_{\theta z} h^2}{c}$$

$$\Delta = \frac{t_2 h}{t_1 + t_2}$$

Integrally Stiffened Waffle Shells

The approximate elastic constants for shells with closely spaced integral ribs running in a wafflelike pattern were obtained from Ref. 11-2. Figure 11-4 shows the type of construction being considered and also defines the geometrical parameters.

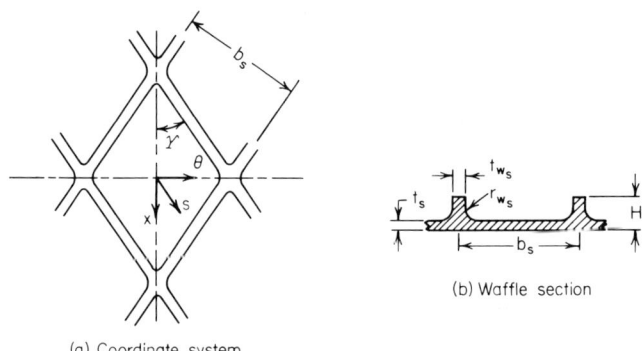

(a) Coordinate system

(b) Waffle section

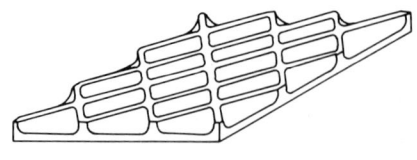

(c) Typical pattern

figure 11-4 *Waffle construction.*

The definitions of the material properties are:

E = Young's modulus
G = shear modulus
μ = Poisson's ratio

The elastic constants for integrally stiffened waffle construction are

$$B_x = \frac{EH}{1 - \mu_x'\mu_\theta'} \frac{\bar{A}_s^2}{A_\theta}$$

$$B_\theta = \frac{EH}{1 - \mu_x'\mu_\theta'} \frac{\bar{A}_s^2}{A_x}$$

$$D_x = EH^3 \left[I_x - \frac{A_s^2 A_x}{\bar{A}_s^2} (\bar{k}_x - \bar{k}_s)^2 \right]$$

$$D_\theta = EH^3 \left[I_\theta - \frac{A_s^2 A_\theta}{\bar{A}_s^2} (\bar{k}_\theta - \bar{k}_s)^2 \right]$$

$$D_{x\theta} = EH^3 \left(\frac{I_{x\theta}}{2} \right)$$

$$D_{Q_x} = \infty$$

$$D_{Q_\theta} = \infty$$

$$G_{x\theta} = EH(A_{x\theta})$$

$$\mu_x = \frac{\bar{I}_s^2}{I_\theta \bar{A}_s^2 - A_s^2 A_\theta (\bar{k}_\theta - \bar{k}_s)^2}$$

$$\mu_\theta = \frac{\bar{I}_s^2}{I_x \bar{A}_s^2 - A_s^2 A_x (\bar{k}_x - \bar{k}_s)^2}$$

$$\mu_x' = \frac{A_s}{A_\theta}$$

$$\mu_\theta' = \frac{A_s}{A_x}$$

where

$$\bar{A}_s^2 = A_x A_\theta - A_s^2$$

$$\bar{I}_s^2 = I_s \bar{A}_s^2 + A_s A_x A_\theta (\bar{k}_x - \bar{k}_s)(\bar{k}_\theta - \bar{k}_s)$$

$$A_x = \frac{1}{1 - \mu^2} \frac{t_s}{H} + \frac{A_{w_s}/b_s}{H} \cos^4 Y$$

$$A_\theta = \frac{1}{1 - \mu^2} \frac{t_s}{H} + \frac{A_{w_s}/b_s}{H} \sin^4 Y$$

$$A_s = \frac{\mu}{1-\mu^2}\frac{t_s}{H} + \frac{A_{w_s}/b_s}{H}\sin^2 Y \cos^2 Y$$

$$A_{x\theta} = \frac{1}{2(1+\mu)}\frac{t_s}{H} + \frac{A_{w_s}/b_s}{H}\sin^2 Y \cos^2 Y$$

$$\bar{k}_x = \frac{1}{A_x}\frac{A_{w_s}/b_s}{H}\bar{k}_{w_s}\cos^4 Y$$

$$\bar{k}_\theta = \frac{1}{A_\theta}\frac{A_{w_s}/b_s}{H}\bar{k}_{w_s}\sin^4 Y$$

$$\bar{k}_s = \frac{1}{A_s}\frac{A_{w_s}/b_s}{H}\bar{k}_{w_s}\sin^2 Y \cos^2 Y$$

$$\bar{k}_{x\theta} = \frac{1}{A_{x\theta}}\frac{A_{w_s}/b_s}{H}\bar{k}_{w_s}\sin^2 Y \cos^2 Y$$

$$I_x = \frac{1}{12(1-\mu^2)}\left(\frac{t_s}{H}\right)^3 + \frac{I_{w_s}/b_s}{H^3}\cos^4 Y + \frac{1}{1-\mu^2}\frac{t_s}{H}(\bar{k}_x)^2$$
$$+ \frac{A_{w_s}/b_s}{H}[(\bar{k}_{w_s} - \bar{k}_x)^2 \cos^4 Y]$$

$$I_\theta = \frac{1}{12(1-\mu^2)}\left(\frac{t_s}{H}\right)^3 + \frac{I_{w_s}/b_s}{H^3}\sin^4 Y + \frac{1}{1-\mu^2}\frac{t_s}{H}(\bar{k}_\theta)^2$$
$$+ \frac{A_{w_s}/b_s}{H}(\bar{k}_{w_s} - \bar{k}_\theta)^2 \sin^4 Y$$

$$I_s = \frac{\mu}{12(1-\mu^2)}\left(\frac{t_s}{H}\right)^3 + \frac{I_{w_s}/b_s}{H^3}\sin^2 Y \cos^2 Y + \frac{\mu}{1-\mu^2}\frac{t_s}{H}(\bar{k}_s)^2$$
$$+ \frac{A_{w_s}/b_s}{H}(\bar{k}_{w_s} - \bar{k}_s)^2 \sin^2 Y \cos^2 Y$$

$$I_{x\theta} = \frac{1}{6(1+\mu)}\left(\frac{t_s}{H}\right)^3 + 4\frac{I_{w_s}/b_s}{H^3}\sin^2 Y \cos^2 Y + \frac{2}{1+\mu}\frac{t_s}{H}(\bar{k}_{x\theta})^2$$
$$+ 4\frac{A_{w_s}/b_s}{H}(\bar{k}_{w_s} - \bar{k}_{x\theta})^2 \sin^2 Y \cos^2 Y$$

$$\frac{A_{w_s}/b_s}{H} = 2\left\{1 - \left[1 - 0.43\left(\frac{r_{w_s}}{t_s}\right)^2\frac{t_s}{t_{w_s}}\right]\frac{t_s}{H}\right\}\frac{t_{w_s}}{t_s}\frac{t_s}{b_s}$$

$$\bar{k}_{w_s} = \frac{2}{\dfrac{(A_{w_s}/b_s)}{H}}\left[\frac{1}{2}\left(1 - \frac{t_s}{H}\right)^2 + 0.14\left(\frac{r_{w_s}}{t_s}\right)^3\frac{t_s}{t_{w_s}}\left(\frac{t_s}{H}\right)^2\right]\frac{t_{w_s}}{t_s}\frac{t_s}{b_s} + \frac{1}{2}\frac{t_s}{H}$$

$$\frac{I_{w_s}/b_s}{H^3} = 2\left[\frac{1}{12}\left(1 - \frac{t_s}{H}\right)^3 + \left(1 - \frac{t_s}{H}\right)\left(\frac{1}{2} - \bar{k}_{w_s}\right)^2 + 0.01\left(\frac{r_{w_s}}{t_s}\right)^4 \frac{t_s}{t_{w_s}}\left(\frac{t_s}{H}\right)^3\right.$$

$$\left. + 0.43\left(\frac{r_{w_s}}{t_s}\right)^2 \frac{t_s}{t_{w_s}} \frac{t_s}{H}\left(\bar{k}_{w_s} - \frac{1}{2}\frac{t_s}{H} - 0.218\frac{r_{w_s}}{t_s}\frac{t_s}{H}\right)^2\right]\frac{t_{w_s}}{t_s}\frac{t_s}{b_s}$$

The special case of $Y = 90°$ corresponds to a shell with circumferential stiffeners $2t_{w_s}$ wide and b_s apart. The case of $Y = 0$ corresponds to a shell with longitudinal or meridional stiffeners only, depending on the type of shells considered, $2t_{w_s}$ wide and b_s apart.

For more exact formulas of the elastic constants for this type of construction, or if a more complex type of construction is used which contains stiffeners in the x and θ directions as well as skewed stiffeners, Ref. 11-2 may be used.

Special Cases

If the elastic constants for the shell wall can be expressed in the following manner:

$$B = B_x = B_\theta \qquad D = D_x = D_\theta \qquad \mu = \mu_x = \mu_\theta = \mu_x' = \mu_\theta'$$

$$\frac{G_{x\theta}}{B_x} = \frac{1 - \mu}{2} \qquad \frac{D_{x\theta}}{D_x} = 1 - \mu$$

this shell can be treated as a pseudo unstiffened isotropic shell of thickness t with a Young's modulus E where

$$t = \sqrt{\frac{12D}{B}}$$

$$E = \frac{B(1 - \mu^2)}{t}$$

Using these expressions to obtain t and E, the equations which are presented for the stress analysis and stability analysis of thin unstiffened isotropic shells may be used directly without resorting to the considerably more complicated orthotropic-shell equations. The elastic constants for several types of composite construction which may be treated in this manner will be presented.

If an integrally stiffened waffle shell has the stiffeners in the pattern shown in Fig. 11-5 and the Poisson's ratio of the material is 1/3, the necessary elastic constants are

$$\mu = \frac{1}{3}$$

$$B = \frac{9}{8} Et(1 + a_1)$$

$$D = \frac{9}{8} \frac{Et^3}{1 + a_1} [3a_1(1 + a_2)^2 + (1 + a_2)(1 + a_1 a_2^2)]$$

where

$$a_1 = \frac{t_{w_s}(H - t_s)}{t_s b_s} \qquad a_2 = \frac{H - t_s}{t_s}$$

and where E is Young's modulus of the material.

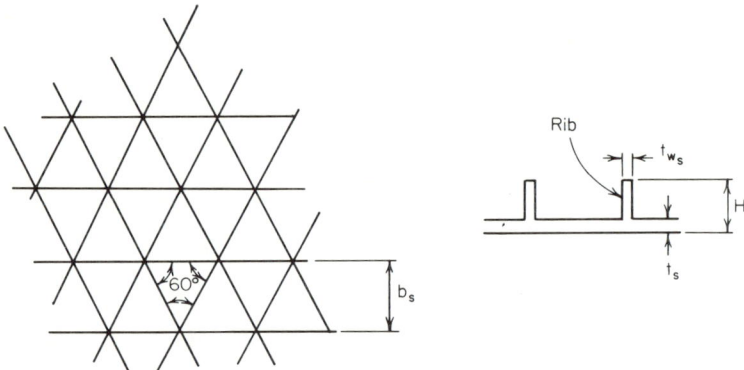

figure 11-5 *Special waffle construction.*

If the facing sheets of a sandwich-type construction (see Fig. 11-3) are isotropic and have the same Poisson's ratios and if in addition the core of the sandwich resists very little bending or stretching but is relatively stiff in transverse shear, the necessary elastic constants are

$$B = \frac{E_1 t_1 + E_2 t_2}{1 - \mu^2}$$

$$D = \frac{E_1 t_1 E_2 t_2 h^2}{(1 - \mu^2)(E_1 t_1 + E_2 t_2)}$$

where E = Young's modulus of elasticity of the facing-sheet material
μ = Poisson's ratio of the facing-sheet material

The subscripts 1, 2, represent the facing sheet to which the material corresponds.

A layered shell may be considered pseudo-isotropic if (1) each layer is made from the same orthotropic material, (2) each layer is the same thickness, (3) there are many layers, and (4) the natural axes of the orthotropic layers (see Fig. 11-2) are oriented in the following manner:

$$Y = +30°, -30°, 90°, +30° - 30°, 90°, +30°, \text{etc.}$$

The necessary elastic constants are

$$B = \frac{E_L t_L}{1 - \mu_L^2}$$

$$D = \frac{E_L t_L^3}{12(1 - \mu_L^2)}$$

where

$$\mu_L = \frac{E_\alpha + 2[3\mu_\alpha E_\beta - 2G_{\alpha\beta}(1 - \mu_\alpha \mu_\beta)] + E_\beta}{3E_\alpha + 2[\mu_\alpha E_\beta + 2G_{\alpha\beta}(1 - \mu_\alpha \mu_\beta)] + 3E_\beta}$$

$$E_L = \frac{3}{8} \frac{1 - \mu_L^2}{1 - \mu_\alpha \mu_\beta} \{3E_\alpha + 2[\mu_\alpha E_\beta + 2G_{\alpha\beta}(1 - \mu_\alpha \mu_\beta)] + 3E_\beta\}$$

t_L is the total thickness of the shell and E_α, E_β, μ_α, μ_β, and $G_{\alpha\beta}$ are the material properties of each layer in the direction of the natural axis as previously defined.

11-3 Cylinders

Axial Compression, Orthotropic Cylinders

The following stability analysis for simply supported orthotropic cylinders subjected to axial compression is based primarily on the theory from Refs. 11-3 and 11-4. This analysis may be used for layered construction (such as filament-wound) and for stiffened construction (such as integrally stiffened waffle construction) if the stiffeners are very close together. The definition of the elastic constants used in this section and the formulas for computing the elastic constants for typical types of construction are given in Sec. 11-2. The design-allowable buckling load per unit width N_x for moderate-length orthotropic cylinders is

$$N_x = \gamma \frac{2}{R} [B_\theta D_x (1 - \mu_x' \mu_\theta')]^{1/2} U$$

and the allowable compressive load for the cylinder is

$$P_{cr} = N_x 2\pi R$$

R and L are the cylinder radius and length, respectively. U is the buckling parameter, which may be obtained from the subsequent procedure.

The following parameters are defined in terms of the elastic constants and geometry:

$$\bar{G} = \frac{B_\theta(1 - \mu_x'\mu_\theta')}{2G_{x\theta}} - \mu_\theta'$$

$$\bar{D} = \frac{D_{x\theta}}{D_x} + \mu_\theta$$

$$Z^2 = \frac{B_\theta}{12 D_x} \frac{1 - \mu_x'\mu_\theta'}{R^2} L^4$$

$$\omega_2 = \frac{B_x D_\theta}{B_\theta D_x}$$

$$\omega_3 = \frac{\bar{D}}{\bar{G}}$$

For $\omega_2 \geqslant 1$, $\omega_3 \geqslant 1$, $U = 1$.
For $\omega_2 \geqslant \omega_3$, $\omega_3 < 1$, $U = U_1$.
For $\omega_2 < \omega_3$, $\omega_2 < 1$, $U = 1$ if $U_2 \geqslant 1$; $U = U_2$ if $U_2 < 1$.
The parameter γ can be obtained from Fig. 11-6.

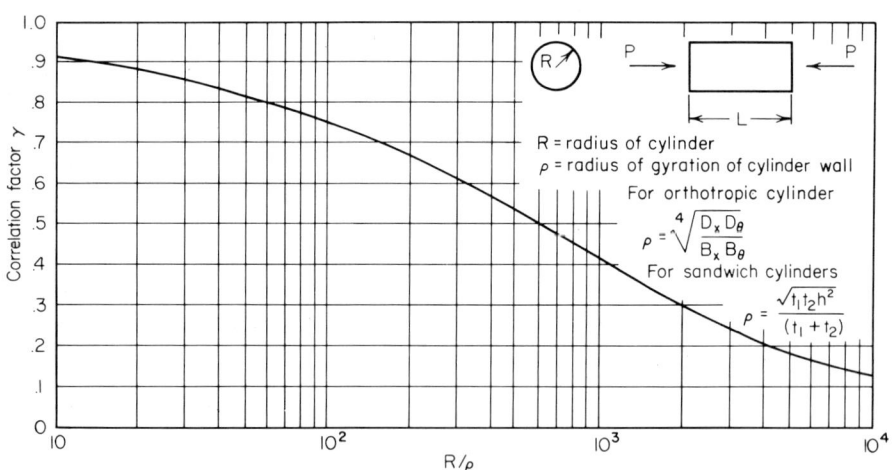

figure 11-6 *Design-correction coefficient for cylinders subjected to axial compression.*

The parameter U_1 is obtained from the formula

$$U_1 = \left[\frac{1 + 2\bar{D}\psi^2 + (D_\theta/D_x)\psi^4}{1 + 2\bar{G}\psi^2 + (B_\theta/B_x)\psi^4} \right]^{1/2}$$

for
$$\psi^2 = S_1 \pm \sqrt{S_1^2 + S_2}$$

$$S_1 = \frac{1}{2\bar{G}} \frac{1 - \omega_2}{\omega_2 - \omega_3}$$

$$S_2 = \frac{B_x}{B_\theta} \frac{1 - \omega_3}{\omega_2 - \omega_3}$$

Only values of $\psi^2 > 0$ may be used to compute U_1. If both values of ψ^2 are greater than zero, then U_1 must be computed for each ψ^2 and the smallest U_1 must be used to compute N_x.

The parameter U_2 can be obtained from the formula

$$U_2 = \frac{\pi^2}{4\sqrt{3}\,Z\gamma}\left[1 + 2\bar{D}\psi^2 + \frac{D_\theta}{D_x}\psi^4 + \frac{12Z^2\gamma^2/\pi^4}{1 + 2\bar{G}\psi^2 + (B_\theta/B_x)\psi^4}\right]$$

where
$$m = 1 \qquad \psi = \frac{L}{2\pi R}\frac{n}{m}$$

The value of n must be varied until the minimum value of U_2 is found. The quantity n is the number of half waves of the buckles in the circumferential direction. Therefore, n is restricted to even positive integers greater than 4. The quantity m is the number of half waves in the axial direction and for this case is restricted to 1.

The correlation factor γ was introduced to allow for the discrepancy between test data and the buckling theory. If $\gamma = 1$, the classical theoretical buckling load is obtained. The curve for γ given in Fig. 11-6 was obtained by replotting the curve given in Fig. 10-9 as a function of R/ρ, where ρ is the radius of gyration of the shell wall, as suggested in Ref. 11-5. This in effect uses the test data for isotropic cylinders to obtain γ for orthotropic cylinders. For radii of gyration which differ in the axial and circumferential direction, Ref. 11-5 recommends using the geometric mean of the two radii of gyrations; therefore,

$$\rho = \sqrt[4]{\frac{D_x}{B_x}\frac{D_\theta}{B_\theta}}$$

The method of obtaining γ for orthotropic cylinders has been verified by only a limited amount of test data, and caution should be used in applying it to a design. Some test results, such as the tests described in Refs. 11-4 and 11-6, indicate that this method may be conservative, but the specimens tested were not typical of large-production parts.

The preceding analysis is good only for moderate-length cylinders;

therefore γZ must be greater than about $2U_3$, where, if $U = 1$ or $U = U_2$, then $U_3 = 1$, and if $U < 1$ or $U \neq U_2$, then

$$U_3 = \left(1 + 2\bar{D}\psi^2 + \frac{D_\theta}{D_x}\psi^4\right)^{1/2}\left(1 + 2\bar{G}\psi^2 + \frac{B_\theta}{B_x}\psi^4\right)^{1/2}$$

For short cylinders, $\gamma Z < 2U_3$ the parameter U is given by the equation

$$U = \frac{\pi^2}{4\sqrt{3}\,Z\gamma}\left[m^2\left(1 + 2\bar{D}\psi^2 + \frac{D_\theta}{D_x}\psi^4\right) + \frac{12Z^2\gamma^2}{\pi^4 m^2}\frac{1}{1 + 2\bar{G}\psi^2 + (B_\theta/B_x)\psi^4}\right]$$

where
$$\psi = \frac{L}{2\pi R}\frac{n}{m}$$

The value of m and n must be varied until the minimum value of U is found. The quantities n and m are the number of half waves of the buckles in the circumferential and axial directions, respectively. Therefore, n is restricted to even positive integers greater than 4 and m is restricted to positive integers.

For axial-stiffened cylinders ($D_x > 300D_\theta$) in the short-cylinder range, the following formula is recommended in Ref. 11-7:

$$N_x = \frac{c\pi^2 D_x}{L^2} + \frac{2\gamma}{R}\sqrt{B_x D_\theta}$$

Coefficient c is equivalent to the column-fixity coefficient in Euler's column formula. It is recommended that this formula be restricted to geometry where $\gamma\sqrt{B_x D_\theta}/R < c\pi^2 D_x/L^2$. Verification of this formula has been limited, and the range of validity is not well defined; therefore, it should be used with caution.

If a cylinder is stiffened with stringers and frames, it is recommended that Chap. 12 be used unless the stringers and frames are very close together. The test results of Ref. 11-8 for cylinders with light frames and heavy stringers indicate that the analysis presented in this section may be unconservative for cylinders with large frame spacing.

Plasticity may be considered by modifying Young's modulus in the stiffness constants (Ref. 11-3) or may be accounted for with a plasticity factor as in Chap. 10.

Torsion, Orthotropic Cylinders

The curves in Ref. 11-9 will be used to determine the buckling load for orthotropic cylinders subjected to torsion. The design-allowable shear load per unit length of circumference is

$$N_{x\theta} = \frac{\gamma K_s \pi^2 D_x}{L^2}$$

and the design-allowable torque for the cylinder is

$$T_{cr} = N_{x\theta} 2\pi R^2$$

The buckling coefficient K_s may be obtained from Fig. 11-7 for the elastic constants as given in Table 11-1, and γ may be obtained from

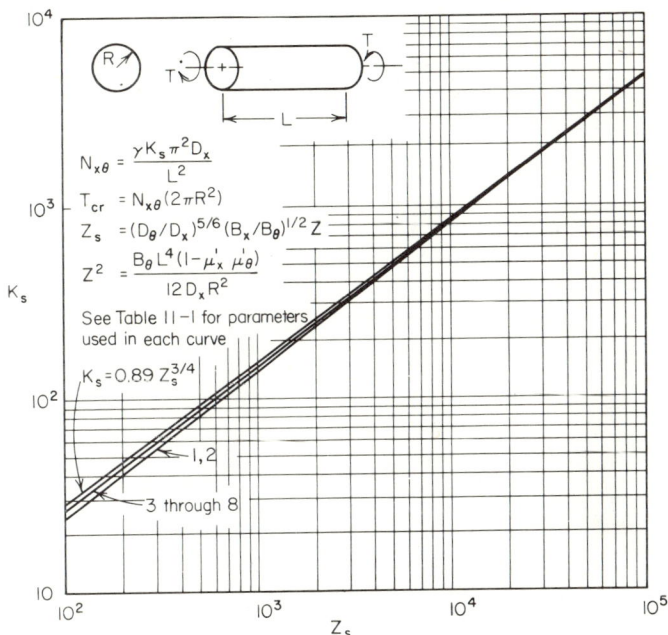

figure 11-7 *Buckling coefficient for orthotropic cylinders subjected to torsion.*

TABLE 11-1 Cylinder Parameters used for K_s - Z Plots

Curve number	$\dfrac{D_{x\theta}}{D_x} + \mu_\theta$	$\dfrac{D_\theta}{D_x}$	$\dfrac{B_x}{B_\theta}$	$\dfrac{B_\theta(1-\mu'_x \mu'_\theta)}{2G_{x\theta}} - \mu'_\theta$
1	0	1/8	4	1
2	0	1/2	1	1
3	0	2	1/4	1
4	1	1/8	4	4
5	1	1/2	1	4
6	1	2	1/4	4
7	1	4	1/8	4
8	1	8	1/16	4

298   **Structural Analysis of Shells**

Fig. 11-8. The definition of the elastic constants used in this section and formulas for computing the elastic constants for typical types of construction are given in Sec. 11-2.

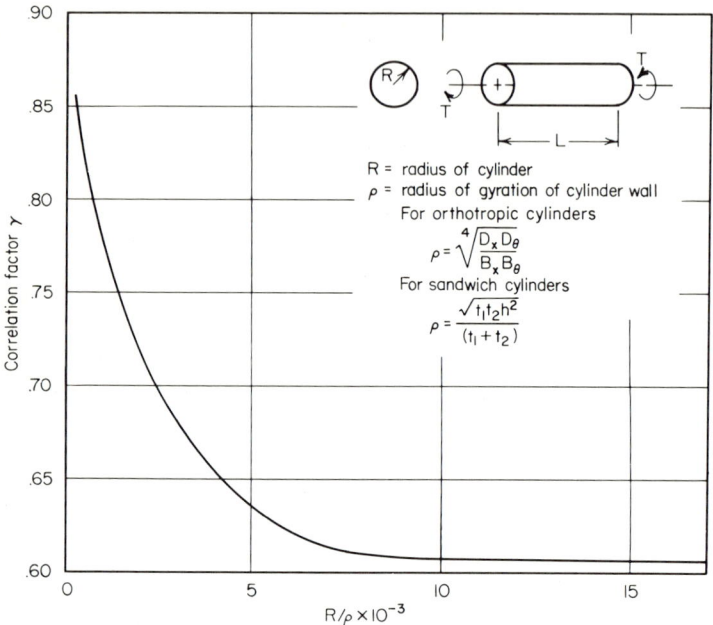

figure 11-8 *Design-correction coefficient for cylinders subjected to torsion.*

The coefficient γ reduces the theory presented in Ref. 11-9 by the same percentage as the theory for homogeneous isotropic cylinders was reduced to obtain the curve presented in Fig. 10-11.

The method of analysis presented may be used for cylinders with simply supported edges, although a small rotational restraint at the edge of the shell is included in the results presented in Fig. 11-7. It can be seen from Fig. 11-7 that for large values of Z_s, all curves merge into one line. The equation of this line as given in Ref. 11-9 is

$$K_s = 0.89 Z_s^{3/4}$$

where

$$Z_s = \left(\frac{D_\theta}{D_x}\right)^{5/6} \left(\frac{B_x}{B_\theta}\right)^{1/2} Z$$

$$Z^2 = \frac{B_\theta(1 - \mu_x' \mu_\theta') L^4}{12 D_x R^2}$$

If the value of Z_s is large enough, this formula can be used to estimate the critical torque for geometries other than the ones given in Table 11-1.

This section should not be used for any type of cylinder with skin that buckles between stiffeners prior to general instability failure. The test data presented in Ref. 11-8 indicate that the results may be unconservative for this case.

Bending, Orthotropic Cylinders

The buckling formulas presented for orthotropic cylinders subjected to axial compression may be used to determine the design-allowable buckling stress for orthotropic cylinders subjected to bending if γ is obtained from Fig. 11-9. Figure 11-9 was obtained from Fig. 10-13 in the same manner as Fig. 11-6 was obtained from Fig. 10-9.

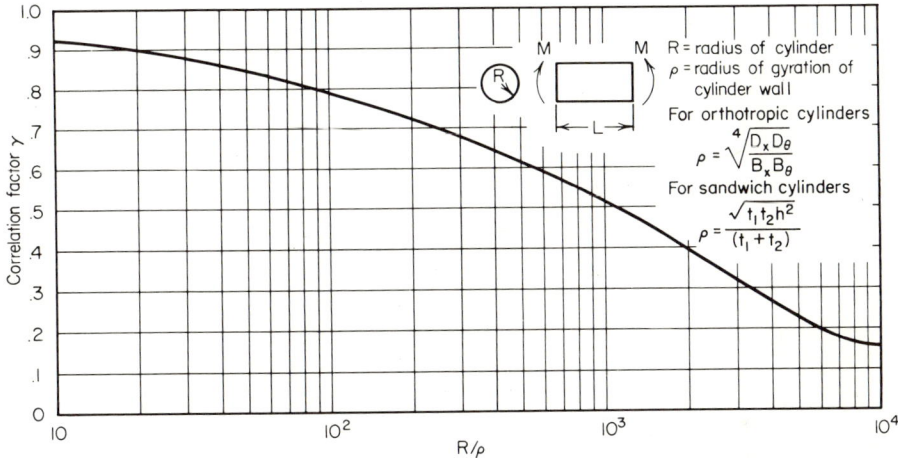

figure 11-9 *Design-correction coefficient for cylinders subject to bending.*

For bending, N_x is the maximum compressive load per unit length due to the bending moment (e.g., outer-fiber load). If the stresses are elastic, the design-allowable bending moment may be obtained from

$$M_{cr} = \pi N_x R^2$$

The analysis presented for bending of orthotropic cylinders assumes that the theoretical buckling load for bending is the same as the theoretical buckling load for axial compression. While this is true for isotropic cylinders, the theoretical buckling loads for bending can be greater than for axial compression if the cylinder is orthotropic. Therefore, this approach will be conservative for some geometries.

External Pressure, Orthotropic Cylinders

The curves presented in Ref. 11-9 will be used to determine the buckling pressure of orthotropic cylinders with simply supported edges subjected to lateral external pressure (no pressure acting in the longitudinal direction on bulkheads). The design-allowable pressure is

$$p_{cr} = \frac{K_p \pi^2 D_x}{RL^2}$$

The buckling coefficient K_p may be obtained from Fig. 11-10 or 11-11 for several ratios of the elastic constants. The definition of the elastic

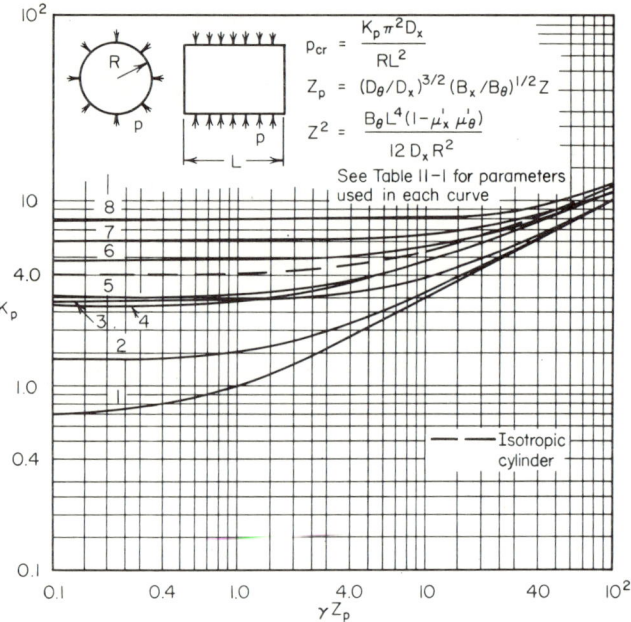

figure 11-10 *Buckling coefficient for short orthotropic cylinders subjected to lateral external pressure.*

constants and formulas for computing the elastic constants for typical types of construction are given in Sec. 11-2.

The coefficient γ was introduced to reduce the theory presented in Ref. 11-9 by the same percentage as the theory for homogeneous isotropic cylinders was reduced to obtain the curve presented in Fig. 10-15. Therefore, $\gamma = 0.8$. There is very little test data to verify this choice of γ.

It can be seen from Fig. 11-11 that for large values of Z_p all curves merge into one line. The equation of this line is $K_p = 1.039 Z_p^{1/2}$, and the design-allowable pressure is

$$p_{cr} = \frac{5.5 D_\theta^{3/4} \gamma^{1/2} [B_x(1 - \mu_x' \mu_\theta')]^{1/4}}{LR^{3/2}} \tag{11-1}$$

If the value of Z_p is large enough, this formula can be used to estimate the critical pressure of geometries other than the ones given in Table 11-1.

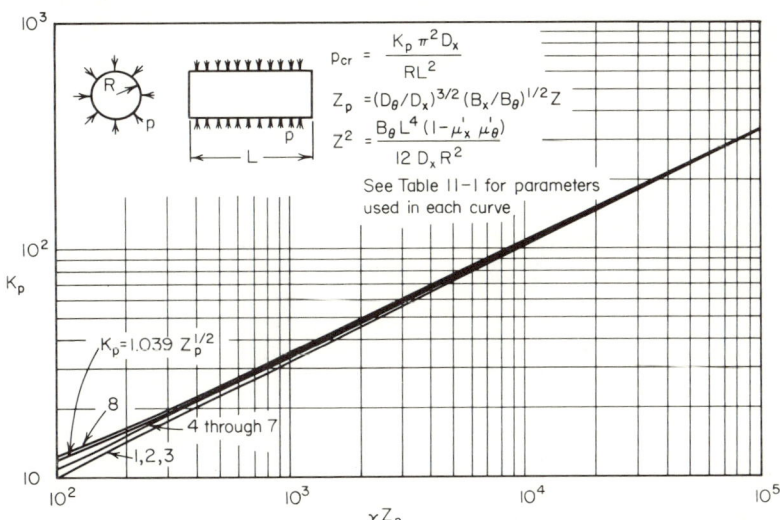

figure 11-11 *Buckling coefficient for long orthotropic cylinders subjected to lateral external pressure.*

For very long cylinders,

$$\frac{L}{R} > 1.83 \left[\frac{B_x(1 - \mu_x' \mu_\theta') R^2}{D_\theta} \right]^{1/4}$$

the buckling pressure becomes independent of length and may be computed from the equation

$$p_{cr} = \frac{3 \gamma^{1/2} D_\theta}{R^3} \tag{11-2}$$

If the stiffness parameters are not in the range given in Table 11-1 and Z_p is not large, K_p may be determined from

$$K_p = \frac{1}{\psi^2} \left(1 + 2\bar{D}\psi^2 + \frac{D_\theta}{D_x} \psi^4 \right) + \frac{12 Z^2 \gamma^2}{\pi^4} \left[\frac{1}{1 + 2\bar{G}\psi^2 + (B_\theta/B_x)\psi^4} \right]$$

where $\bar{G}$ and $\bar{D}$ are defined as before.

$$m = 1$$

$$\psi = \frac{nL}{2\pi Rm}$$

$$Z^2 = \frac{B_\theta(1 - \mu_x'\mu_y')L^4}{12D_x R^2}$$

The value of n must be varied until the minimum value of K_p is found. The quantity n is restricted to even positive integers greater than or equal to 4.

If an orthotropic cylinder is subjected to lateral and axial external pressure (see Fig. 11-12, for definition of load), the design-allowable buckling coefficient is

$$K_p = \frac{m^2}{\frac{1}{2} + \psi^2}\left(1 + 2\bar{D}\psi^2 + \frac{D_\theta}{D_x}\psi^4\right) + \frac{12Z^2/\pi^4 m^2}{(\frac{1}{2} + \psi^2)[1 + 2\bar{G}\psi^2 + (B_\theta/B_x)\psi^4]}$$

where

$$\psi = \frac{nL}{2\pi Rm} \quad \text{and} \quad p_{cr} = \frac{\gamma K_p \pi^2 D_x}{RL^2}$$

The value of m and n must be varied until the minimum value of K_p is found. The parameter n is restricted to even positive integers greater

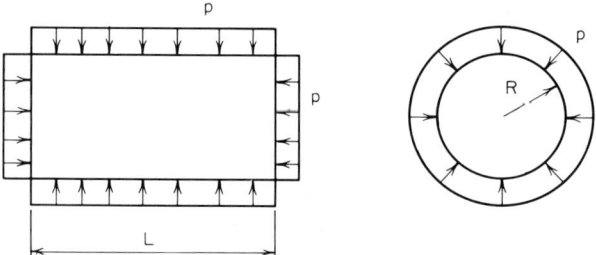

figure 11-12 *Cylinder subjected to lateral and axial external pressure.*

than or equal to 4, and m is restricted to positive integers. The case of $n = 0$ must also be checked. If $n \neq 0$, Ref. 11-10 recommends the correlation factor $\gamma = 0.75$. If $n = 0$, γ may be obtained from Fig. 11-6. If Z_p is large or if the cylinder is very long, the buckling coefficient for lateral and axial external pressure may be obtained using Eq. 11-1 or 11-2 provided that the correlation factor $\gamma = 0.56$ is used. If the buckling pressure obtained from Eq. 11-1 or 11-2 is greater than the buckling pressure for the case of $n = 0$, the buckling pressure is given by

$$p_{cr} = \frac{4\gamma}{R^2} [B_\theta D_x (1 - \mu_x' \mu_\theta')]^{1/2}$$

where γ is obtained from Fig. 11-6. This corresponds to the case of $n = 0$ and is independent of the lateral pressure. Therefore, the axial compressive correlation factor is recommended. The range of validity of the formulas for large Z or very long cylinders is unknown for the cases of axial and lateral pressure and should be used with caution if unusual stiffness ratios are used. However, they are very useful in obtaining quick estimates of the buckling pressure.

The design-allowable pressure is for complete buckling of the shell (e.g., when buckles have formed all the way around the cylinder). If single buckles are not allowable for a particular design, the pressure computed by the preceding formulas may be unconservative.

Shells that are relatively stiff in the circumferential direction and relatively free of initial imperfection will be less likely to have single isolated buckles at pressures less than the design-allowable pressures which have been given.

If the stresses are in the plastic range, a reduced modulus must be included in the stiffness constants (Ref. 11-7) or a plasticity correction factor such as that given in Chap. 10 should be used.

11-4 Cones

Axial Compression, Orthotropic Cones

The limited amount of information available on orthotropic cones is not in a form suitable for design analysis until additional information is available. The equivalent-cylinder approach recommended in Chap. 10 should be used. The cone shown in Fig. 11-13a can be analyzed as a cylinder with a radius $R_e = R_1/\cos \alpha$ and length L. The design-allowable load per inch N_x for the equivalent cylinder can be obtained from Sec. 11-3. The design-allowable total compressive load for the cone can be obtained from

$$P_{cr} = 2\pi R_e N_x \cos^2 \alpha$$

This method of analysis should be used with caution and should be limited to cones with $\alpha < 30°$.

304 *Structural Analysis of Shells*

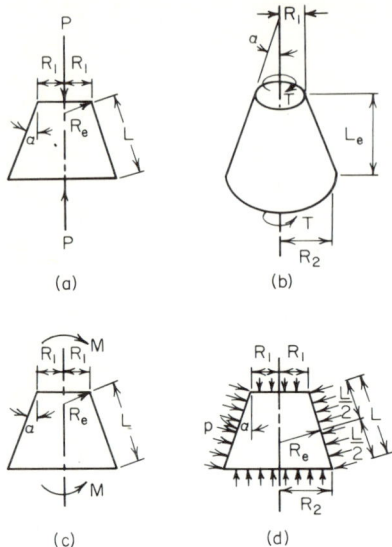

figure 11-13 *Cones subjected to various loadings.*

Torsion, Orthotropic Cones

Until additional information is available, the equivalent-cylinder approach recommended in Sec. 10-3 should be used for orthotropic cones subjected to torsion. The cone shown in Fig. 11-13b (with the corresponding nomenclature) can be analyzed as a cylinder with a radius

$$R_e = \left[1 + \left(\frac{1 + R_2/R_1}{2}\right)^{1/2} - \left(\frac{1 + R_2/R_1}{2}\right)^{-1/2}\right] R_1 \cos \alpha$$

and length $L_e = L/\cos \alpha$.

The design-allowable shear per unit length $N_{x\theta}$ for the equivalent cylinder can be obtained from Sec. 11-3. The design-allowable torque for the cone can be obtained from

$$T_{cr} = 2\pi R_e^2 N_{x\theta}$$

The design-allowable shear stress for the cone should be based on T_{cr}.

This method should be used with caution and should be limited to cones with $\alpha < 30°$.

Bending, Orthotropic Cones

Until additional information is available, the equivalent-cylinder approach recommended in Sec. 10-3 should be used for orthotropic cones subjected to bending. The cone shown in Fig. 11-13c (with corresponding nomenclature) can be analyzed as a cylinder with a radius $R_e = R_1/\cos\alpha$ and length L. The design-allowable load per inch N_x for the equivalent cylinder can be obtained from Sec. 11-3. If the stresses are elastic, the design-allowable moment for the cone can be obtained from

$$M_{cr} = \pi R_1^2 N_x \cos\alpha$$

This method of analysis should be used with caution and should be limited to cones with $\alpha < 30°$.

Lateral and Axial External Pressure, Orthotropic Cones

Until additional information is available, the equivalent-cylinder approach recommended in Sec. 10-3 should be used for an orthotropic cone subjected to lateral and axial pressure as shown in Fig. 11-13d. The cone can be analyzed as a cylinder with a radius

$$R_e = \frac{R_1 + R_2}{2 \cos\alpha}$$

and length L. The design-allowable pressure can be obtained from Sec. 11-3. This method of analysis should be used with caution and should be limited to cones with $\alpha < 30°$.

REFERENCES

11-1. Ambartsumyan, S. A.: *Theory of Anisotropic Shells*, NASA F-118, May, 1964.
11-2. Dow, N., C. Libove, and R. E. Hubka: *Formulas for the Elastic Constants of Plates with Integral Waffle-like Stiffening*, NACA RM L53 E13a, 1953.
11-3. Gerard, G.: Compressive Stability of Orthotropic Cylinders, *J. Aerospace Sci.*, October, 1962.
11-4. Milligan, R., G. Gerard, and C. Lakshmikantham: *General Instability of Orthotropically Stiffened Cylinders under Axial Compression*, AIAA Paper 66-139, 3rd Aerospace Sciences meeting, January, 1966.
11-5. Peterson, J.: *Weight-strength Studies of Structures Representative of Fuselage Construction*, NACA TN 4114, 1957.
11-6. Tasi, J., A. Feldman, and D. Stang: *The Buckling Strength of Filament-wound Cylinders under Axial Compression*, NASA CR-266, July, 1965.
11-7. Peterson, J., R. Whitley, and J. Deaton: *Structural Behavior and Compressive Strength of Circular Cylinders with Longitudinal Stiffening*, NASA TN D-1251, 1962.
11-8. Dunn, L.: *Some Investigations of the General Instability of Stiffened Metal Cylinders*, IX, *Criterions for the Design of Stiffened Metal Cylinders Subjected to General Instability Failures*, NACA TN 1198, 1947.
11-9. Becker, H., and G. Gerard: Elastic Stability of Orthotropic Shells, *J. Aerospace Sci.*, vol. 29, no. 5, May, 1962.
11-10. *Buckling of Thin-walled Circular Cylinders*, NASA SP-8007, August, 1968.

Chapter 12

STABILITY OF STIFFENED SHELLS

12-1 General

The stiffened shells which are discussed in the following sections are cylinders which consist of a thin metal sheet stiffened by frames (circumferential stiffening elements) and stringers (longitudinal stiffening elements). In general, this type of shell should be analyzed for three modes of failure: (1) material failure, (2) buckling between stiffeners, and (3) general-instability failure.

If the frames and stiffeners are close together, and the stiffener eccentricities are small, the procedures presented in Chap. 11 may be useful for the general-instability analysis, but the range of applicability of the method is not well defined. In this section, a different method of analysis is presented for frame- and stringer-stiffened cylinders subjected to compressive loads in the axial direction as well as for frame-stiffened cylinders subjected to lateral and axial external pressure.

12-2 Frame- and Stringer-stiffened Cylinders

Axial Compression, Frame- and Stringer-stiffened Cylinders

If a cylindrical shell having both frames (circumferential stiffeners) and stringers (longitudinal stiffeners) is subjected to axial compression, it may fail in one of three distinct ways. The types of failure are classified as (1) material failure, (2) buckling between frames, and (3) general instability.

Material Failure: The stress caused by the applied axial load should be compared with the material's allowable stress. Methods of obtaining stresses in shells due to various types of loading are discussed in previous chapters.

Buckling between Frames: Buckling between frames will occur in a cylinder that has relatively heavy frames and light stringers; the cylinder tends to act as a number of isolated axially stiffened cylinders each of which is one frame spacing long. Failure will occur by some form of instability of the stringers, modified by the effect of the attached sheet. The frames will remain circular in cross section. The only function of the frames in this case will be to determine the end-fixity coefficient of the stringers. The four forms of instability which must be investigated for this type of failure are as follows:

1. Buckling of the sheet between stringers and frames
2. Crippling of the stringers
3. Torsional instability of the stringers
4. Lateral buckling of the sheet stringer panel between frames

Although there are four distinct instability modes, the ultimate buckling failure between frames of a stiffened cylinder subjected to stress is usually a combination of these modes.

Buckling of the sheet between the stringers and frames does not necessarily constitute an ultimate failure of the structure; however, the buckling stresses σ_{cs} of the sheet must be known to determine the stress distribution in the cylinder. The buckling stresses for this mode may be calculated by the methods presented for axial compression of curved panels in Sec. 10-2.

Crippling is a local instability failure of the elements of the stringers and is defined as any type of failure in which the cross sections of the stringers are distorted in their own plane but not translated or rotated. The length of the buckle involved in a crippling failure is of the same order of magnitude as its cross-sectional dimensions. A typical beginning of failure by crippling is shown in Fig. 12-1.

Crippling generally occurs in stringers having wide, thin flanges.

308 Structural Analysis of Shells

The crippling stress is defined as σ_{cc} and may be determined by the usual methods of analysis for columns.

One of the procedures for computing the allowable crippling stress of stringers with simple cross sections will be presented. This method is primarily from Ref. 12-1. The stringer is broken up into parts as shown in Fig. 12-2.

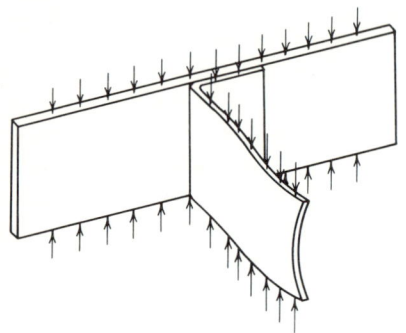

figure 12-1 *Typical failure in crippling.*

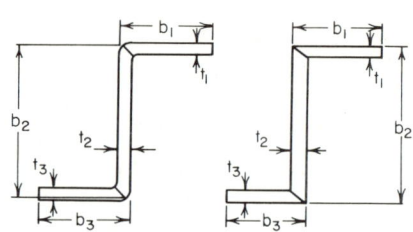

figure 12-2 *Separation of sections into subsections.*

The allowable crippling stress for the total stringer is

$$\sigma_{cc} = \frac{\sum_{i=1}^{m} \sigma_{cc_i} A_i}{\sum_{i=1}^{m} A_i}$$

$$\sigma_{cc_i} = C_{e_i} \sqrt{\sigma_{cy_i} E_i \eta_i} \left(\frac{t_i}{b_i}\right)^{3/4}$$

where
- m = number of parts the stringer is cut in
- i = subscript referring to part number
- C_e = material and shape parameter constant derived from test specimens (see Table 12-1)
- $C_e = C_{e_1}$ = if one edge is free such as parts 1 and 3 of Fig. 12-2
- $C_e = C_{e_2}$ = if both edges are attached to adjacent parts such as part 2 in Fig. 12-2
- σ_{cy} = compressive yield stress of material
- E = compressive modulus of elasticity of material
- A_i = area of part
- η = plasticity correction given by curve A if C_{e_1} is used and curve C if C_{e_2} is used (see Sec. 10-7)

TABLE 12-1 Material and Shape Parameter Constants

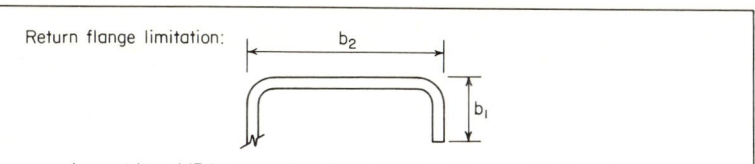

b_1 must be $\geq 1/3\, b_2$ to consider the b_2 leg as continuous at both ends

Material	C_{e_1}	C_{e_2}	Maximum element stress σ_{cy}, psi	Modulus $E \times 10^{-6}$, psi
24S–T4 Bare, Extruded	0.312	0.590	40,000	10.5
24S–T4	0.312	0.590	36,000	10.5
75S–T6 Bare	0.312	0.590	71,000	10.5
75S–T6 Clad	0.312	0.590	66,000	10.5
75S–T6 Extruded	0.312	0.590	70,000	10.5
Ti 6AL–4V Annealed (Formed)	0.304	0.771	126,000	15.8
Ti 6AL–4V Annealed (Extruded)	0.304	0.771	120,000	15.8
Ti 6AL–4V Heat Treated (Formed and Extruded)	0.304	0.771	150,000	17.0
4130 Formed, 4340 Extruded	0.312	0.735	175,000	29.0
1/2 Heat (301 Formed) (303 Extruded)	0.365	0.800	85,000	26.0
17–7PH Formed	0.333	0.631	158,000	30.0
17–7PH Extruded	0.333	0.631	176,000	30.0
AM350 Formed, AM355 Extruded	0.333	0.631	165,000	28.6
Inconel X Formed	0.300	0.750	105,000	31.0
H-11 Formed and Extruded	0.296	0.704	280,000	30.9
PH15–7Mo (cres) Formed and Extruded	0.300	0.700	220,000	30.0
René 41 Formed	0.300	0.700	130,000	31.0
Ti 4AL–3Mo–IV Formed	0.304	0.771	160,000	16.0
Ti 5AL–2.5 (Formed and Extruded)	0.304	0.771	115,000	15.5

If σ_{cc_i} is below the proportional limit, then $\eta_i = 1$. At higher stresses, η_i must be compatible with σ_{cc_i}. The stress σ_{cc_i} must be less than or equal to σ_{cy_i}. The amount of sheet to include with the part of the stringer attached to the skin depends on the method of attachment.

Torsional instability occurs when the cross section of the stringer rotates but does not distort or translate in its own plane. Typical torsional modes of instability are shown in Fig. 12-3.

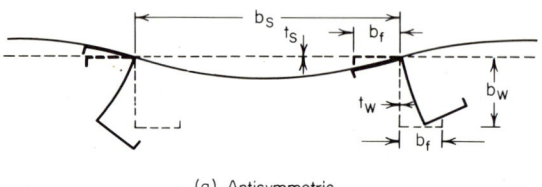

(a) Antisymmetric

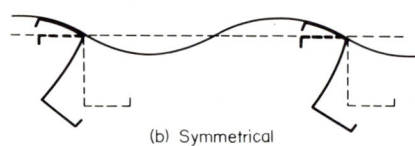

(b) Symmetrical

figure 12-3 *Torsional mode of instability.*

The methods of analysis of torsional instability of stiffeners attached to sheets, as suggested in Ref. 12-2, will be described. For the case of cylinders with typical ring spacing $[d > \pi(E\eta_G \Gamma/k)^{1/4}]$, the allowable torsional instability stress σ_{ct} for the mode shown in the preceding figure is

$$\sigma_{ct} = G\eta_A \frac{J}{I_p} + 2 \frac{\sqrt{\Gamma}}{I_p} \sqrt{E\eta_G k}$$

where G = elastic shear modulus for stringer material
η_A = plasticity correction given by curve A in Sec. 10-7
E = Young's modulus for stringer material
η_G = plasticity correction given by curve G in Sec. 10-7
I_p = polar moment of inertia of section about center of rotation, in.^{4}
J/I_p and $\sqrt{\Gamma}/I_p$ = values that may be obtained from Figs. 12-4 and 12-5 for two commonly used types of stringers
J = torsion constant of the stringer

$$GJ = \frac{\text{torque}}{\text{twist per unit length}}$$

Γ = torsional bending constant, in.^{6}
k = rotational spring constant
b, d = stringer and frame spacing, respectively
$1/k = 1/k_{\text{web}} + 1/k_{\text{sheet}}$
$k_{\text{web}} = Et_w^3/(4b_w + 6b_f)$
$k_{\text{sheet}} = \lambda_1(Et_s^3/b)$
$\lambda_1 = 1$ for the symmetric mode
$\lambda_1 = \frac{1}{3}[1 + 0.6(\sigma_{ct} - \sigma_{cs})/\sigma_{cs}]$ for the antisymmetric mode
σ_{cs} = compressive buckling stress of the sheet (see Sec. 10-2)

If $\sigma_{ct} < 4.33\,\sigma_{cs}$, the antisymmetric mode is critical. If $\sigma_{ct} > 4.33\sigma_{cs}$, the symmetric mode of failure is critical. Since λ_1, η_A, and η_G depend on σ_{ct}, the solution for σ_{ct} is in general a trial-and-error procedure. Start with the assumption that $\lambda_1 = 1$, $\eta_A = \eta_G = 1$, calculate σ_{ct}, and correct for plasticity if required. Correct λ_1 if required, and repeat the procedure until the desired convergence is obtained. Then check to see if $d > \pi(E\eta_G\Gamma/k)^{1/4}$.

If $d < \pi(E\eta_G\Gamma/k)^{1/4}$, the allowable torsional instability stress is

$$\sigma_{ct} = G\eta_A\frac{J}{I_p} + \frac{E\eta_G\Gamma}{I_p}\left(\frac{\pi}{d}\right)^2 + \frac{k(d/\pi)^2}{I_p}$$

where

$$\Gamma = \left(\frac{\sqrt{\Gamma}}{I_p}\right)^2 I_p^2$$

and J/I_p and $\sqrt{\Gamma}/I_p$ may be obtained from Fig. 12-4 or 12-5.

The formulas which have been presented may be used for stringers with sections other than those shown in Figs. 12-4 and 12-5 if the values of I_p, J, and Γ are known.

Lateral buckling of the sheet-stringer panel between frames is essentially a column instability in which the cross section of the stringer translates. It is customary to idealize the sheet stringer as a column with length equal to the frame spacing d. The lateral buckling stress of the sheet-stringer panel is defined as σ_{cp}. If no restraint exists normal to the sheet-stringer panel, it is free to buckle in alternate in and out waves in which the frames are nodes. In such a case, the panel will act as a pinned-end column and the effective end fixity c is equal to 1.0. Because of the curvature of the shell, a certain restraint to the outward buckle mode will exist because that deflection mode will involve some stretching of the sheet in the hoop direction. This restraint as well as the torsional restraint of the frame will tend, in general, to provide an effective end-fixity coefficient in the sheet-stringer column that is somewhat greater than 1.0.

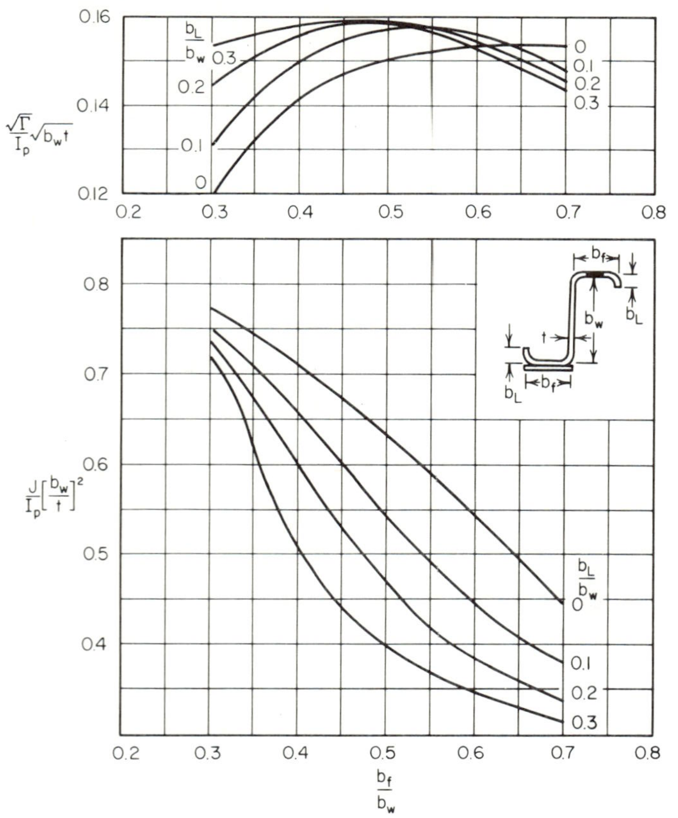

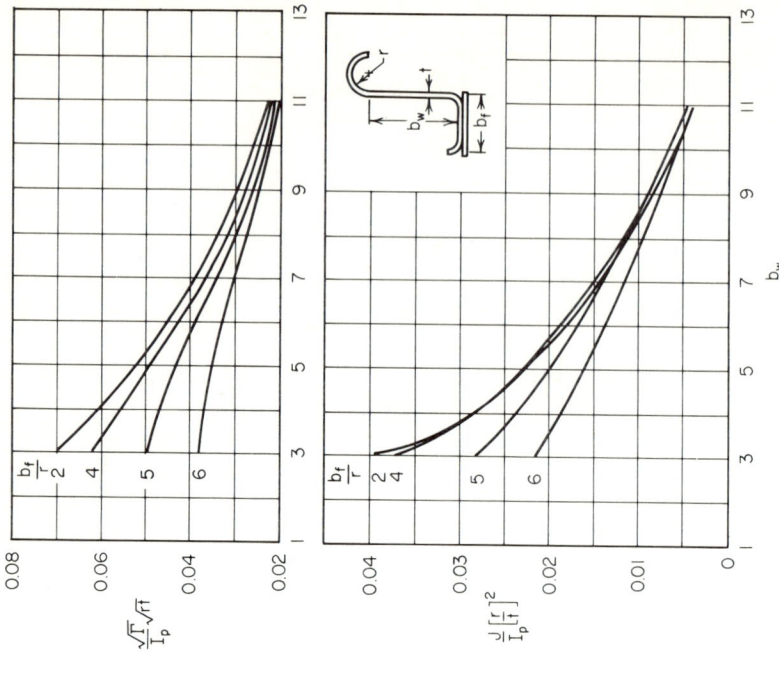

figure 12-5 Torsional-section properties for J stringer—sheet panels.

figure 12-4 Torsional-section properties for lipped Z stringer—sheet panels.

A satisfactory means of determining c is not available at the present time, and the use of fixity coefficients greater than 1.0 must be substantiated by tests.

The buckling stress of the sheet-stringer panel depends on the effective width of skin w_e which is acting with the stringer. Two common examples of effective width of sheet are shown in Fig. 12-6.

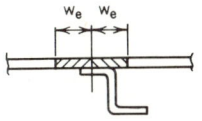

(a) Single line attachment

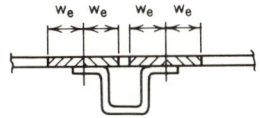

(b) Double line attachment

figure 12-6 *Examples of effective width.*

For calculating the effective width of sheet acting with a stiffener, the following equation has been found to give results consistent with tests:

$$w_e = 0.85 t_s \sqrt{\frac{E}{\sigma_{cr}/\eta}}$$

where $\sigma_{cr} = \sigma_{cp}$
σ_{cp} = lateral buckling stress of sheet-stringer panel

When σ_{cr} is known, σ_{cr}/η may be found using curves C in Sec. 10-7.

Other methods of obtaining w_e are available (Ref. 12-3), but in general, the difference will be small if the skin does not carry a large percentage of the load. σ_{cp} depends on the radius of gyration ρ of the sheet-stringer column, which in turn depends on w_e; therefore, an iteration procedure is needed to determine σ_{cp}. In addition, there may be an interaction between the lateral buckling of the panel and the torsional buckling or crippling of the stringer. The following procedure is recommended for determination of the allowable stress for the sheet-stringer panel:

1. Determine the radius of gyration of the stiffener cross section about the centroidal axis.

2. Determine the effective slenderness ratio L'/ρ of the stiffener alone

$$\frac{L'}{\rho} = \frac{d}{\rho \sqrt{c}}$$

3. Determine the crippling stress σ_{cc} and the torsional instability stress σ_{ct}. The lower of these two stresses σ_0 determines the intercept $L'/\rho = 0$ of a modified Johnson parabola and defines a complete column curve, as can be seen in Fig. 12-7.

4. Using the column curve and the slenderness ratio determined in steps 1, 2, and 3, record the value of σ_{cp}.

5. Compute the effective width of sheet acting with the stiffener.

6. Use the curves in Fig. 12-8 to compute ρ of the stiffener plus effective sheet. A_{st} and ρ_{st} are the area and radius of gyration of the stringer, respectively.

7. Compute L'/ρ for the new value of ρ.

8. Enter the column curve a second time with the new L'/ρ and record the corrected value of σ_{cp}.

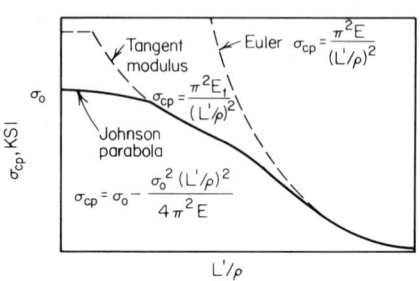

figure 12-7 *Critical stress for columns.*

9. Repeat steps 5, 6, and 7 until satisfactory convergence to a final stress σ_{cp} is obtained. Convergence generally occurs after two or three iterations.

10. The allowable load P_{cr} for the skin-stringer combination with single line attachment is

$$P_{cr} = \sigma_{cp}(A_{st} + 2w_e t_s) + \sigma_{cs}(b - 2w_e) t_s$$

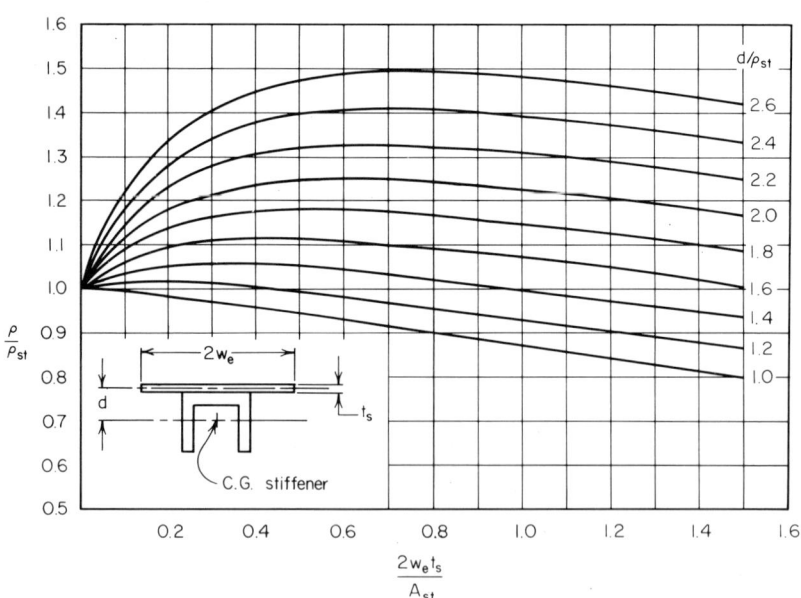

figure 12-8 *Variation of radius of gyration for sheet-stiffener combinations.*

Stability of Stiffened Shells

The preceding analysis assumes that the skin-stringer panel between the frames buckles as a column and neglects the stiffening effects of the curvature. If the stiffeners are close together, the formula given in Sec. 11-3 for axially stiffened cylinders may be used to account for the effects of curvature.

General Instability: The general-instability type of failure will occur in a structure which has frames and stringers that fail simultaneously under the critical load; that is, collapse takes place in a manner so as to destroy the load-carrying properties of all three structural elements—sheet, frames, and stringers. A buckling equation for stiffened cylinders in compression (Ref. 12-4) is given by

$$N_x = \gamma \left(\frac{L}{m\pi}\right)^2 \frac{\begin{vmatrix} A_{11} & A_{12} & A_{13} \\ A_{21} & A_{22} & A_{23} \\ A_{31} & A_{32} & A_{33} \end{vmatrix}}{\begin{vmatrix} A_{11} & A_{12} \\ A_{21} & A_{22} \end{vmatrix}} \quad \text{for } (n \geq 4)$$

where N_x is the buckling load per unit width of circumference, and the remaining parameters are defined as follows:

$$A_{11} = \bar{B}_x \left(\frac{m\pi}{L}\right)^2 + \bar{G}_{x\theta} \left(\frac{n}{R}\right)^2$$

$$A_{22} = \bar{B}_\theta \left(\frac{n}{R}\right)^2 + \bar{G}_{x\theta} \left(\frac{m\pi}{L}\right)^2$$

$$A_{33} = \bar{D}_x \left(\frac{m\pi}{L}\right)^4 + \bar{D}_{x\theta} \left(\frac{m\pi}{L}\right)^2 \left(\frac{n}{R}\right)^2 + \bar{D}_\theta \left(\frac{n}{R}\right)^4$$
$$+ \frac{\bar{B}_\theta}{R^2} + \frac{2\bar{C}_\theta}{R}\left(\frac{n}{R}\right)^2$$

$$A_{12} = A_{21} = (\bar{B}_{x\theta} + \bar{G}_{x\theta}) \frac{m\pi}{L} \frac{n}{R}$$

$$A_{23} = A_{32} = \frac{\bar{B}_\theta}{R} \frac{n}{R} + \bar{C}_\theta \left(\frac{n}{R}\right)^3$$

$$A_{31} = A_{13} = \frac{\bar{B}_{x\theta}}{R} \frac{m\pi}{L} + \bar{C}_x \left(\frac{m\pi}{L}\right)^3$$

$$\bar{B}_x = \frac{Et_s}{1-\mu^2} + \frac{E_s A_s}{b}$$

$$\bar{B}_\theta = \frac{Et_s}{1-\mu^2} + \frac{E_f A_f}{d}$$

$$\bar{B}_{x\theta} = \frac{\mu E t_s}{1 - \mu^2}$$

$$\bar{G}_{x\theta} = \frac{E t_s}{2(1 + \mu)}$$

$$\bar{D}_x = \frac{E t_s^3}{12(1 - \mu^2)} + \frac{E_s I_s}{b} + \tilde{z}_s^2 \frac{E_s A_s}{b}$$

$$\bar{D}_\theta = \frac{E t_s^3}{12(1 - \mu^2)} + \frac{E_f I_f}{d} + \tilde{z}_f^2 \frac{E_f A_f}{d}$$

$$\bar{D}_{x\theta} = \frac{E t_s^3}{6(1 + \mu)} + \frac{G_s J_s}{b} + \frac{G_f J_f}{d}$$

$$\bar{C}_x = \tilde{z}_s \frac{E_s A_s}{b}$$

$$\bar{C}_\theta = \tilde{z}_f \frac{E_r A_r}{d}$$

where A_s, A_f = stiffener and frame area, respectively
b = stiffener spacing
d = frame spacing
E_f, E_s = Young's modulus of frame and stiffeners, respectively
I_f, I_s = moment of inertia of frame and stiffeners, respectively, about their centroid
J_f, J_s = beam torsion constant of frame and stiffeners, respectively
G_s, G_f = shear modulus of stiffeners and frames, respectively

$\tilde{z}_s$ and $\tilde{z}_f$ is the distance from the middle of the thin sheet to the centroid of the stiffener and frame, respectively. $\tilde{z} > 0$ when stiffener or frame is outside; $\tilde{z} < 0$ when stiffener or frame is inside.

Prebuckling deformations are not taken into account in the derivation of the equation. The cylinder edges are assumed to be supported by rings that are rigid in their own plane but offer no resistance to rotation or bending out of their plane. For given cylinder and stiffener dimensions, the values of m and n to be used are those which minimize $\bar{N}_x$. The quantity n is the number of half waves of the buckles in the circumferential direction and should be restricted to even positive integers greater than or equal to 4. The quantity m is the number of half waves of the buckles in the longitudinal direction and should be limited to positive integers.

The half wavelengths of the buckles which give the lowest $\bar{N}_x$ should be greater than the stiffener spacings to obtain reasonable accuracy. The allowable compressive load for cylinder is $P_{cr} = N_x 2\pi R$.

The equation presented can be specialized for various types of cylinders which have been treated separately in the literature (see Refs. 12-5 to 12-11). The unusually large number of parameters does not permit any definitive numerical results to be shown. For combinations of parameters representative of stiffened shells, calculations indicate that external stiffening, whether stringers or rings or both, can be more effective than internal stiffening for axial compression. Generally, calculations neglecting stiffener eccentricity yield unconservative values of the buckling load of internally stiffened cylinders and conservative values of the buckling load of externally stiffened cylinders (Ref. 12-7). An extensive investigation of the variation of the buckling load with various stiffener parameters is reported in Ref. 12-9. The limited experimental data (Refs. 12-12 to 12-19) for stiffened shells are in reasonably good agreement with the theoretical results for the range of parameters investigated.

If $\gamma = 1$ is used, the theoretical buckling load is obtained. On the basis of available data, it is recommended that $\gamma = 0.75$ be used to obtain the buckling loads of cylinders with closely spaced, moderately large stiffeners. Correlation coefficients covering the transition from unstiffened cylinders to stiffened cylinders with closely spaced stiffeners have not been fully investigated. While theory and experiment (Ref. 12-18) indicate that restraint against edge rotation and longitudinal movement significantly increases the buckling load, not enough is known about the edge restraint of actual cylinders to warrant taking advantage of these effects unless substantiated by tests.

A quick estimate of the buckling load can be obtained by neglecting eccentricities of the stiffeners and using Sec. 11-3.

Bending, Frame- and Stringer-stiffened Cylinders

Theoretical and experimental results (Refs. 12-8, 12-13, and 12-20 to 12-22) indicate that the critical maximum load per unit circumference of a stiffened cylinder in bending can exceed the critical unit load in axial compression. In the absence of an extensive investigation, it is recommended that the critical maximum load per unit circumference of a cylinder with closely spaced stiffeners be taken as equal to the critical load in axial compression, which may be calculated from the previous paragraphs. If the stresses are elastic the design-allowable bending moment is $M_{cr} = \pi N_x R^2$.

12-3 Frame-stiffened Cylinders

Lateral and Axial External Pressure, Frame-stiffened Cylinders

If a cylindrical shell with frames (circumferential stiffeners) is subjected to lateral and axial external pressure, it may fail in three distinct ways. The types of failure will be classified as material failure, buckling between frames, and general instability. A brief discussion of these failure modes is presented here. A more detailed discussion of frame-stiffened cylinders subjected to external pressure is presented in Ref. 12-23.

Material Failure: For purposes of analysis, the stress and deflection distributions of a cylinder prior to buckling can be assumed to be axisymmetric. Therefore, the analysis methods of previous chapters can be used to determine the stresses in the cylinder. A more detailed procedure is discussed in Ref. 12-24. The actual stresses must be compared with the material's allowable stresses. In addition, the compressive stress in the frame must be compared with the local buckling stresses of the frame.

Buckling between Frames: This failure will occur in a cylinder having relatively heavy frames. The sheet will buckle between frames and the frames will remain circular in cross section. The allowable buckling stress for this mode of failure can be obtained from Sec. 10-3, using a cylinder length equal to the frame spacing. A more detailed method of analyzing this mode of failure is presented in Ref. 12-25.

General Instability: General-instability failure will occur when the frame buckles with the sheet at the critical load. The design-allowable general instability pressure p_{cr} for a frame-stiffened cylinder subjected to lateral and axial external pressure can be obtained from

$$p_{cr} = \frac{\gamma E}{30 \times 10^6} (p_s + p_f)$$

where E = Young's modulus of the material (lb/in.^{2}) (the frame and sheet must be made from the same material)

p_s can be obtained from Fig. 12-9 if $(I_e/dR^3) \times 10^6 < 10$ and from Fig. 12-10 if $(I_e/dR^3) \times 10^6 \geq 10$

p_f can be obtained from Fig. 12-11 or 12-12

The quantity p_{cr} must be computed for $n = 2, 3, 4,$ and 5; its lowest value is the critical allowable buckling pressure. This graphical method for determining p_s and p_f was obtained from Refs. 12-26 and 12-27, which are based on Ref. 12-28. Reference 12-28 analyzes simply supported cylinders. The test results of Ref. 12-29 indicate an appreciable increase in buckling pressure due to edge fixity, but a method of including these effects for design purposes has not been developed.

The parameter γ is introduced to reduce the theoretical results of Ref. 12-28 to a value which may be used for design purposes. It is a function of many variables such as initial out of roundness of the cylinder. For the test data given in Refs. 12-29 and 12-30, $\gamma = 1$ is adequate.

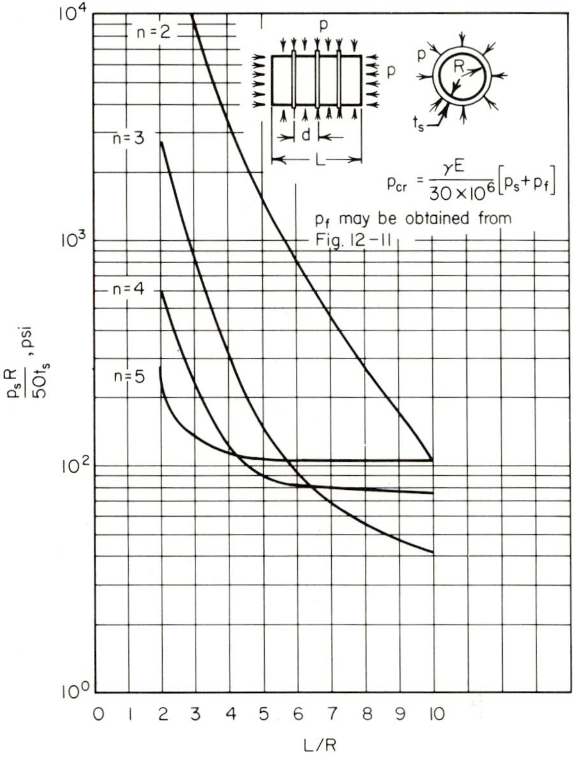

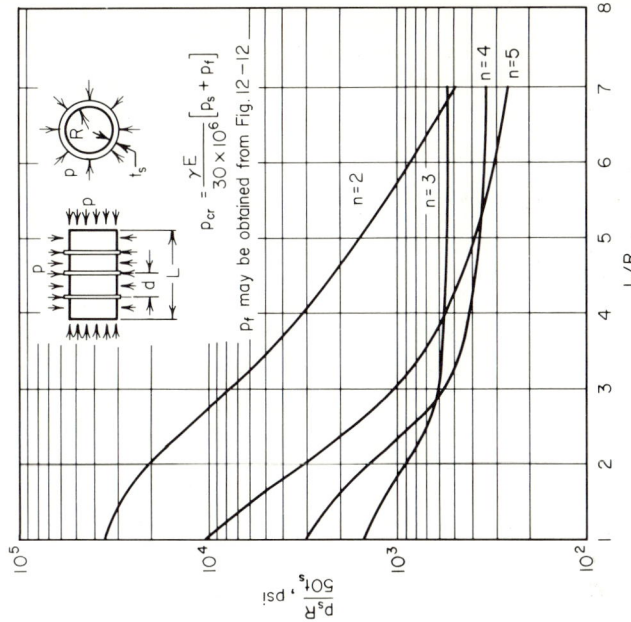

figure 12-9 *Shell pressure factor p_s as a function of bulkhead spacing L/R for $(I_e/dR^3) \times 10^6 \leq 10$.*

figure 12-10 *Shell pressure factor p_s as a function of bulkhead spacing L/R for $(I_e/dR^3) \times 10^6 \leq 10$.*

319

However, the cylinders in Refs. 12-29 and 12-30 were machined to very close tolerances. Presently, it is not known what value of γ to use, but for $(12I_e/dt_s)^{3/2}(L^2/t_s^4 R) > 4 \times 10^3$, $\gamma = 0.9$ is probably reasonable.

The value of I_e required in Fig. 12-11 or 12-12 can be calculated from

$$I_e = \frac{A_f e^2}{1 + A_f/d_e t_s} + I_f + \frac{d_e t_s^3}{12}$$

where A_f, I_f = area of frame and moment of inertia of frame about its own neutral axis, respectively
e = distance from middle surface of sheet to centroid of frame
$d_e = (d - t_w)F_1 + t_w$
t_w = frame web thickness

The value of F_1 can be obtained from Fig. 12-13. Parameter λ_2 in Fig. 12-13 is a function of p_{cr}, which is unknown. A good approximation can be obtained by using $\lambda_2 = 0$. For cases in which $\lambda_2 < 2$, this approximation gives results within 5 percent accuracy.

For the loading case of external pressure, Ref. 12-31 states that a cylinder with frames on the inside will be stronger than a cylinder with frames on the outside if $Z > 500$, where $Z = (1 - \mu^2)^{1/2} L^2/Rt_s$. If the frames are on the inside and $Z > 500$, the results will be slightly conservative because the curves presented are for external frames. Therefore, the curves presented will probably be conservative if the frames are located on the outside because most stiffened cylinders fall in the range $Z > 500$.

If the parameters for a particular design do not fail in the range of parameters presented in Figs. 12-11 and 12-12, the following formula (given in Ref. 12-32) can be used to estimate the design-allowable external pressure:

$$p_{cr} = \frac{5.5\gamma E(I_e/d)^{3/4} t_s^{1/4}}{LR^{3/2}}$$

provided

$$\left(\frac{12I_e}{dt_s}\right)^{3/2} \frac{1}{t_s^4} \frac{L^2}{R} > 4 \times 10^3$$

The theoretical results of Ref. 12-28 were reduced 10 percent for design purposes, and it has been found that the theory of Ref. 12-28 predicts buckling pressures that are, in some cases, 80 percent of the theoretical results given in Ref. 12-32; therefore, $\gamma = 0.8 \times 0.9 = 0.72$ can be used in the preceding formula to obtain the design-allowable buckling pressure.

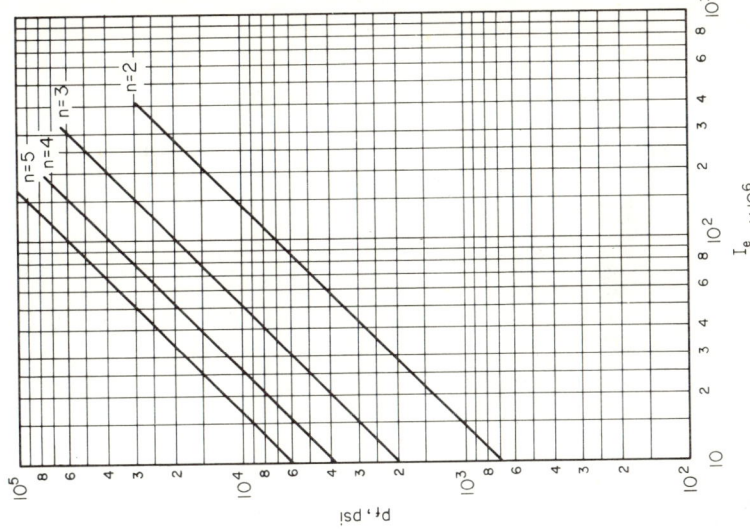

figure 12-12 *Frame pressure factor p_f as a function of frame stiffness I_e/dR^3.*

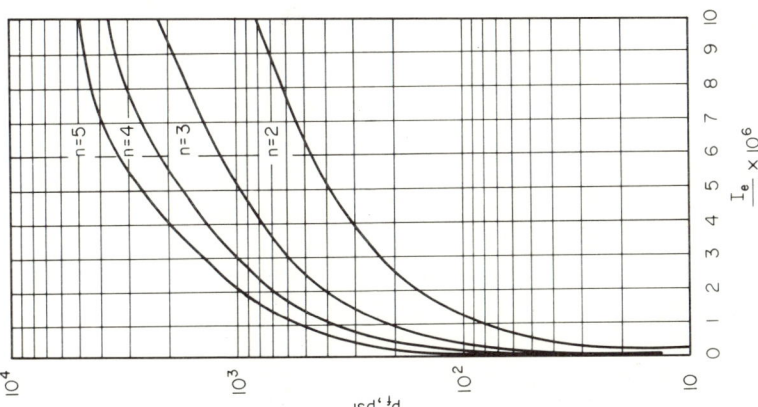

figure 12-11 *Frame pressure factor p_f as a function of frame stiffness I_e/dR^3.*

322 *Structural Analysis of Shells*

A simple method of including plasticity in the preceding formulas is not available, but an estimate can be obtained by using the plasticity

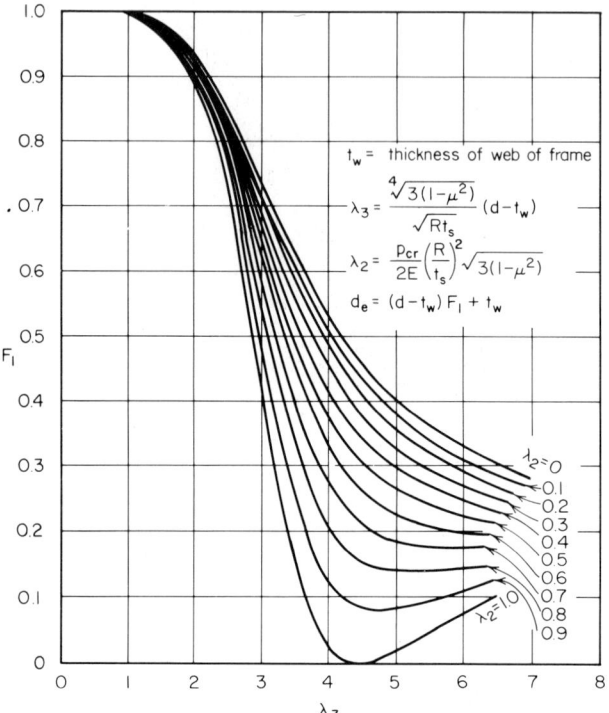

figure 12-13 *Stress function for determining effective length.*

correction factor suggested in Sec. 10-3 for external pressure of unstiffened cylinders. A more detailed procedure is presented in Ref. 12-33.

REFERENCES

12-1. *Structures Manual*, NAA Los Angeles Division, NA-52-400.
12-2. Argyris, J.: Flexure-torsion Failure of Panels, *Aircraft Eng.*, June–July, 1954.
12-3. Peterson, J., R. Whitley, and J. Deaton: *Structural Behavior and Compressive Strength of Circular Cylinders with Longitudinal Stiffening*, NASA TN D-1251, 1962.
12-4. Jones, R. M.: Buckling of Circular Cylindrical Shells with Multiple Orthotropic Layers and Eccentric Stiffeners, *Rept.* TR-0158(S3820-10)-1, *Air Force Rept.* SAMSO-TR-67-29, Aerospace Corp., September, 1967.
12-5. Stein, M., and J. Mayers: *Compressive Buckling of Simply Supported Curved Plates and Cylinders of Sandwich Construction*, NACA TN 2601, 1952.

Stability of Stiffened Shells

12-6. Becker, H., and G. Gerard: Elastic Stability of Orthotropic Shells, *J. Aeron. Sci.*, vol. 29, no. 5, pp. 505–512, 520, May, 1962.

12-7. Block, D. L., M. F. Card, and M. M. Mikulas, Jr.: *Buckling of Eccentrically Stiffened Orthotropic Cylinders*, NASA TN D-2960, August, 1965.

12-8. Hedgepeth, J. M., and D. B. Hall: Stability of Stiffened Cylinders, *J. Am. Inst. Aeronautics Astronautics*, vol. 3, no. 12, pp. 2275–2286, December, 1965.

12-9. Singer, J., M. Baruch, and O. Harari: On the Stability of Eccentrically Stiffened Cylindrical Shells under Axial Compression, *Intern. J. Solids Structures*, vol. 3, no.4, pp. 445–470, July, 1967.

12-10. Tasi, J.: Effect of Heterogeneity on the Stability of Composite Cylindrical Shells under Axial Compression, *J. Am. Inst. Aeronautics Astronautics*, vol. 4, no. 6, pp. 1058–1062, June, 1966.

12-11. Van der Neut, A.: The General Stability of Stiffened Cylindrical Shells under Axial Compression, National Aerospace Research Institute (Amsterdam), *Rept.* S-314, 1947.

12-12. Dickson, J. N., and R. H. Brolliar: *The General Instability of Ring-stiffened Corrugated Cylinders under Axial Compression*, NASA TN D-3089, January, 1966.

12-13. Meyer, R. R.: Buckling of Ring-stiffened Corrugated Cylinders Subjected to Uniform Axial Load and Bending, Douglas Aircraft Co., *Rept.* DAC-60698, July, 1967.

12-14. Peterson, J. P., and M. B. Dow: *Compression Tests on Circular Cylinders Stiffened Longitudinally by Closely Spaced Z Section Stringers*, NASA Memo 2-12-59L, 1959.

12-15. Peterson, J. P., R. Q. Whitely, and J. W. Deaton: *Structural Behavior and Compressive Strength of Circular Cylinders with Longitudinal Stiffening*, NASA TN D-1251, 1962.

12-16. Becker, H., G. Gerard, and R. Winter: Experiments on Axial Compressive General Instability of Monolithic Ring-stiffened Cylinders, *J. Am. Inst. Aeronautics Astronautics*, vol. 1, no. 7, pp. 1614–1618, July, 1963.

12-17. Card, M. F., and R. M. Jones: *Experimental and Theoretical Results for Buckling of Eccentrically Stiffened Cylinders*, NASA TN D-3639, October, 1966.

12-18. Milligan, R., G. Gerard, and C. Lakshinikantham: General Instability of Orthotropically Stiffened Cylinders under Axial Compression *J. Am. Inst. Aeronautics Astronautics*, vol. 4, no. 11, pp. 1906–1913, November, 1966.

12-19. Singer, J.: *The Influence of Stiffener Geometry and Spacing on the Buckling of Axially Compressed Cylindrical and Conical Shells*, Preliminary Preprint Paper, Second IUTAM Symposium on the Theory of Thin Shells, Copenhagen, September, 1967.

12-20. Block, D. L.: *Buckling of Eccentrically Stiffened Orthotropic Cylinders under Pure Bending*, NASA TN D-3351, March, 1966.

12-21. Peterson, J. P., and J. K. Anderson: *Bending Tests of Large Diameter Ring-stiffened Corrugated Cylinders*, NASA TN D-3336, March, 1966.

12-22. Card, M. F.: *Bending Tests of Large Diameter Stiffened Cylinders Susceptible to General Instability*, NASA TN D-2200, April, 1964.

12-23. Pulos, J.: *Structural Analysis and Design Considerations for Cylindrical Pressure Hulls*, David Taylor Model Basin Report 1639, April, 1963.

12-24. Pulos, J., and V. Salerno: *Axisymmetric Elastic Deformations and Stresses in Ring-stiffened Perfectly Circular Cylindrical Shell under External Hydrostatic Pressure*, David Taylor Model Basin Report 1497, September, 1961.

12-25. Reynolds, T. E.: *Elastic Lobar Buckling of Ring-supported Cylindrical Shells under Hydrostatic Pressure*, David Taylor Model Basin Report 1614, September, 1962.

12-26. Reynolds, T. E.: *A Graphical Method for Determining the General-instability Strength of Stiffened Cylindrical Shells*, David Taylor Model Basin Report 1106, September, 1957.
12-27. Ball, W.: *Formulas and Curves for Determining the Elastic General-instability Pressures of Ring-stiffened Cylinders*, David Taylor Model Basin Report 1570, January, 1962.
12-28. Kendrick, S.: *The Buckling under External Pressure of Circular Cylindrical Shells with Evenly-spaced, Equal Strength Circular Ring-frames*, part III, Naval Construction Research Establishment Report R-244, 1953.
12-29. Reynolds, T., and W. Blumenberg: *General Instability of Ring-stiffened Cylindrical Shells Subjected to External Hydrostatic Pressure*, David Taylor Model Basin Report 1324.
12-30. Galletly, G. et al.: General Instability of Ring-stiffened Cylindrical Shells Subjected to External Hydrostatic Pressure—A Comparison of Theory and Experiment, *J. Appl. Mech.*, vol. 25; *Trans. ASME*, vol. 80, June, 1958.
12-31. Singer, J., M. Baruch, and O. Harari: *Further Remarks on the Effects of Eccentricity of Stiffeners on the General Instability of Stiffened Cylindrical Shells*, Technion-Israel Institute of Technology, TAE Report 42, August, 1965.
12-32. Becker, H.: *Handbook of Structural Stability*, part VI, *Strength of Stiffened Curved Plates and Shells*, NASA TN 3786, 1956.
12-33. Lunchick, M.: *Plastic General-instability Pressure of Ring-stiffened Cylindrical Shells*, David Taylor Model Basin Report 1587.

Chapter 13

STABILITY OF SANDWICH SHELLS

13-1 General

Sandwich-type construction is a composite construction consisting of three integrally attached layers. The middle layer of the sandwich is the core; the outer two layers are the facing sheets. Generally, the facing sheets are very thin relative to the overall thickness of the sandwich and the elastic modulus of the facing-sheet material is much larger than the corresponding effective modulus of the core. The primary difference between sandwich shells and orthotropic or isotropic shells is the relatively low transverse shear stiffness of the sandwich construction; therefore, the transverse shear stiffness must be included in the analysis.

Generally, sandwich construction should be analyzed for three modes of failure: (1) material failure, in which the applied stresses exceed the material allowable stresses (see Chap. 7); (2) general-instability failure, in which the whole shell fails with the core and facings acting together; and (3) local-instability failure, in which the facing sheet fails because it is not sufficiently stabilized by the core. (The forms of local instability for sandwich construction with honeycomb core are intracell buckling and

326   Structural Analysis of Shells

wrinkling. Design formulas for these two modes of local instability are given in Sec. 13-2.)

Although a considerable amount of theoretical information is available concerning the general instability of sandwich shells, not enough test data are available to obtain design curves directly. Therefore, the design curves for homogeneous isotropic shells are used to reduce the theoretical buckling loads for sandwich shells to design-allowable buckling loads.

13-2   Local Instability

Intracell Buckling

If the core of a sandwich is constructed of cellular (honeycomb) material, it is possible for the facings to buckle or dimple into the spaces between core walls. Dimpling of the facings may not lead to failure unless the amplitude of the dimples becomes large and causes the dimples to grow across the core cell walls, resulting in wrinkling of the facings. Dimpling that does not cause total structural failure may be sufficiently severe that permanent dimples remain after removal of the load. Reference 13-10 presents analysis methods for intracell buckling which will be updated as more information becomes available.

An approximation of the design-allowable uniaxial intracell buckling-failure stress σ_{cr} can be obtained from the formulas given in Ref. 13-1.

$$\frac{\sigma_{cr}}{\eta} = \frac{K_L E}{1 - \mu^2}$$

where E = Young's compression modulus of elasticity of the facing sheet
μ = Poisson's ratio of the facing sheet

The coefficient K_L can be obtained from Fig. 13-1.

For elastic buckling, the plasticity correction term $\eta = 1$ is used. In the inelastic range, Ref. 13-2 recommends

$$\eta = \frac{2E_t}{E + E_t}$$

See Sec. 10-7 for the definition of η and E_t.

For most materials, the curves E_1 of Sec. 10-7 are a sufficiently accurate representation of this plasticity correction.

It should be noted that the formula for obtaining σ_{cr} is based primarily on test data from brazed flat honeycomb sandwich panels with PH 15-7 Mo core and facings. Limited test results indicate that the formula may be used for other types of materials and bonding methods.

It can be seen that the formula for computing σ_{cr} is independent of the foil thickness of the core and does not include any interaction between a wrinkling failure and an intercell-buckling failure. Until an adequate

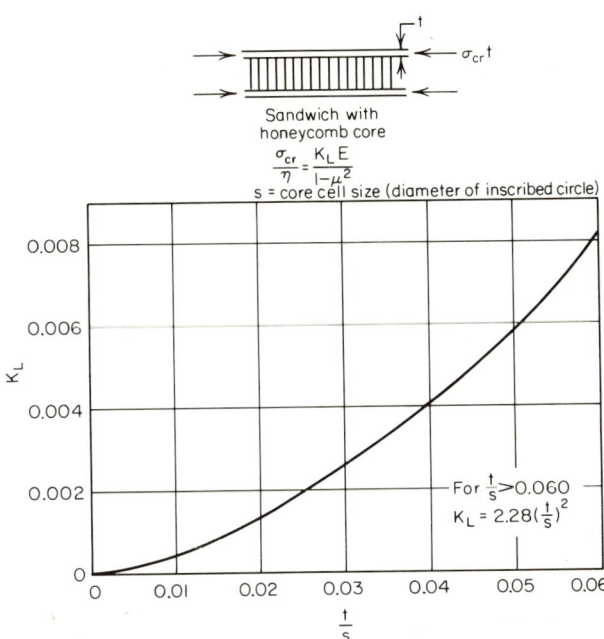

figure 13-1 *Intracell-buckling coefficient.*

method of analysis is developed which includes all the important variables, it is recommended that a limited number of compression tests be conducted to verify a design that may be critical in intercell buckling.

If an element of a sandwich shell is subjected to inplane shear or combined loadings, the maximum principal compressive stress in the facing sheet should be compared with σ_{cr} to determine if intracell buckling may occur. If the stress normal to the maximum principal stress is a tensile stress, the preceding formula for σ_{cr} is probably adequate. If this stress is a compressive stress, Ref. 13-2 indicates that the following reduction of σ_{cr} is necessary:

$$\frac{\sigma_{cr}}{\eta} = \frac{K_L}{\sqrt[3]{1+S^3}} \frac{E}{1-\mu^2}$$

where

$$S = \frac{\text{min principal compressive stress in facing}}{\text{max principal compressive stress in facing}}$$

Dimples may occur at stress levels less than σ_{cr}, but the sandwich will carry more load. An estimate of the stress σ_0 at which initial dimpling may occur is given in Ref. 13-3 as

$$\frac{\sigma_0}{\eta} = \frac{2E}{1-\mu^2}\left(\frac{t}{s}\right)^2$$

where t = thickness of facing sheet
s = cell size of core

For elastic buckling, the plasticity correction term, $\eta = 1$ is used. In the inelastic range, the stress σ_0 can be found by using curves G in Sec. 10-7.

It should be mentioned, however, that for sandwich construction with very thin facing sheets, dimples in the facings can sometimes be observed before the sandwich is loaded. These dimples are the result of the manufacturing procedure.

Although the intracell-buckling formulas are based on data for flat panels, they are adequate for curved panels because the cell size is usually much less than the radius of curvature.

Face-sheet Wrinkling

When a facing sheet of a sandwich element is subjected to axial compression, face-sheet wrinkling may occur. This failure is similar to buckling of a plate on an elastic foundation. Typical buckling modes for wrinkling are shown in Fig. 13-2.

The resulting failure mode will depend on the flatwise compressive strength of the core relative to the flatwise tensile strength of the bond between the facing cores. Typical wrinkling failures are shown in Fig. 13-3.

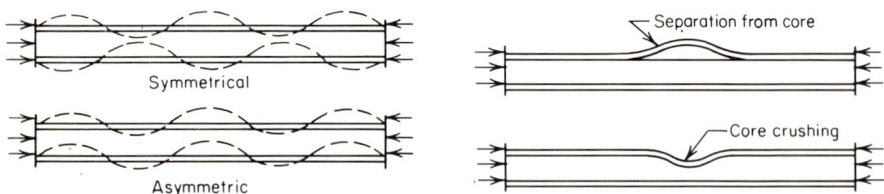

figure 13-2 *Buckling modes for wrinkling.* **figure 13-3** *Wrinkling failures.*

If the bond between the facing and the core is strong, the facing can still wrinkle outward by causing tension failure of the core.

It can be seen that the wrinkling load depends on the stiffness and strength of the foundation system. Since the facing is never flat, the wrinkling load will also depend on the initial facing eccentricity or original waviness. A method of analysis which includes these variables is given in Ref. 13-4. One method of determining the initial waviness parameter needed in Ref. 13-4 is presented in Ref. 13-5.

Reference 13-10 presents the most up to date analysis methods for wrinkling which will be updated as more information becomes available. The wrinkling analysis from Ref. 13-2 will be presented in the following paragraphs because it is easy to apply.

The design-allowable uniaxial compressive wrinkling stress σ_{cr} can be obtained from

$$\frac{\sigma_{cr}}{\eta} = K_L \sqrt[3]{EE_c G_c}$$

where $K_L = 0.43$
$E =$ Young's compressive modulus of elasticity of the facing sheet
$E_c =$ flatwise compression modulus of the core (in a direction normal to the surface of the shell)
$G_c =$ transverse shear modulus of the core in the direction of the maximum compressive stress

For elastic buckling, the plasticity correction term $\eta = 1$ is used. In the inelastic range, Ref. 13-2 recommends

$$\eta = \sqrt[3]{\frac{3E_t + E_s}{4E}}$$

E_t and E_s are defined in Sec. 10-7. For most materials, curves C of Sec. 10-7 are sufficiently accurate representations of this plasticity correction coefficient. A limited number of tests should be conducted to verify the analysis procedure for a particular sandwich configuration.

The information on wrinkling of sandwich facings pertains primarily to flat panels. However, it is adequate for shells because the wavelength of the buckle is small relative to the radius of curvature.

If an element of a sandwich shell is subjected to inplane shear or combined loadings, the maximum principal compressive stress in the facing sheet should be compared with σ_{cr} to determine whether wrinkling will occur. Theoretically, it has been shown that the wrinkling stress is unaffected by the stresses normal to the maximum compressive principal stress. However, Ref. 13-2 has indicated that if the principal stresses are both compression, the following reduction in K is necessary:

$$K_L = \frac{0.43}{\sqrt[3]{1 + S^3}}$$

where

$$S = \frac{\text{min principal compressive stress in facing}}{\text{max principal compressive stress in facing}}$$

For a sandwich with orthotropic core, wrinkling should also be checked in the direction in which the shear modulus of the core is lowest.

13-3 Cylinders

Axial Compression, Sandwich Cylinders

The curve presented in Ref. 13-6 will be used to determine the buckling stress for sandwich cylinders subjected to axial compression. The results are applicable to a sandwich with unequal homogeneous isotropic facing sheets with an orthotropic core that does not carry inplane loads. The design-allowable buckling stress is

$$\frac{\sigma_{cr}}{\eta} = \gamma C_c E \frac{h}{R} \frac{2\sqrt{t_1 t_2}}{\sqrt{1 - \mu^2}(t_1 + t_2)}$$

where E = Young's modulus of the facing-sheet material
μ = Poisson's ratio of the facing-sheet material
G_{xz}, $G_{\theta z}$ = transverse shear moduli of the core in the longitudinal and circumferential directions, respectively

The buckling coefficient C_c and a definition of the geometrical parameters are given in Fig. 13-4. The correction factor γ was introduced to reduce the theoretical results of Ref. 13-6 to values that can be used for design purposes. The factor γ can be obtained from Fig. 11-6, where

$$\rho = \frac{\sqrt{t_1 t_2 h^2}}{t_1 + t_2}$$

Existing test data show that the value of γ may be conservative for some values of the parameters, but a sufficient number of tests have not been conducted to justify increasing the value of γ.

The method for obtaining γ is discussed in Sec. 11-3 for orthotropic cylinders. This method will be consistent if the shear stiffness of the core is large (i.e., when V_c is small). The formula for V_c is given in Fig. 13-4. Because of manufacturing limitations and local instability problems, V_c is usually small. However, if V_c is large (i.e., when $V_c > 0.5$), γ will approach 1. There is not enough information available to obtain γ as a function of V_c; therefore, Fig. 11-6 should be used to obtain γ for all values of $V_c < 0.5$.

The curve presented in Fig. 13-4 is for a sandwich with very thin facing sheets ($c/h \approx 1$). For small values of c/h, the results of Fig. 13-4

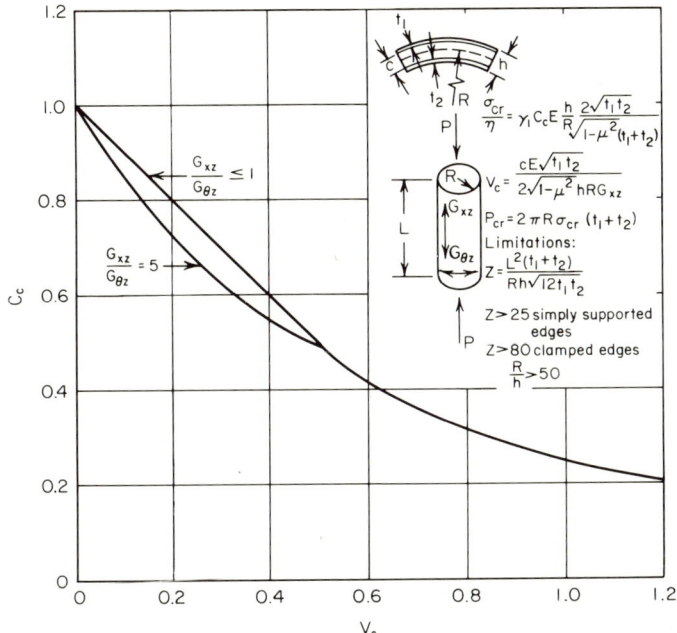

figure 13-4 *Classical buckling coefficient for sandwich cylinders subjected to axial compression.*

may yield excessive error for certain values of the parameters. For values of $c/h > 0.9$, Fig. 13-4 should be adequate for $V_c > 0.5$. Sandwich cylinders subjected to axial compression must be analyzed for local instability, as discussed in Sec. 13-2.

For elastic buckling, the plasticity correction term $\eta = 1$ is used. In the inelastic range, an estimate of the stress σ_{cr} can be found by using curves E_1 in Sec. 10-7. The parameter V_c is a function of the stress level for stresses above the proportional limit. By assuming V_c is independent of the stress level, the results will be conservative. For most practical designs, the difference will be very small.

References 13-11 and 13-12 may be used to obtain buckling loads for sandwich cylinders with a corrugated core subjected to axial compression.

Shear or Torsion, Sandwich Cylinders

The curves presented in Ref. 13-7 will be used to determine the buckling stress for sandwich cylinders subjected to torsion. The results are applicable to a sandwich with unequal homogeneous isotropic facing sheets and an orthotropic core which does not carry inplane loads. The design-allowable shear buckling stress is

$$\frac{\tau_{cr}}{\eta} = \gamma C_s E \frac{h_1}{R}$$

where E = Young's modulus of the facing-sheet material
$G_{xz}, G_{\theta z}$ = transverse shear modulus of the core in the longitudinal and circumferential directions, respectively

A definition of the geometrical parameters and C_s can be obtained from Figs. 13-5 through 13-10. The correction factor γ was introduced

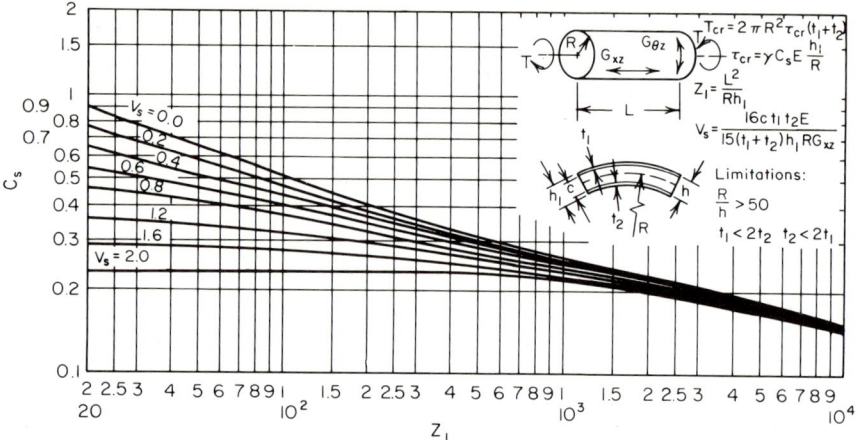

figure 13-5 Buckling coefficients for cylinders having isotropic facings $c/h = 1.0$, $G_{xz}/G_{\theta z} = 0.4$.

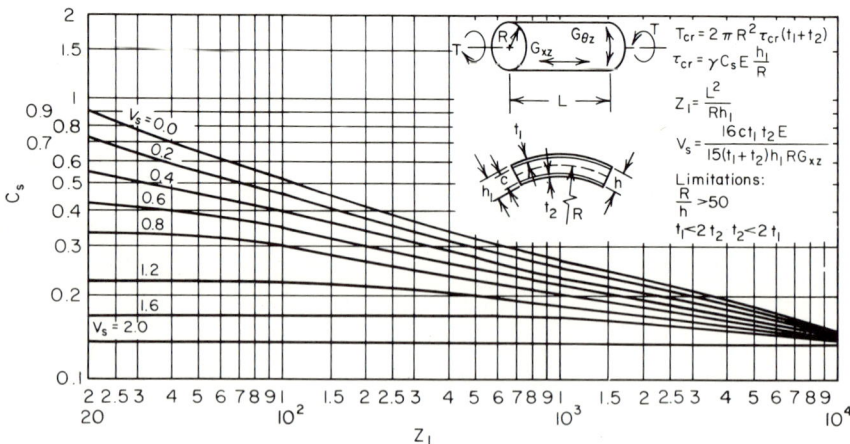

figure 13-6 Buckling coefficients for cylinders having isotropic facings $c/h = 1.0$, $G_{xz}/G_{\theta z} = 1.0$.

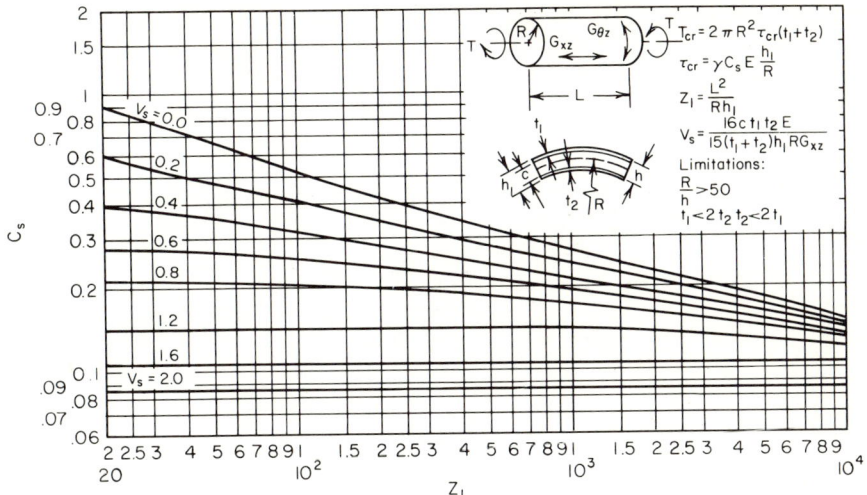

figure 13-7 *Buckling coefficients for cylinders having isotropic facings* $c/h = 1.0$, $G_{xz}/G_{\theta z} = 2.5$.

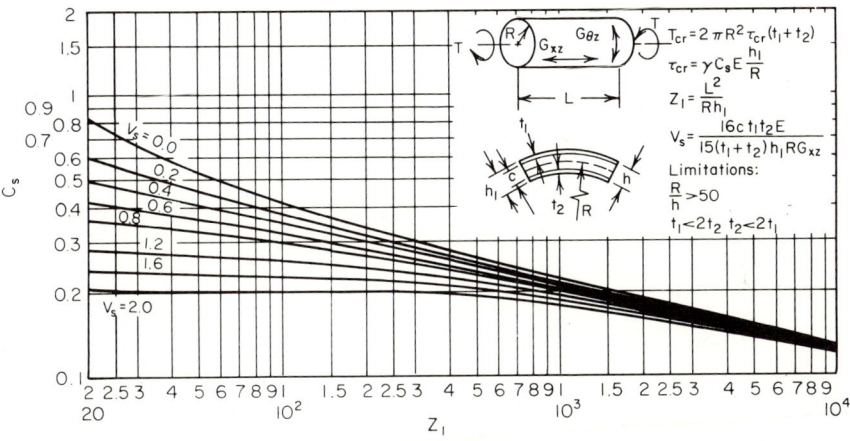

figure 13-8 *Buckling coefficients for cylinders having isotropic facings* $c/h = 0.82$, $G_{xz}/G_{\theta z} = 0.4$.

to reduce the theoretical results of Ref. 13-7 to values that can be used for design. γ can be obtained from Fig. 11-8, where

$$\rho = \frac{\sqrt{t_1 t_2 h^2}}{t_1 + t_2}$$

The method used to obtain the curve shown in Fig. 11-8 is discussed in in Sec. 11-3 for torsion for orthotropic cylinders. The correction factor

334 Structural Analysis of Shells

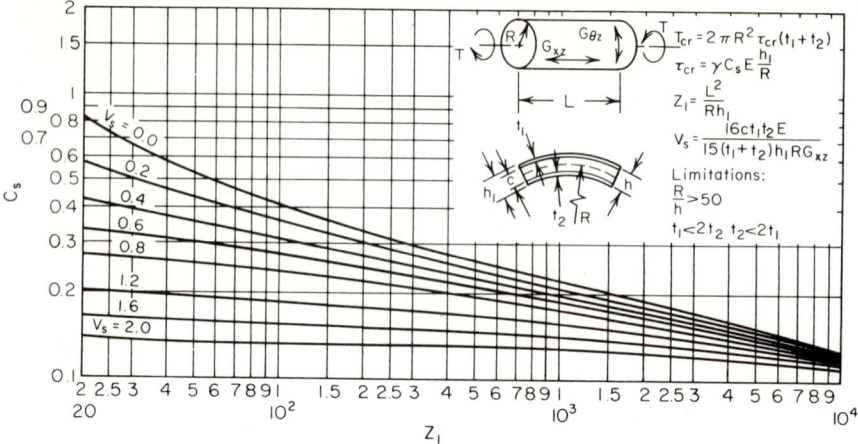

figure 13-9 *Buckling coefficients for cylinders having isotropic facings* $c/h = 0.82$, $G_{xz}/G_{\theta z} = 1.0$.

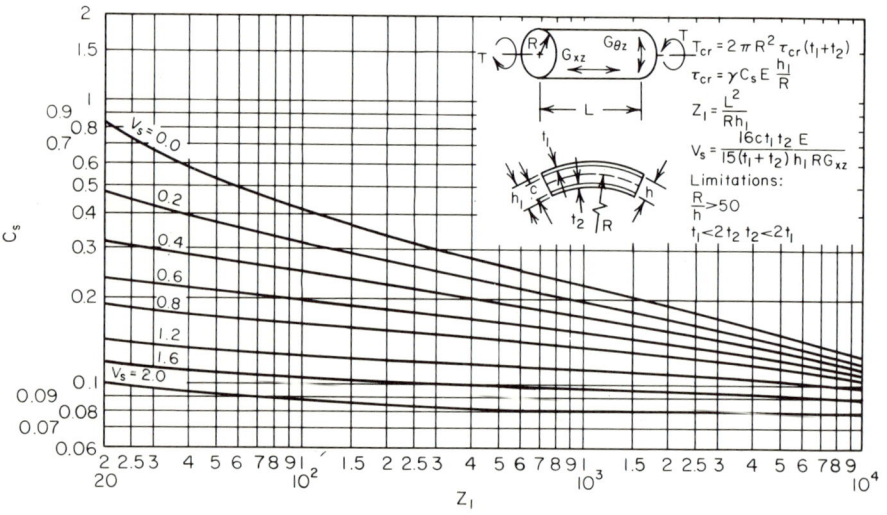

figure 13-10 *Buckling coefficients for cylinders having isotropic facings* $c/h = 0.82$, $G_{xz}/G_{\theta z} = 2.5$.

used in Sec. 11-3 and this section will be consistent if V_s is small (i.e., a core with relatively large shear stiffness). Because of manufacturing limitations and local instability considerations, V_s is usually small. However, if V_s is large, γ will approach 1 for all values of R/ρ. There is no information available to obtain γ as a function of V_s.

Figure 13-6 predicts results that are, at most, about 6 percent higher than the results of Ref. 13-8 for isotropic sandwich cylinders with simple supports. Therefore, the results presented in Figs. 13-5 through 13-10 are probably sufficiently accurate for cylinders with simply supported edges.

Sandwich cylinders subjected to torsion must be analyzed for local instability, as discussed in Sec. 13-2. For elastic buckling, the plasticity correction term $\eta = 1$ is used. If the stresses are above the proportional limit, the procedure discussed in Sec. 10-2 for shear of curved panels can be used. The parameter V_s is also a function of the stress level for stresses above the proportional limit. By assuming that V_s is independent of the stress level, the results will be conservative. For most practical designs, the differences will be small.

References 13-11 and 13-12 can be used to obtain the buckling load for sandwich cylinders with a corrugated core subjected to torsion.

Bending, Sandwich Cylinders

The formula presented for axial compression of sandwich cylinders can be used to determine the design-allowable buckling stress for unpressurized sandwich cylinders subjected to bending if γ is obtained from Fig. 11-9.

$$\rho = \frac{\sqrt{t_1 t_2 h^2}}{t_1 + t_2}$$

For bending, σ_{cr} is the maximum compressive stress due to the bending moment (outer fiber stress). Figure 11-9 is based on the correction factors used to modify the small-deflection theoretical results for homogeneous isotropic cylinders subjected to bending. If the stresses are elastic, the design-allowable bending moment may be obtained from $M_{cr} = \pi R^2 \sigma_{cr} (t_1 + t_2)$.

Existing test data show that the value of γ may be conservative for some values of the parameters, but not enough tests have been conducted to justify increasing the value of γ.

External Pressure, Sandwich Cylinders

The curves presented in Ref. 13-9 will be used to determine the buckling stress for sandwich cylinders with simply supported edges subjected to lateral external pressure. The results are applicable for a sandwich with isotropic facings of equal or unequal thickness and of the same or different materials. The core material may be orthotropic or isotropic but may not carry inplane loads. The design-allowable buckling pressure is

$$p_{cr} = \frac{\gamma C_p}{1 - \mu^2} \frac{E_1 t_1 + E_2 t_2}{R}$$

336 *Structural Analysis of Shells*

where E_1, E_2 = Young's moduli of the outer and inner facing sheets, respectively

G_{xz}, $G_{\theta z}$ = transverse shear moduli of the core in the longitudinal and circumferential directions, respectively

$\mu = (\mu_1 E_1 t_1 + \mu_2 E_2 t_2)/(E_1 t_1 + E_2 t_2)$

μ_1, μ_2 = Poisson's ratios of the outer and inner facing sheets, respectively

The buckling coefficient C_p and a definition of the geometrical parameters may be obtained from Figs. 13-11 through 13-14. G_{xz} does not

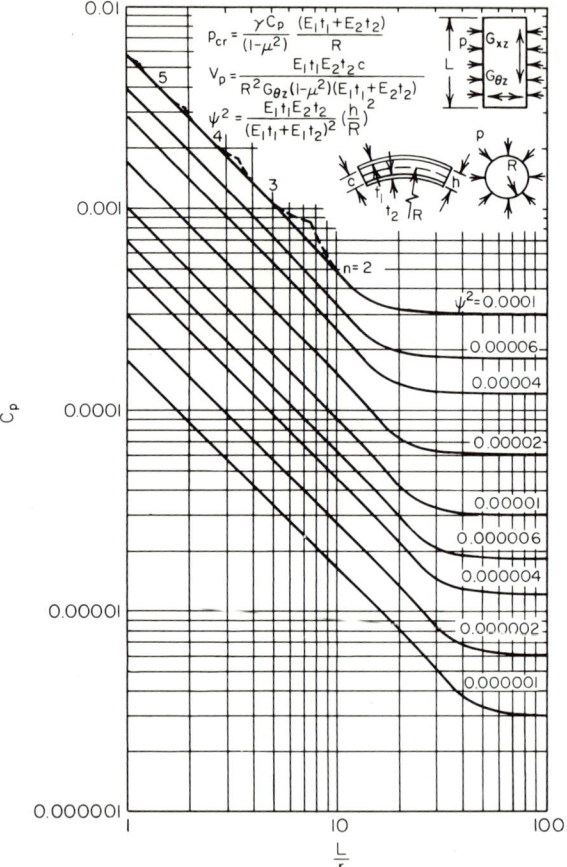

figure 13-11 *Buckling coefficient C_p for sandwich cylinders under external radial pressure. Isotropic facings; isotropic or orthotropic core; $V_p = 0$.*

enter into the analysis for most practical designs. The coefficient γ was introduced to reduce the theory presented in Ref. 13-9 by the same percentage that the theory for homogeneous isotropic cylinders was reduced to obtain the curve presented in Fig. 10-15. Therefore, if

$$Z_p = \frac{E_1 t_1 + E_2 t_2}{\sqrt{12 E_1 t_1 E_2 t_2}} \frac{L^2}{Rh} (1 - \mu^2)^{1/2}$$

$$\gamma = 0.9 \quad \text{for} \quad Z_p > 10^2$$

$$\gamma = 1 - Z_p \times 10^{-3} \quad \text{for} \quad Z_p < 10^2$$

For large values of L/R, it can be seen that C_p becomes independent of L/R. The design-allowable buckling pressure can then be obtained from

$$p_{cr} = \frac{3\gamma D_\theta}{R^3 [1 + 4 D_\theta / (R^2 G_{\theta z} h)]}$$

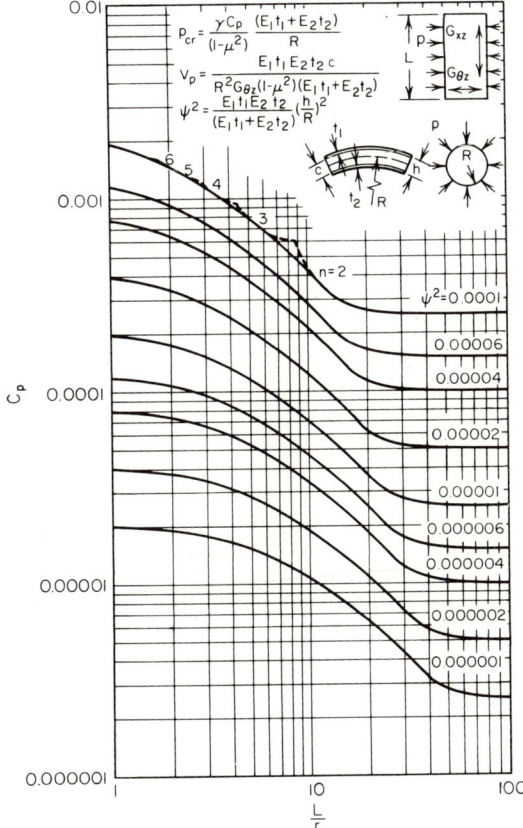

figure 13-12 *Buckling coefficient C_p for sandwich cylinders under external radial pressure. Isotropic facings; isotropic or orthotropic core; $V_p = 0.05$.*

where
$$D_\theta = \frac{E_1 t_1 E_2 t_2 h^2}{(1-\mu^2)(E_1 t_1 + E_2 t_2)}$$

The procedure presented for lateral external pressure may be used to estimate the design buckling load for sandwich cylinders subjected to lateral and axial external pressure (see Fig. 11-12) if $Z_p > 100$ and $\gamma = 0.75$ is used.

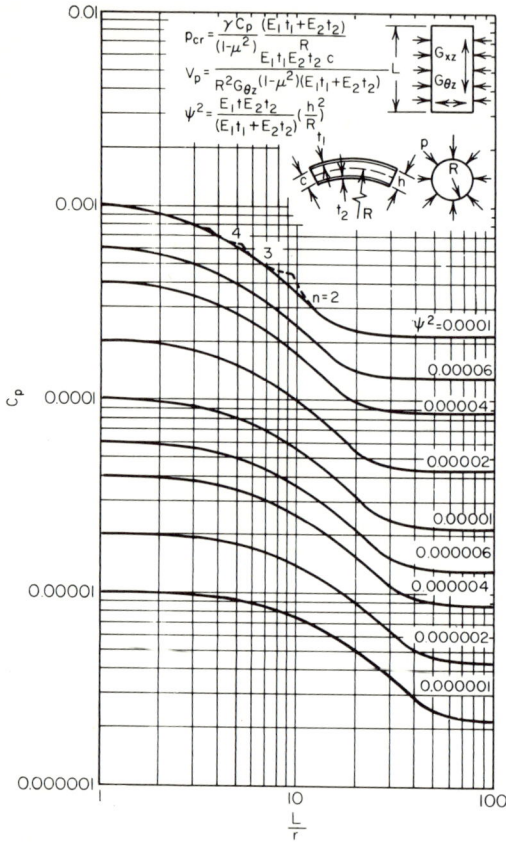

figure 13-13 *Buckling coefficient C_p for sandwich cylinders under external radial pressure. Isotropic facings; isotropic or orthotropic core; $V_p = 0.10$.*

The pressure p_{cr} is the design-allowable pressure for complete buckling of the cylinder (e.g., when buckles have formed all the way around the cylinder). For some values of the parameters, single buckles will occur at pressure less than p_{cr}; therefore, if single buckles are not allowable for a particular design, the results of this section may be unconservative.

Stability of Sandwich Shells   339

In addition, if single buckles occur at pressures less than p_{cr}, the resulting stresses may fail the core, causing a complete collapse of the cylinders. Sandwich shells, however, are generally relatively stiff with only small initial imperfection; therefore, for most cases, sandwich shells will be considerably less likely to have single isolated buckles at pressures less than the design-allowable pressures which have been given.

The curves presented in Figs. 13-11 through 13-14 are for a sandwich construction with very thin facing sheets ($c/h \approx 1$). For small values of c/h, the results of Figs. 13-11 through 13-14 may yield excessive errors for certain values of the parameters. For values of $c/h > 0.9$, the curves should be adequate.

Sandwich cylinders subjected to lateral pressure must be analyzed for local instability as described in Sec. 13-2. If the stresses are above the proportional limit and both facing sheets are made of the same material, the procedure discussed in Sec. 10-3 for external pressure, unstiffened cylinders, can be used. The parameter V_p is also a function of the stress level for stresses above the proportional limit. By assuming that V_s is independent of the stress level, the results will be conservative. For most practical designs, the difference will be small.

References 13-11 and 13-12 may be used to obtain buckling loads for sandwich cylinders with a corrugated core subjected to external pressure.

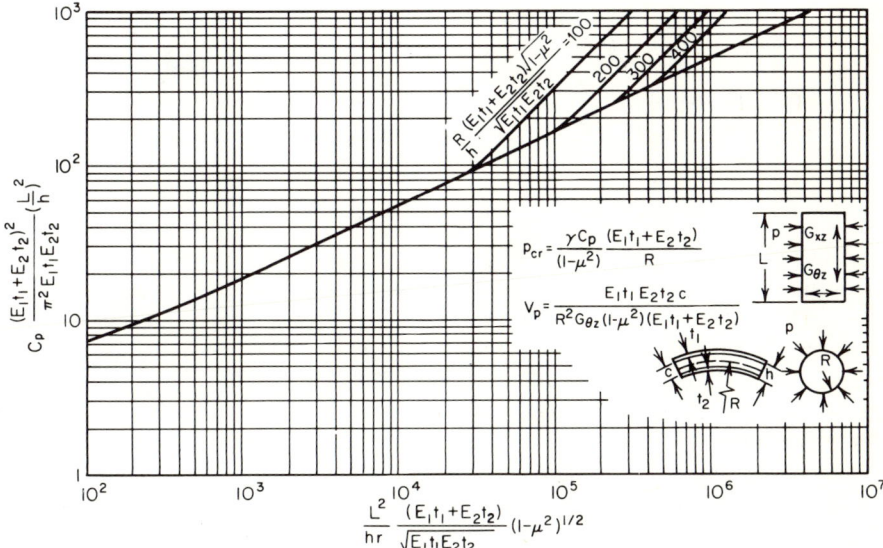

figure 13-14 Modified buckling coefficient for sandwich cylinders under external radial pressure. Isotropic facings; isotropic or orthotropic core; $V_p = 0$.

13-4 Cones

Axial Compression, Sandwich Cones

Until additional information is available, the equivalent-cylinder approach recommended in Sec. 10-4 for axial compression of unstiffened cones may be used. The cone shown in Fig. 11-13a can be analyzed as a cylinder with a radius $R_e = R_1/\cos \alpha$ and length L. The design-allowable stress σ_{cr} for the equivalent cylinder can be obtained from Sec. 13-3 for axial compression of a sandwich cylinder. The design-allowable total compressive load for the cone may be obtained from

$$P_{cr} = 2\pi R_e \sigma_{cr}(t_1 + t_2) \cos^2 \alpha$$

This method of analysis should be used with caution and should be limited to cones with $\alpha < 30°$.

Torsion, Sandwich Cones

Until additional information is available, the equivalent-cylinder approach recommended in Sec. 10-4 for torsion of unstiffened cones can be used for sandwich cones subjected to torsion. The cone shown in Fig. 11-13b can be analyzed as a cylinder with a radius

$$R_e = \left[1 + \left(\frac{1 + R_2/R_1}{2}\right)^{1/2} - \left(\frac{1 + R_2/R_1}{2}\right)^{-1/2}\right] R_1 \cos \alpha$$

and length $L_e = L/\cos \alpha$.

The design-allowable shear stress τ_{cr} for the equivalent cylinder can be obtained from Sec. 13-3 for torsion of a sandwich cylinder. The design-allowable torque for the cone can be obtained from the equation

$$T_{cr} = 2\pi R_e^2 \tau_{cr}(t_1 + t_2)$$

The design-allowable shear stress for the cone should be based on T_{cr}.

This method of analysis should be used with caution and should be limited to cones with $\alpha < 30°$. For inelastic stresses, the reduction of τ_{cr} due to plasticity should be based on the stresses at the smaller end of the cone and not on the stress of the equivalent cylinder.

Bending, Sandwich Cones

Until additional information is available, the equivalent-cylinder approach recommended in Sec. 10-4 for bending of unstiffened cones can be used for sandwich cones subjected to bending. The cone shown in Fig. 11-13c can be analyzed as a cylinder with a radius $R_e = R_1/\cos \alpha$ and length L. The design-allowable stress σ_{cr} can be obtained from

Sec. 13-3 for bending of sandwich cylinders. If the stresses are elastic, the design-allowable moment for the cone can be obtained from

$$M = \pi R_1^2 \sigma_{cr}(t_1 + t_2) \cos^2 \alpha$$

This method of analysis should be used with caution and should be limited to cones with $\alpha < 30°$.

Lateral and Axial External Pressure, Sandwich Cones

Until additional information is available, the equivalent-cylinder approach recommended in Sec. 10-4 for lateral and axial pressure on unstiffened cones can be used for sandwich cones subjected to lateral and axial pressure as shown in Fig. 11-13d. The cone can be analyzed as a cylinder with radius $R_e = (R_1 + R_2)/(2 \cos \alpha)$ and length L. The design-allowable buckling pressure p_{cr} can be obtained from Sec. 13-3 for lateral and axial external pressure on cylinders provided $Z_p > 100$.

For inelastic stresses, the reduction of p_{cr} due to plasticity should be based on the stresses at the larger end of the cone and not on the stress of the equivalent cylinder.

This method of analysis should be used with caution and should be limited to cones with $\alpha < 30°$.

13-5 Doubly Curved Shells

If the sandwich core is resistant to transverse shear so that its shear stiffness can be assumed to be infinite, then for unequal-thickness facings, the equivalent unstiffened isotropic material thickness, $\bar{t}$, and modulus of elasticity, $\bar{E}$, are given by

$$\bar{t} = \frac{\sqrt{12}\, h}{\sqrt{E_1 t_1/E_2 t_2} + \sqrt{E_2 t_2/E_1 t_1}}$$

$$\bar{E} = \frac{E_1 t_1 + E_2 t_2}{\bar{t}}$$

$$\mu = \frac{\mu_1 E_1 t_1 + \mu_2 E_2 t_2}{E_1 t_1 + E_2 t_2}$$

where E_1, E_2 = Young's modulus of the inner and outer facing, respectively.
t_1, t_2 = inner and outer facing, sheet thickness, respectively
μ_1, μ_2 = Poisson's ratio of inner and outer facing sheet, respectively

and for equal-thickness facings, with the same modulus of elasticity, by

$$\bar{t} = \sqrt{3}\, h$$

$$\bar{E} = \frac{2Et}{\sqrt{3}\, h}$$

where E is Young's modulus of the facing sheet and t is the facing-sheet thickness. These equivalent properties can be used in conjunction wich the recommended practices in Secs. 10-5 and 10-6 for isotropic doubly curved shells. The buckling pressure or load for an unstiffened, isotropic shell of thickness $\bar{t}$ and modulus $\bar{E}$ will be the same as for the sandwich shell provided the stresses are elastic and the core shear stiffness is large. Most cores fall into this classification. If the stresses are inelastic, the reduction in the buckling load should be based on the stress in the sandwich shell, not the equivalent unstiffened shell.

REFERENCES

13-1. *S & ID Structural Manual*, NAA S & ID, 543-G-11, North American Rockwell Corp., Downey, Calif., 1964.
13-2. Konishi, D., et al.: *Honeycomb Sandwich Structures Manual*, NAA Los Angeles Division, Report NA58-899.
13-3. Norris, C.: *Short-column Compressive Strength of Sandwich Constructions as Affected by Size of Cells of Honeycomb Core Materials*, Forest Products Laboratory, FPL-026, January, 1964.
13-4. Yusuff, S.: Face Wrinkling and Core Strength in Sandwich Construction, *J. Roy. Aeron. Soc.*, vol. 64, March, 1960.
13-5. Harris, B., and W. Crisman: Face-wrinkling Mode of Buckling of Sandwich Panels, *ASCE J. Eng. Mech. Div.*, EM3, June, 1965.
13-6. Zahn, J., and E. Kuenzi: *Classical Buckling of Cylinders for Sandwich Construction in Axial Compression—Orthotropic Cores*, Forest Products Laboratory, FPL-018, 1963.
13-7. March, H. W., and E. W. Kuenzi: *Buckling of Sandwich Cylinders in Torsion*, Forest Products Report 184, January, 1958.
13-8. Kiciman, M. O., and D. Y. Konishi: *Stability of Honeycomb Sandwich Cylinders*, ASME Paper 61-AV-36, 1961.
13-9. Kuenzi, E., B. Bohannan, and G. Stevens: *Buckling Coefficients for Sandwich Cylinders of Finite Length under Uniform External Lateral Pressure*, Forest Product Laboratory Research Note FPL-0104, September, 1965.
13-10. *Structural Sandwich Composites*, Department of Defense, MIL-HDBK-23, 1968.
13-11. Baker, E. H.: Stability of Circumferentially Corrugated Sandwich Cylinders Subjected to Combined Loads, *J. Am. Inst. Aeronautics Astronautics*, vol. 2, no. 12, December, 1964.
13-12. Harris, L. A., and E. H. Baker: *Elastic Stability of Simply Supported Corrugated Core Sandwich Cylinders*, NASA D-1510, 1962.

INDEX

Aluminum-alloy sheet and plate, plasticity correction and, 267–270
Apex, concentrated load at, of spherical caps, 254–256
Approximate method, any meridional shape, 99–104
   cylindrical shells, 100–101
Axial compression:
   buckling stress and: in curved cylindrical panels, 222–225
      in cylinders, 222–232, 238–241
         with elastic core, 241–242, 245–246
      in frame- and stringer-stiffened cylinders, 307–317
      in orthotropic cones, 303–305
      in orthotropic cylinders, 188–189, 293–303
      in sandwich cones, 340
      in sandwich cylinders, 330–331
      in toroidal segments, shallow bowed-out, 261–262

Axial compression (*Cont.*):
   in unstiffened cones, 246–248, 253
Axisymmetric loading:
   arbitrary partial, 44, 45
   bending theory for, 11–14, 19
Axisymmetric membrane shells, 41

Barrel vaults, 205–208
Beam systems, shells of, 199–204
Beam theory, elementary, Navier's hypothesis in, 21–22
Bending:
   axial forces and, 188–189
   buckling stress and: in cones, unstiffened, 249–251, 253
      in curved panels, 227–228
      in cylinders, unstiffened, 234–236, 238–240
      in frame- and stringer-stiffened cylinders, 317
   membrane theory of shells and, 22–23

Index

Bending (*Cont.*):
  in orthotropic cones, 304–305
  in orthotropic cylinders, 299
  in sandwich cones, 340–341
  in sandwich cylinders, 335
Bending moments, 5
Bending stiffness, 180–182
Bending theory:
  for axisymmetrically loaded shells of revolution, 11–14
  force method, 28–35
  unit-loads solution, 30
Best-fit curves, 221
Boundaries:
  elastically built-in or elastically restricted, 32
  fixed, restraints for, 32
  free or unrestrained, restraints of, 32
  restraints of, in force method, 32
  [*See also* Edge(s)]
Boundary disturbances:
  matrix of unknowns and, 159, 162–164
  unknowns and, 159–161
  [*See also* Edge(s)]
Buckling:
  between frames, 307, 308
  intracell, in sandwich shells, 326–328
  in orthotropic composite shells, 279–280
  of sheet stringer, 311, 313, 315–317
  wrinkling, 328
Buckling coefficients:
  for orthotropic cylinders, 297, 300–302
  for sandwich cylinders, 330–339
Buckling load, 221–222
  definition of, 220
Buckling stress:
  for cones, 246–253
  for curved cylindrical panels, 222–229
  for cylinders, 229–246
  elastic, 265
  inelastic, 265–276
    combined loadings and, 275–276
    plasticity correction factor and, 265–274
  for other shapes, 256–264
  for spherical shells, 253–256
Bulkhead(s):
  combined, approximate method, 104
  ellipsoidal and toroidal, internal pressure, 258–260

Bulkhead(s) (*Cont.*):
  interaction between two shell elements, 33–36
    compound and spherical, 33
  not approximated with sphere alone, 104
Bulkhead spacing, shell pressure factor as function of, 319
Byrne, Ralph, Jr., 22

Cantilever cylindrical shell, 199–200
Cassini shells, 61–65
Cassini's curve, 62
Circular plates, 141–150
  with central hole, 145, 146, 148
  with clamped edges, 149
  simply supported, 142, 144
    with central hole, 145, 146
  stresses in, due to edge elongation, 147
Circular rings, 151
  end deformations for, 147
  symmetrically loaded, 152
Circumferential load, 57, 107, 109
Clapeyron, E., 23
Collapse-loads, 255
Columns:
  critical stress for, 314
Composite shells (*see* Orthotropic shells, composite)
Composite walls, shells with, 181–195
Compression (*see* Axial compression)
Conical segment, edge distortion in, 169–171
Conical shells:
  buckling stress in unstiffened, 246–253
    under axial compression, 246–248
    in bending, 249–251
    combined loading in, 253
    lateral and axial external pressure on, 251–253
    shear in, 248–249
  closed, 86–89
  F factors for, in Hampe's solution, 126–130
  with free edges, edge distortion in, 168
  loaded by wind loading, 197
  membrane solutions for, 53–54
  open, 86–94
  orthotropic, 303–305
  orthotropic composite, 189–190

Conical shells (*Cont.*):
  primary solutions for, 50–54
  sandwich, 340–341
  secondary solutions for, 86–94
Core, sandwich, 194
Corrective loadings, 34–35
Crippling in stringers, 307–308
Critical stress for columns, 313–314
Curved meridian, shells of revolution with, 69–70
Curved panels, 205–208
  cylindrical, 205, 222–229
  elliptical and cycloidal, 208
Cylindrical panels, buckling stress for, 222–229
Cylindrical shells:
  approximate method for, 100–101
    maximum stresses in, 135–140
  buckling stress in, 229–246
    under axial compression, for thin-walled cylinders, 229–232
    bending and, 234–236
    in combined loading, 238–241
    under external pressure, 236–238
    shear in, in unstiffened cylinders, 232–234
  cantilevered and free, 199–200
  continuous, under deadweight, 203
  with elastic core, buckling in, 241–246
  F factors for, in Hampe's solution, 127
  frame- and stringer-stiffened, 307–317
  free and cantilevered, 199–200
  with free edges, edge distortion in, 165–166
  loaded as a beam, 202, 203
  loaded by wind loading, 197, 201
  long, 95–96
  membrane solutions for, 58
  multilayered filament-wound, 279
  orthotropic, 293–303
  orthotropic composite, 185–187
  primary solutions for, 55–58
  with rotationally symmetric discontinuities in geometry or loading, 121–123
  sandwich (*see* Sandwich shells, cylinders)
  secondary solutions for, 95–99
  short, 96
  sign convention for, 96, 99
  simple and fixed-beam, 201–204

Cylindrical shells (*Cont.*):
  with uniform thickness, Hampe's solutions for, 107–120
  with various edges, and loadings, 176, 177, 179

Deadweight loading, 3
  cylindrical shell under axisymmetrical, 57
  cylindrical shell under unsymmetrical, 203
  loading components from, 4
Deflection line in Hampe's solution, 109
Deformations:
  geometrical relations between, and elastic laws, 14–17
  of shell elements, equation of, 154–155
  unit, 34
  (*See also* Edge deformations; End deformations for circular plates and rings; Stresses and deformations)
Design-allowable buckling load, 221–222
Design-correction coefficients for orthotropic cylinders, 294, 298, 299
Differential element of shell, 7
Differential equations for equilibrium, 6–14
Dimpling of sandwich shell, 326, 368
Discontinuities, rotationally symmetric, cylinders with, 121–123
Displacement and rotation for Cassini domes, 63, 64
Displacements:
  approximate method for, 100
  for ellipsoidal shells, 61, 62
  in irregular shell, 71–73
  membrane, 42–44
    geometric relations between, 43
Distortions in spherical shell, 82–85

Edge(s):
  built-in, 32, 106, 124–126
  clamped, circular plate with, 149–150
  of cylindrical shells, 176–177, 179
  elongation of, in circular plates, 147
  fixed, 32, 106
  fixed or pinned, in spherical shell, Hampe's solution for, 124–126
  free, 32, 106

346 **Index**

Edge(s) (*Cont.*):
  in Hampe's solutions, 110
  pinned, 106, 124–126
  of spherical shell, 83, 85, 124–126
Edge deformations, secondary solutions for, 97, 98, 103
  (*See also* Deformations)
Edge distortions:
  in conical segment, 169–171
  in conical shells, 168
  in cylindrical shell, 165–167
  due to secondary loadings, 178
  in spherical segments, 173
  in spherical shell, 172–174
Edge loadings (*see* Loadings, edge)
Edge restraints, definition of, 26
Elastic constants in orthotropic composite shells, 279–293
Elastic core, cylinders with, 241–246
Elastic laws, 14–20
  Kirchhoff-Love hypothesis based on, 21, 22
Elastically restrained boundaries, 32
Elasticity equations in linear shell theory, 2
Elements (*see* Shell-element equations)
Ellipsoidal bulkheads (*see* Bulkheads)
Ellipsoidal shells, complete, under uniform external pressure, 256
Elliptical shells, 59–60
End deformations for circular plates and rings, 147
Equations:
  junction, 157
  sets of, for corrective loadings, 162–164
  shell-element, 154–156
Equilibrium:
  condition of, 5–6
Equilibrium equations, 6–13
Equivalent-cylinder approach, 246, 248
Exponential-loading case, 56
Extensional stiffness, 180–182
External loadings, 3
External pressure (*see* Pressure)

F factors, definition of, in Hampe's solutions, 126–134
Face-sheet wrinkling, 328–330
Failure theories, 210–213
First-order-approximation shell theory, 21–22
Fixity moment in cylinders, 140
Flügge, W., 22
Force method, 25–39
  definition of, 26
  procedure for, 30–32
  of solution, 29–30
  in two shell elements, 33–36
  unit edge, 30, 31
Forces:
  internal, 3–5
    elastic laws and, 18–20
  membrane (*see* Membrane forces)
  transverse shear, 4–5
Frame pressure factors, 318, 321
Frame-stiffened cylinders, 307–322
  under axial compression, 307–317
  under lateral and axial external pressure, 317–322
Frame stiffness, 321

Geckler's assumption, 104
General-instability failure in stiffened cylinders, 315–322
Geometry of shells, 2–3, 69–70
  complicated, breakdown for, 33
  elastic laws, 14–17
  various, interaction between, 32–36

Hampe, E., 96, 135
Hampe's solution, 106–140
  approximate method for location and maximum stresses in cylinders, 135–140
  for cylinders, 107–123, 135–140
    approximate method for stresses in, 135–140
    with rotationally symmetric discontinuities in geometry or loading, 121–123
    with uniform thickness, 107–120
  definition of F factors in, 126–135
  for spherical shell, any fixity at lower boundary, 124–126
Hooke's law, 17
  for layered shells, 282
Hydrostatic-pressure loading over portion of spherical shell, 49

Inelastic buckling (*see* Buckling)
Instability:
  local, of sandwich shells, 326–330
  of stiffened cylinders: general, 315–322
    torsional, of stringer, 310–312
Interaction:
  method of, in force method, 33–36
  in multishell structure, 158–162
  theoretical approach to, 217–219
Interaction analysis in approximate method, 100–101
Interaction curves and stress ratios, 214–217
Interaction relations, 216–217
Intracell buckling, 326–328
Isotropic facings:
  sandwich with, 287–288
  sandwich cylinders with, 332–339

Junction equations, 157
Junctions, equilibrium of, 156–157

Kempner, J., 22
Kirchhoff-Love hypothesis, 21, 22

Lamé, Gabriel, 23
Lateral pressure (*see* Pressure)
Layered shells, orthotropic, elastic constants for, 282–285
Length, effective, stress function for, 322
Linear loadings, 56
Linear shell theory, 2
Load(s):
  at apex, concentrated, in spherical caps, 254–255
  axisymmetrical, bending theory for, 11–20
  bending and shear, 12
  buckling, 221–222
    definition of, 220
  collapse, 220
  concentrated, and membrane theory, 28
  in cylinders in Hampe's solution, 135–140
    circumferential, 136
    linear, 137–140
    internal, 18–20

Load(s) (*Cont.*):
  in stresses and deformations, determining, 36–39
  vertical, shell of revolution under, 71–73
Loading(s):
  axisymmetric, arbitrary partial, 44
  circumferential, 57
  combined: in curved panels, 229
    inelastic buckling and, 275–276
    in unstiffened cones, 253
    in unstiffened cylinders, 238–241
  in conical segment, 169–171
  in conical shells, 50–52, 168
  corrective, final sets of equations for, 162–164
  corrective edge, definition of, 26
  in cylindrical shells, 55–57, 165–167, 201–204
    Hampe's solutions and, 107, 108, 110, 121–123
    with various edge conditions, 176–177, 179
  deadweight, 57
  distortions due to secondary, 178
  edge: in conical shell, open, 87–94
    in cylindrical shells, 95–99
    in irregular shells, solutions for, 100
    in open spherical shells, 79–81
    in orthotropic cylinder, 186, 187
  external, 3
  hydrostatic-pressure, over portion of spherical shell, 49
  linear, 56
  primary: in spherical shell, 172–174
    tables on, 173
  secondary: in spherical segment, 175
    tables on, 175
  in spherical shells, 45–46, 170–175
  statically indeterminate and corrective, in multishell structures, 158, 159
  trigonometrical, 56
  unit, 34
  unit edge, 31
    definition of, 26
  unsymmetrical, 196–208
  wind, shells loaded by, 197, 201
Loading case, exponential-, 56
Location of maximum stresses in Hampe's solutions, 135–140
Love, A. E. H., 21, 22

Margin of safety, 210, 214, 215
Material and shape parameter constants, 309
Material failure in stiffened cylinders, 307, 318
Maximum moment in Hampe's solutions, 137–139
Maximum normal stress theory, 217–218
Maximum shear stress theory, 218
Maximum stresses in Hampe's solutions, 135–140
Membrane displacements, 42–44
Membrane forces, 4
  for Cassini shells, 63
  internal, 40–42
Membrane shells, axisymmetric, determination of forces of, 41
Membrane solutions:
  for Cassini shells, 65
  for conical shells, 53–54
  for cylindrical shells, 57–58
  spherical shell, stress and deformation in, 47–48
Membrane theory, 22–23, 27–28
  for elliptical shells, 59
  for shells of revolution, 6–11, 45
Meridian, 3
  curved, other shells of revolution with, 70
  any shape of, 44–45
Meridian plane, 3
Moment(s):
  bending and twisting, 5
  unit, 75
Momentless state of stress, 23
Multiaxial failing stress, 209–211
Multishell structures, 153–180

Navier's hypothesis in elementary beam theory, 21–22
Novoshilov, V. V., 22, 27

Orthotropic facings, sandwich with, 286–287
Orthotropic shells, 182–185
  composite, stability of, 279–305
  cones, 303–305
  cylinders, 185–188, 293–303
  elastic constants in, 279–293

Orthotropic shells (*Cont.*):
  cylinders, stability of, 293–303
  spheres and cones, 189–190

Panels:
  curved, 205–208
  cylindrical (*see* Cylindrical panels)
Parallels, 3
Plasticity correction curves, 267–274
Plasticity correction factor, plasticity coefficient, 265–274
Plates:
  circular (*see* Circular plates)
  flat, buckling load in, 221
Poisson's ratio, 17
Pressure:
  external: in curved panels, 228
    in orthotropic cylinders, 300–303
    in sandwich cylinders, 335–339
    in unstiffened cylinders, 236–238, 240–241
    with elastic core, 241–244
  hydrostatic, on toroidal segments, 263, 264
  internal, in ellipsoidal and toroidal bulkheads, 258–260
  lateral, on toroidal segments, 263, 264
    axial external and: in frame-stiffened cylinders, 317
    for orthotropic cones, 305
    on sandwich cones, 341
    in unstiffened cones, 251–253
    in cylinders with elastic core, 245–246
  uniform, in Cassini shells, 61–65
  uniform external: in complete circular toroidal shells, 260–261
    in complete ellipsoidal shells, 256
    of shallow toroidal segments, 263–264
    in spherical caps, 253–254
  uniform internal, in complete oblate spheroidal shells, 258
Pressure factor, 319, 321
  frame, 321
  shell, 319
Pressurized shells and buckling, 224–228, 231–238, 247–251
Primary solutions, 40–74
Principal-stress equations, 217

Radius of curvature for various types of shells, 72, 73
Radius of gyration, 313, 314
Restraints of boundaries, 32
Revolution:
  shells of (*see* Shells of revolution)
  surfaces of, 3
Rigidity:
  extensional, 19
  flexural, 19
Rings, circular (*see* Circular rings)
Rotational parameters and displacement for ellipsoidal shells, 61

Safety:
  factor of, 215, 216, 219
  margin of, and allowable stresses, 209–219
Salvadori, M. G., 71
Sandwich cores, 194
Sandwich shells, 192–195
  cones, 340–341
  cylinders, 330–339
  doubly curved shells and, 341–342
  elastic constants for, 286–288
  face-sheet wrinkling in, 328–330
  honeycomb, 193
  intracell buckling in, 326–328
  with isotropic facings, 287–289
  with orthotropic facings, 286–287
  stability of, 325–342
Second-order-approximation theories for shells of revolution, 22
Secondary solutions, 31, 75–105
  approximate method in, 99–102
  definition of, 26
  orthotropic composite, 184–186
  practical considerations in, 102–105
Shear:
  in cones, unstiffened, 248–249, 253
  in curved panels, 225–227
  in cylinders: in Hampe's solutions, 139, 140
    unstiffened, 232–234, 238–240, 244–245
  in orthotropic cones, 304
  in orthotropic cylinders, 296–299
  in sandwich cones, 340
  in sandwich cylinders, 331–336

Shear distortion, transverse, in cylinder of sandwich construction, 186–187
Shear stress theory, maximum, 218
Shears, unit, 75
Sheet stringer, 306–317
Shell(s):
  of beam systems, 199–204
  Cassini, 61–65
  combined, 153
  with composite or stiffened walls, 181–196
  conical (*see* Conical shells)
  cylindrical (*see* Cylindrical shells)
  deformations of (*see* Deformations)
  doubly curved, 341–342
  elliptical (*see* Elliptical shells)
  geometries of, 2–3, 14–17, 32–36, 69–70
  irregular: displacements in, 71
    edge-loading solutions for, 100
    primary solutions for, 71–73
  membrane (*see* Membrane shells)
    theory for (*see* Membrane theory)
  monocoque, 27
  nonshallow, 26, 27, 29
    conical, 50
    spherical, 45
  sandwich (*see* Sandwich shells)
  spherical (*see* Spherical shells)
  statically analogous, 102
  statically determined, 23
  stiffened, 190–192, 306–324
  symmetrical deformation of, 42
  theory of, 1–24
    bending, 11–20
    classification of, 20–23
    first-order-approximation, 21–22
    general, 1–2
    large or finite-deflection, 24
    linear shell, 1, 2
    membrane, 6–11, 22–23
      (*See also* Membrane theory)
    nonlinear, 23–24
    second-order-approximation, 21–22
    special, 22
  thick, 27
  thin, 27
  toroidal (*see* Toroidal shells)
  two, interaction between, 33–36
  unstiffened, stability of, 220–278

350 *Index*

Shell(s) (*Cont.*):
  unsymmetrically loaded, 196–208
  (*See also* Multishell structures)
Shell-element equations, 154–156
Shell pressure factor and bulkhead spacing, 319
Shells of revolution, 4
  axisymmetrically loaded, 11–14, 41, 45
  membrane theory for, 6–11
  unsymmetrically loaded shells and, 196–198
Skin-stringer, 306–317
Small-deflection theory for buckling load, 221
Snap-through, axisymmetric, 253–255
Solutions:
  force method of, 29–30
  membrane (*see* Membrane solutions)
  primary, 30–31, 40–74
    definition of, 26
  secondary (*see* Secondary solutions)
  special (*see* Special solutions)
  unit-loads, 30
Special solutions, 106–152
  circular plates, 141–150
  circular rings, 151
  Hampe's solutions, 106–140
Spherical caps:
  buckling pressure of, 253–256
  concentrated load at apex of, 254–256
Spherical segment with free edges, edge distortion in, 173
Spherical shells:
  any fixity at lower boundary, Hampe's solutions for, 124–126
  buckling of, 253–256
  F factors for, in Hampe's solutions, 132
  with free edges, edge distortion in, 172–174
  loaded by wind loading, 197
  orthotropic composite, 189–191
  primary solutions for, 45–49
    hydrostatic-pressure loading over portion of, 49
    stress and deformation in, 47–48
  secondary solutions for, 76–85, 103–104
  with unsymmetrical boundaries, 198
Spheroidal shells:
  complete oblate, uniform internal pressure in, 258

Spheroidal shells (*Cont.*):
  oblate, buckling under external pressure, 256
  prolate, buckling under external pressure, 256, 257
Stability of unstiffened shells, 220–278
Statically determinate loadings, 158, 159, 162
Statically indeterminate elements:
  in free body diagram, 163
  structural system combined from, 163–164
Statically indeterminate loadings, 158, 159
Steel, alloy and stainless, and plasticity correction, 271–273
Stiffened shells, 190–192
  stability of, 306–324
Stiffened walls, shells with composite or, 179–195
Stiffness:
  bending, 19, 180–182
  extensional, 19, 180–182
  frame, and frame pressure factor, 321
  in orthotropic composite shells, 279
Strain and elastic laws, 14–15
Stress(es):
  allowable, and margin of safety, 209–219
  critical, for columns, 314
  displacement and, of spherical shells, 82
  Hooke's law and, 17
  internal, 3–5
  maximum, 135–140
  momentless state of, 23
  in plates due to edge elongation, 147
  in shell elements, 18, 19
  (*See also* Buckling stress)
Stress-ratio interaction curves, 214–217, 238
Stress theory:
  maximum normal, 217–218
  maximum shear, 218
Stresses and deformations:
  of cylinders, Hampe's formulas for, 107
  determination of, at any point on shell, 36–39
  in ellipsoidal shells, 61
  in elliptical shells, 59–61
  in force method, 31–32

Stresses and deformations (*Cont.*):
   in spherical shell membrane solution, 47–48
Stringer-stiffened cylinders, 306–317
Structural system:
   from statically determinate elements, 162–166
   from statically indeterminate elements, 163–164

Tests:
   multiaxial, 209–211
   uniaxial, 209, 210
Theories (*see* Shells, theories of)
Titanium-alloy sheet, plasticity correction and, 273–274
Torispherical closure, 259–260
Toroidal bulkheads, 258–260
Toroidal segments, 261–264
Toroidal shells, 65–68
   complete circular, 260–261
   primary solutions for, 65–68
Torsion (*see* Shear)
Torsional-buckling coefficients for cylinders with elastic core, 245
Torsional instability of stiffeners, 310–313
Transverse shear forces, 4–5
Trigonometrical loadings, 56
Twisting moments, 5

Uniaxial tests, 209–210
Unit deformations, 34
Unit loadings, 31, 34–35
   secondary solutions and, 75
Unit-loads solution, substitution for bending theory by, 30
Unit moments, 75
Unit shears, 75
Unknowns, matrix of: boundary disturbances and, 159, 162–164
   in multishell structures, 159, 160, 164
Unpressurized shells and buckling, 222–235, 238–240, 246–250
Unstiffened shells, 220–278
Unsymmetrical loading, 196–208

Vlasov, V. Z., 22

Waffle shells, integrally stiffened, elastic constants for, 288–292
Walls, composite or stiffened, 181–195
Width, effective, of sheet with stiffener, 313
Wrinkling, face-sheet, 328–330
Wrinkling failures, 328

Young's modulus of elasticity, 17